ENGINEERING THERMODYNAMICS

ENGINEERING THERMODYNAMICS

R.K. Singal
Ph.D., FIE
Professor and Consultant
rk.singal@gmail.com

Mridul Singal
B.Tech (IIT, Kanpur), M.Tech (IISc, Bangalore)
I/T Architect, IBM India Pvt. Ltd.

Rishi Singal
B.Tech (PEC Chandigarh), Ph.D. (PMI, USA)
Project Manager, Accentire India Pvt. Ltd.

TECHSAR™
NEW DELHI

Published and Printed by
TechSar Pvt. Ltd.
4435-36/7, Ansari Road, Daryaganj
New Delhi–110 002 (India)
E-mail: tech_info@techsarworld.com
Website: www.techsarworld.com

ISBN 978-93-80026-51-0

Reprint 2025

Preface

Engineering Thermodynamics has been designed to serve as a textbook for undergraduate and postgraduate students of engineering. The book has been written in simple language duly supplemented by solved numerical problems, diagrams and mathematical analysis. The book is also aimed to serve as reference manual for practising engineers and those opting for competitive examinations. A large number of objective-type questions have been included as appendix.

Suggestions and criticisms for the improvement of the book are welcome with gratitude.

Authors

Contents

PART I

THERMODYNAMIC LAWS AND RELATIONS

CHAPTER 1

Basic Concepts

1.1 INTRODUCTION

Thermodynamics is the basic science that deals with the conversion of heat into mechanical work. The application of basic thermodynamics for the solution of engineering problems is called engineering thermodynamics or applied thermodynamics. Engineering thermodynamics plays a vital role in the design and development, analysis and optimization of various thermal machines, plants and systems connected with power generation and power consumption. The design, operation and performance of electric-generating power-plants are largely dependent upon the science of thermodynamics. Thermodynamics laws have been formulated based on observations of common experience. Steam and nuclear power plants, internal combustion engines, gas turbines, jet propulsion, compressors, chemical plants, heat transfer, air-conditioning and refrigeration plants are based on thermodynamic laws and principles.

The whole of heat supplied (Q_1) to a heat engine (Fig. 1.1) from a heat source cannot be converted into mechanical energy or work (W). Some portion of heat

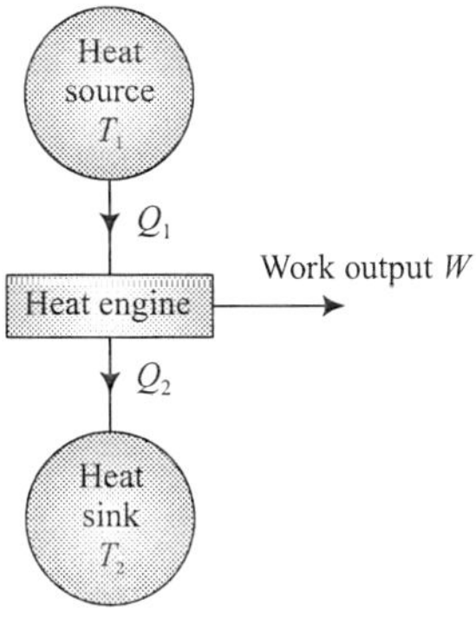

Fig. 1.1 Heat engine

(Q_2) has to be degraded at a low temperature T_2 and rejected to the environment or a heat sink. The part of heat which is available for conversion into maximum work is called *exergy*. The portion of heat energy which is not available for conversion into work is measured by a thermodynamic property called *entropy*. Thus, thermodynamics is a science which deals with the computation of energy, exergy and entropy.

Thermodynamics is based on four laws:

1. Zeroth law deals with *equality of temperature* and introduces the concept of *thermal equilibrium*. It helps in the *measurement of temperature*.
2. First law is the law of conservation of energy and introduces the *concept of internal energy*. The techniques of mass and energy balance based on first law are used for the performance analysis and measurement of energy efficiency of different machines and plants.
3. Second law was formulated to overcome the inherent limitations of first law of thermodynamics. It puts limit on the amount of conversion of *heat into work* and introduces the *principle of increase of entropy*.
4. Third law defines the *absolute zero of entropy*. Entropy has zero value at absolute zero temperature.

These laws are based on experimental observations and have no mathematical proof. Their validity stems from the fact that these laws have not been violated.

1.2 SCOPE OF THERMODYNAMICS

Thermodynamics has very wide applications.

1. *Associate Branch of Engineering:* Thermal engineering is a very important associate branch of mechanical, chemical, metallurgical, automobile, aerospace, marine, textile, energy technology and environmental engineering. Thermodynamics is the basic science of thermal engineering.
2. *Applications of Thermal Engineering:* Whenever there is combustion, heating or cooling, exchange of heat for carrying out chemical reactions, propulsion of rockets, railway engines, ships, production of mechanical or electrical power, etc., application of thermal engineering is required.
3. *Process Engineering:* Process engineering of pharmaceutical, refinery, fertilizer, food processing, sugar, paper and pulp, organic and inorganic chemical plants is based on laws and principles of thermodynamics.
4. *Different Sectors:* All process and engineering industries, agriculture, transport, commercial and domestic activities use thermal engineering for their design and operation.
5. *Power and Energy Technology:* The power sector and energy technology are purely dependent on the laws of thermodynamics. Central power

plants, captive power plants based on coal, nuclear power plants, gas turbine power plants, engines for automobiles, ships, railways, spacecraft are designed using laws of thermodynamics. Similarly, direct energy conversion devices such as fuel cells, thermionic, thermoelectric, solar cells, magneto-hydrodynamic plants are based on thermodynamics.

6. *Power Consumers:* Refrigeration and air-conditioning, cooling, heating and ventilation plants, domestic, commercial and industrial lighting, heating and cooling, agriculture, transport and industrial machines use laws of thermodynamics for their design and operation.

It can be safely concluded that there is hardly any human activity which does not involve thermodynamics. Modern industrial society depends upon thermodynamics for its existence and growth.

1.3 SPECIAL TECHNIQUES AND ASSUMPTIONS

Special techniques are used for the study of engineering thermodynamics. Highly simplified substitutes and idealizations are made to replace actual physical materials and actions.

1.3.1 System

A region or a quantity of matter is identified as system for the study and analysis of a problem. The system is isolated from the surrounding through a real or imaginary boundary. A system is a specified region, not necessarily of constant volume or fixed boundaries, where transfers and conversions of energy and mass are to be studied. Both energy and mass can cross the boundaries of an open system. If the open system is in the transient state, the mass inflow and outflow are not equal or vary with time. The mass within the system changes with time. For example, a pressurizer in a nuclear pressurized-water reactor is a transient open system.

The steady-state open system or steady-state steady-flow (SSSF) system has constant mass and mass and energy flows across the boundaries do not vary with time. Most of the engineering devices like turbines, compressors, boilers, heat exchangers, furnaces, etc. are examples of steady-state open system.

In a closed system, only energy and not mass may cross the boundaries. Steam engine, internal combustion engines, bomb calorimeter are analysed as closed systems.

A third system of some interest is the isolated system, a special example of the closed system. It is one in which neither mass nor energy crosses the boundaries but in which energy transformation may take place within the boundaries.

The comparison of open and closed system is illustrated in Table 1.1.

Table 1.1 Comparison of open and closed system

Open system	*Closed system*
1. Both mass and energy can cross the boundary.	1. Only energy and not mass can cross the boundary.
2. Mass of the system can be constant or vary.	2. Mass of the system is constant.
3. Flow processes take place in open system.	3. Non-flow processes take place in closed system.
4. Special case: Control volume system with control surface.	4. Special case: Isolated system with no mass or energy interaction with the surrounding.
5. Examples: Turbines, pumps, compressors.	5. Examples: Piston-cylinder engine, bomb calorimeter.

1.3.2 System Matter

Fluids in the form of gas, vapour, liquid or mixtures are used in thermodynamic devices to transport energy between the system and its surrounding. If the system matter is homogeneous (uniform) in physical structure and chemical composition throughout, it is called one *phase* system, i.e., single gas or a mixture of gases. A system consisting of liquid and gas is two-phase system. Water at triple point exists as water, ice and steam simultaneously forms a three-phase system. A *pure* substance has homogeneous invariable chemical composition even though there is change of phase. Liquid water, wet steam, mixture of ice and water are examples of pure substance.

1.3.3 Continuum

Thermodynamics makes no hypothesis about the structure of matter of the system. The system is regarded as a *continuum*, i.e., continuous distribution of matter.

Matter consists of large number of molecules which are in a random state of motion. They experience inter-molecular forces of attraction and repulsion. They are separated from each other by a distance of the order of mean-free path. Thus, matter is neither continuous nor homogeneous from microscopic point of view.

However, engineering systems working at high pressure contain large number of molecules in comparatively small volumes. The linear dimensions of the system are quite large as compared to inter-molecular distance, the latter can be neglected. It is assumed that the system contains continuous distribution of matter with no voids or cavities.

As a consequence of continuum hypothesis, each intensive property like pressure, temperature, density are regarded as definite value at a point.

1.3.4 Macroscopic Approach

Based on continuum hypothesis, the system matter is described *macroscopically* based on time-averaged measured values of the properties of the system. In the microscopic approach, the overall behaviour of the matter is predicted by statistically averaging the behaviour of individual molecules. It is rather complex, cumbersome and time-consuming.

Engineering thermodynamic analysis is macroscopic and the results obtained are of sufficient accuracy and validity.

1.3.5 Thermodynamic Properties

Engineering thermodynamics deals with heat and work and also properties of substances that transport heat and work between system and surrounding. The properties are grouped as *measurable* properties in the laboratory with the help of instruments such as pressure, temperature, volume, mass, etc., and *derived* properties which can be calculated from experimental data using thermodynamic relations such as internal energy, enthalpy, specific heat, etc.

1.3.6 Thermodynamic Processes

Properties are defined only when the system is in thermodynamics equilibrium. However, all real processes occur when there is no equilibrium as some potential difference must exist within the system and between a system and surrounding to promote a change of state during a thermodynamic process.

In the analysis of real processes, it is assumed that the system deviates from the equilibrium state by a very small margin. All the intermediate states during a process are assumed as equilibrium state. Such a process in which the system is always very near to the equilibrium state throughout the process is called *quasistatic process.* In engineering thermodynamics, quasistatic processes are considered extensively. It is an idealized process in which many real process closely approach.

1.4 CLASSIFICATION OF THERMODYNAMICS

The field of study of thermodynamics is very wide. It can be studied conveniently under two parts:

1. Basic Thermodynamics for the study of laws of thermodynamics, processes, properties and their relations.
2. Engineering Thermodynamics for the study of thermal machines, plants and systems connected with production and consumption of power.

The thermal machines for power generation include steam turbines, gas turbines, steam engines, I.C.engines, jet propulsion units. Thermal machines for power consumption include refrigeration, cryogenics, air-conditioning, heating, cooling and ventilation, liquid pumping and gas compression plants.

The behaviour of matter can be studied macroscopically in *macroscopic* thermodynamics which is based on continuum hypothesis. In *microscopic* thermodynamics, microscopic approach is used for the overall behaviour of the matter which is predicted by statistically averaging the behaviour of individual molecules. The macroscopic thermodynamic is also called *classical* thermodynamics and microscopic thermodynamics is called statistical thermodynamics. Engineering thermodynamics is classical thermodynamics. The main differences between classical thermodynamics and statistical thermodynamics are discussed in Table 1.2.

Table 1.2 Comparison of classical and statistical thermodynamics

Classical Thermodynamics	*Statistical Thermodynamics*
1. The system is considered a continuum, i.e., continuous distribution of matter without voids and cavities.	1. The system is considered to contain a large number of molecules moving randomly in chaotic fashion.
2. It is also called macroscopic thermodynamics.	2. It is also called microscopic thermodynamics.
3. The thermodynamic properties are measured values which are time-averaged variables.	3. The thermodynamics properties are found out by statistically averaging the behaviour of individual molecules.
4. The results obtained are of sufficient accuracy and validity.	4. The results obtained are very accurate.
5. The study is simple and quick.	5. It is rather complex, cumbersome and time-consuming.
6. Engineering devices are designed and analysed macroscopically. Engineering thermodynamics is classical thermodynamics.	6. For research and study of laws of thermodynamics deeply, statistical thermodynamics is used.

1.5 THERMODYNAMIC PROPERTIES

Property is a quantity to describe the state of a system. The number of properties required to describe a system depends upon the nature of the system. However, each property has a single value at each state. Each state can be represented by a point on a graph with any two properties as coordinates. All properties are point functions.

1.5.1 Characteristics

Special characteristics of thermodynamics properties are:

1. Property is a quantity to describe the state of a system.
2. The number of properties required to describe a system depends upon the nature of the system.
3. Each property has a single value at each state.
4. Each state can be represented by a point on a graph with any two properties as coordinates.
5. All properties are point function.
6. Any operation in which one or more properties of a system change is called *change of state* which may result in a *process* or a *cycle*.

7. Thermodynamic properties are exact differentials.
8. Heat and work are path functions and not properties of the system.
9. A property is a macroscopic characteristic of a system to which numerical value can be assigned.
10. The value of a property is unique and depends on the condition of the system.
11. Thermodynamic properties can be classified as extensive or intensive. Extensive properties depend on the extent or size of the system. Intensive properties do not depend upon the size of the system and have same value throughout a system.
12. There are properties which can be measured directly in the laboratory with the help of instrument, i.e., pressure, temperature, volume, mass, etc. There are properties derived mathematically using the laws of thermodynamics, i.e., internal energy, enthalpy, entropy, specific heat, Helmholz function, Gibb's function, etc.
13. Properties are defined only when a system is in equilibrium.

1.5.2 Extensive and Intensive Properties

There are two types of properties

(a) Intensive property

It does not depend on the extent or size of the system. Examples are pressure, temperature, density, composition, viscosity, etc. It has same value throughout the system.

$$p = p_1 = p_2 = p_3 = \dots$$

where, p = pressure of the total system (Pa)

$p_1, p_2, p_3, \dots$ = pressure of subsystems (Pa)

(b) Extensive property

It depends on the extent or size of the system and is the sum of the properties of various parts of the system. Volume, energy, enthalpy, entropy are the examples of extensive properties.

$$V = V_1 + V_2 + V_3 + \dots$$
$$= \sum V_i$$

where, V = total volume of the system (m^3)

$V_1, V_2, V_3, \dots$ = volume of the subsystems (m^3)

(c) Specific property

It is the extensive property per unit mass of the system. This is an intensive property and is expressed by a lower case letter.

$$\text{Specific volume, } v = \frac{V}{m} \text{ [m}^3\text{/kg]}$$

$$\text{Specific energy, } e = \frac{E}{m} \text{ [kJ/kg]}$$

Like extensive property, specific properties are also additive

$$v = v_1 + v_2 + v_3 + \dots$$

(d) Molar property

It is the extensive property per unit mole.

$$\text{Molar volume} = \frac{\text{Total volume}}{\text{Mole number}}$$

$$\bar{v} = \frac{V}{n} \text{[m}^3\text{/mol]}$$

(e) Properties of continuum

The property of a continuum is the average property at a point within the continuum.

Example 1 Density is an intensive property and may vary from point to point within a system and may also vary with time. Therefore, average density of a system is defined as mass per unit volume of the system.

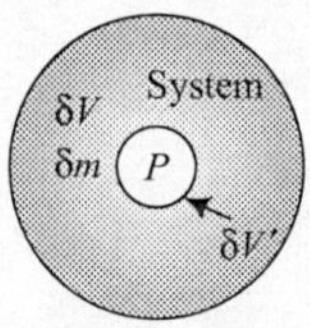

Fig. 1.2 Density of continuum

Consider mass δm of small volume δV of a system. The ratio $\frac{\delta m}{\delta V}$ is the average mass density of the system within the volume δV. If the volume is shrunk to $\delta V'$ about a point P which is regarded as a continuum, the density at a point is defined as:

$$\rho = \underset{\delta V \to \delta V'}{\text{Lt}} \frac{\delta m}{\delta V}$$

Example 2 If δA is a small area of the system on which δF_n is acting which is the normal component of the force. The average pressure of the system,

$$p = \frac{\delta F_n}{\delta A}$$

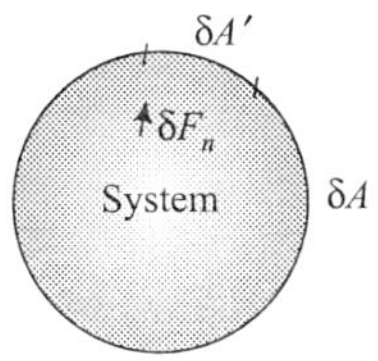

Fig. 1.3 Pressure of continuum

If δA approaches $\delta A'$ which is the value of smallest area of the continuum, the pressure at a point of the continuum is expressed as:

$$p = \underset{\delta A \to \delta A'}{\text{Lt}} \frac{\delta F_n}{\delta A}$$

1.5.3 Measurable Properties

The important thermodynamic properties which can be measured in the laboratory with the help of instruments are mass, volume, pressure and temperature.

1. ***Mass*** is an indication of quantity of matter present in the system and is measured in kg.

A *mole* of a substance has mass numerically equal to the molecular weight of the substance. One mole of oxygen has a mass of 32 g and 1 kg mol (or kmol) of nitrogen has a mass of 28 kg.

2. ***Volume (V)*** is the space occupied by a substance and is measured in m^3. The specific volume may be expressed either on mass basis or mole basis.

Specific volume (v) is the volume of system per unit mass and is measured in m^3/kg.

$$v = \frac{V}{m} [m^3/kg]$$

Molar volume ($\bar{v}$) is the volume per mole and is given as m^3/mol (or $m^3/kmol$).

Specific weight (w) is the force exerted by the system due to gravitational acceleration. The weight (W) of the system is the product of mass (m) and local gravitational acceleration (g) and is measured in newton (N).

$$W = mg\ [N]$$

Specific weight is the weight per unit volume.

$$w = \frac{W}{V} [N/m^3]$$

3. ***Pressure (p)*** is the normal force exerted by a system against the unit area of boundary surface. The pressure at a point is the same in all directions.

$$p = \frac{F}{A}\,[\text{N/m}^2]$$

$$1\ \text{N/m}^2 = 1\ \text{Pa}$$

The pressure is expressed in N/m^2 or pascal (Pa). Pascal is the force of one newton acting on an area 1 m^2.

The unit of pascal is very small. Engineers express pressure in kilopascal (kPa) and megapascal (MPa).

$$1\ \text{kPa} = 10^3\ \text{Pa}$$

$$1\ \text{MPa} = 10^6\ \text{Pa}$$

Bar and standard atmosphere are other units, not within S.I. system widely used.

$$1\ \text{bar} = 10^5\ \text{Pa} = 10^5\ \text{N/m}^2 = 750\ \text{mm Hg}$$

$$1\ \text{atm} = 1.01325\ \text{bar} = 101.325\ \text{kPa} = 760\ \text{mm Hg}$$

The pressure exerted by 1 mm of mercury is called Torr.

$$1\ \text{bar} = 750\ \text{Torr} = 100\ \text{kPa}$$

(i) *Absolute and Gauge Pressure*

The pressure relative to absolute zero pressure is called *absolute* pressure. This is the pressure of the system. In thermodynamic calculations, the value of absolute pressure is considered.

The pressure relative to the atmosphere is called *gauge pressure*. This is the pressure indicated by the instrument.

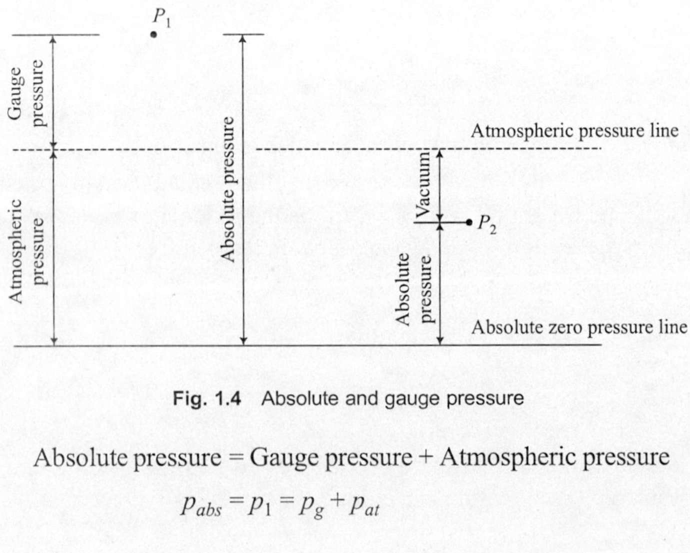

Fig. 1.4 Absolute and gauge pressure

Absolute pressure = Gauge pressure + Atmospheric pressure

$$p_{abs} = p_1 = p_g + p_{at}$$

If the pressure of the system is less than atmosphere, it can be designated as negative gauge pressure or positive pressure called vacuum.

$$p_2 = p_{at} - p_{vac}$$

(ii) *Measurement of Pressure*

There are a number of pressure measuring devices.

1. *Bourdon gauge:* It consists of a metallic bent hollow tube which tries to unbend when there is a change between the inside (system) pressure and outside (atmospheric) pressure. This moves a pointer through a suitable gear-and-lever mechanism against a calibrated scale.

2. *U-tube manometer:* It is the simplest pressure measuring device consisting of a glass *U*-tube filled with a suitable liquid (mercury, coloured water) whose density is known. The height of the balanced column of the liquid is measured and converted into desired pressure units. Manometers are used to measure gauge, atmospheric and absolute pressure.

$$p_{abs} = p_g + p_{at}$$

$$p_g = \rho g h \text{ [N/m}^2\text{]}$$

where, ρ = Density of manometer fluid

$= 13616 \text{ kg/m}^3$ for mercury

$= 1000 \text{ kg/m}^3$ for water

g = Acceleration due to gravity

$= 9.81 \text{ m/s}^2$

h = Height of liquid column in manometer (m)

$$p_{at} = \rho g h_b \text{ [N/m}^2\text{]}$$

where, h_b = Height of liquid column in barometer (m)

Example 1.1 A vessel of cylindrical shape of 50 cm diameter and 75 cm height, contains 4 kg of gas. The pressure gauge mounted on the vessel indicates 620 mm of Hg above atmosphere. If the barometer reading is 760 mm of Hg, calculate the absolute pressure of gas in bar. Also determine the density and specific volume of the gas.

Solution

Vessel diameter, $d = 50 \text{ cm} = \dfrac{50}{100} \text{ m}$

Vessel height, $h = 75 \text{ cm} = \dfrac{75}{100} \text{ m}$

The vessel volume, $V = \dfrac{\pi}{4} d^2 h$

$$V = \frac{\pi}{4}\left(\frac{50}{100}\right)^2 \times \frac{75}{100} = 0.147 \text{ m}^3$$

The mass of gas, $m = 4$ kg

Density of gas, $\rho = \frac{m}{V} = \frac{4}{0.147} = \mathbf{27.21\ kg/m^3}$ **Ans.**

Specific volume, $v = \frac{V}{m} = \frac{0.147}{4} = \mathbf{0.03675\ m^3/kg}$ **Ans.**

The absolute pressure, $p_{ab} = p_{at} + p_g$

$$= \rho g\, h_b + \rho g\, h_g$$

$$= 13.6 \times 10^3 \times 9.81\left(\frac{760}{1000} + \frac{620}{1000}\right) \times 10^{-5}$$

$= 1.84$ bar **Ans.**

4. ***Temperature:*** Temperature may be defined as degree of heat or coldness of a body or environment. If two bodies are brought in contact, heat will flow from a hot body at a high temperature to a cold body at a lower temperature. Temperature is the thermal potential causing the flow of heat energy. It is an intensive thermodynamic property independent of extent or size of the system.

The temperature of a body is measured with the help of an instrument called thermometer.

Thermometers work on different principles using different working substances. These are graduated with different temperature scales.

Two systems have equal temperature if there are no changes in their properties when they are brought in thermal contact with each other.

(i) *Celsius scale* is used in SI system. Triple point of water is used as fixed point. The important temperatures on Celsius scale are:

Triple point of water = 0.01°C

Steam point = 100.00°C

Ice point = 0.00°C

(ii) *Kelvin absolute scale* has fixed point at triple point of water equal to 273.16

$$T, K = t°\text{C} + 273.16$$

(iii) *Thermodynamic temperature scale* is based on the efficiency of an ideal engine operating on the Carnot cycle between any two temperatures.

$$\eta = \frac{T_1 - T_2}{T_1} = \frac{Q_1 - Q_2}{Q_1}$$

where Q_1 is the heat absorbed at temperature T_1 and Q_2 is the heat rejected at temperature T_2.

$$\therefore \qquad \frac{T_1}{T_2} = \frac{Q_1}{Q_2}$$

The ratio of any two temperatures on the thermodynamic scale (Kelvin scale) is equal to the ratio of the heat absorbed to the heat rejected by a Carnot engine operating between these temperatures.

(iv) *Electrical resistance thermometer:* The change of resistance of platinum wire with temperature is calibrated as temperature scale.

$$R = R_i(1 + At + Bt^2)$$

where, R_i = Resistance at ice point

A and B = Constants

(v) *Thermocouple:* An e.m.f. is generated in a circuit due to seeback effect which can be calibrated into a temperature scale.

$$E = a + bt + ct^2 + dt^3$$

where, E = e.m.f. generated

$a, b, c\ldots$ = constants

5. ***Equation of State:*** The functional relationship among the properties, pressure, molar or specific volume and temperature is known as equation of state or characteristic equation of a gas. This can be developed by combining Boyle's Law and Charle's law for a perfect gas.

$$pv = RT$$

$$pV = mRT$$

$$p\bar{v} = MRT$$

$$pV = n(MR)T$$

These equations are called characteristic gas equations.

where, p = pressure of gas [N/m^2]

V = volume of gas [m^3]

m = mass of gas [kg]

v = specific volume

$$= \frac{V}{m} \text{ [m}^3\text{/kg]}$$

$\bar{v}$ = molar volume [m^3/mol]

T = absolute temperature of gas [K]

R = characteristic gas constant [N-m/kgK]

M = molecular mass of gas

n = number of moles of the gas

6. ***Universal Gas Constant:*** The universal gas constant or molar constant ($\overline{R}$) of a gas is the product of the characteristic gas constant (R) and the molecular mass (M) of the gas.

$$\overline{R} = MR$$

The value of $\overline{R}$ is same for all gases and can be derived from Avogadro's law.

where,
$$p = 760 \text{ mm Hg } (1.013 \times 10^5 \text{ N/m}^2)$$
$$T = 273.15 \text{ K}$$
$$\bar{v} = 22.4 \text{ m}^3\text{/kmol}$$

$$\overline{R} = \frac{1.013 \times 10^5 \times 22.4}{273.15} = 8.3143 \text{ kJ/kg mol K}$$

Example 1.2 A tank of 0.35 m^3 capacity contains H_2S gas at 300 K. When 2.5 kg of gas is withdrawn, the temperature in the tank becomes 288 K and pressure 10.5 bar. Calculate the mass of the gas initially present in the tank. Also determine the initial pressure of the gas.

Solution Volume,
$$V_1 = V_2 = 0.35 \text{ m}^3 \text{ (constant)}$$
$$T_1 = 300 \text{ K}$$
$$T_2 = 288 \text{ K}$$
$$p_2 = 10.5 \text{ bar} = 10.5 \times 10^5 \text{ N/m}^2$$

The molecular mass of H_2S gas

$$M = 2 + 32 = 34$$

$$\therefore \quad R = \frac{\overline{R}}{M} = \frac{8314}{34} = 244.53 \text{ J/kg K}$$

Applying equation of state to the final condition,

$$p_2V_2 = m_2RT_2$$

$$\therefore \quad m_2 = \frac{p_2V_2}{RT_2} \frac{10.5 \times 10^5 \times 0.35}{244.53 \times 288} = \mathbf{5.2 \text{ kg}}$$

$$\therefore \quad m_1 = 5.2 + 2.5 = 7.7 \text{ kg} \quad \textbf{Ans.}$$

Applying equation of state to the initial condition

$$p_1V_1 = m_1RT_1$$

$$p_1 = \frac{m_1 RT_1}{V_1} = \frac{7.7 \times 244.53 \times 300}{0.35}$$

$$= \textbf{16.14 bar} \quad \textbf{Ans.}$$

Example 1.3 Determine the size of a spherical balloon filled with hydrogen at 30°C and atmospheric pressure for lifting 400 kg payload. Atmospheric air is at temperature of 27°C and barometer reading is 75 cm of mercury.

Solution Assume size of balloon = $V(\text{m}^3)$

Atmospheric pressure $= p_1 = \rho g h$

$$= 13.6 \times 10^3 \times 9.81 \times \frac{75}{1000} \times 10^{-5}$$

$$= 1 \text{ bar}$$

Gas constant for hydrogen

$$R = \frac{\overline{R}}{M} = \frac{8314}{2} = 4157 \text{ J/kg k}$$

Temperature of atmosphere,

$$T = 27°\text{C} + 273 - 300 \text{ K}$$

Applying equation of state to hydrogen gas in balloon

$$p_1 V_1 = m_1 R T_1$$

$$m_1 = \frac{p_1 V}{RT_1} = \frac{1 \times 10^5 \times V}{4158 \times (273 + 30)}$$

$$= 0.079392 \text{ V (kg)}$$

Mass of atmosphere air displaced by balloon.

$$m_2 = \frac{p_1 V}{RT_1} = \frac{1 \times 10^5 \times V}{287 \times 300} = 1.16144 \text{ V kg}$$

where R for air = 287 J/kg K

$$\text{Payload} = m_2 - m_1$$

$$= (1.16144 - 0.097392) \text{ V}$$

$$= 400$$

$\therefore$

$$V = 369.669 \text{ m}^3$$

Size of spherical balloon

$$\frac{4}{3}\pi R^3 = 369.669$$

$$R = 4.55 \text{ m} \quad \textbf{Ans.}$$

1.5.4 Derived Thermodynamic Properties

These properties are calculated from experimental data by the use of generalized thermodynamic relations.

Important examples are:

1. Internal energy
2. Enthalpy
3. Entropy
4. Specific heat
5. Joule-Thompson Coefficient
6. Helmoltz function
7. Gibb's function

These properties will be discussed in subsequent sections.

1.6 THERMODYNAMIC PROCESSES

The condition or *state* of a system is identified or described by its properties such as pressure, temperature, volume. The number of properties required to describe a system depends upon the nature of the system. Each state can be represented by a point on a graph with any two properties as coordinates.

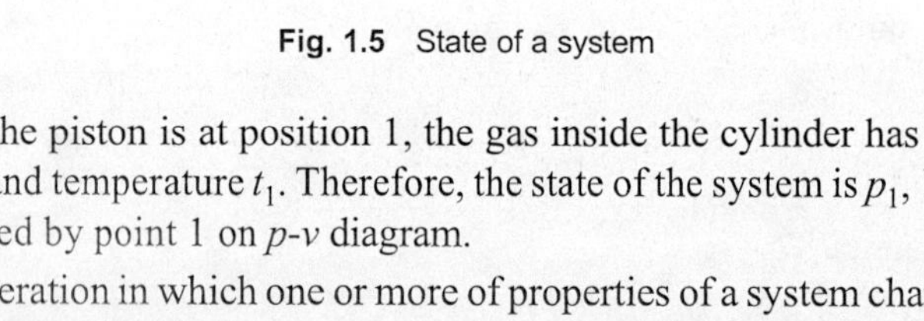

Fig. 1.5 State of a system

When the piston is at position 1, the gas inside the cylinder has pressure p_1, volume V_1 and temperature t_1. Therefore, the state of the system is p_1, V_1 and t_1 and is represented by point 1 on *p-v* diagram.

Any operation in which one or more of properties of a system change is called *change of state.* The piston moves outwards to position 2. The volume increases from V_1 to V_2, pressure decreases from p_1 to p_2 and temperature decreases from t_1 to t_2. The new state is represented by point 2 on the *p-v* diagram. The line joining states 1 and 2 is called path of change of state. When the path of change of state is completely specified, it is called a *process*. The process 1-2 is a constant temperature process if temperature along the path 1-2 remains constant.

The processes during which any one property remains constant are specified by using prefix *iso* before the property.

Isothermal process: temperature remains constant.

Isobaric process: pressure remains constant

Isochoric or isometric process: volume remains constant

When a process or series of processes are performed on a system in such a way that the final state is identical with the initial state, a cyclic process or cycle is said to have been executed. The properties of the system at the end of the *cyclic process* have the same values as those were in the beginning of the cycle.

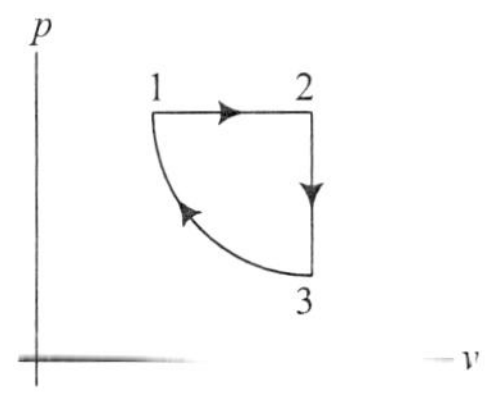

Fig. 1.6 Cyclic process

In Fig. 1.6, process 1-2 is isobaric process, process 2-3 is isochoric process and process 3-1 is isothermal process (say), then 1-2-3-1 is a thermodynamic cycle.

1.6.1 Thermodynamic Equilibrium

The properties of the system are measured and processes are studied when the systems are in thermodynamic equilibrium. A system having same values of properties (pressure, temperature, velocity, elevation, composition, etc.) throughout is in thermodynamic equilibrium. The system is simultaneously in the state of mechanical, thermal, electrical and chemical equilibrium.

1. *Thermal equilibrium* denotes uniformity of temperature or absence of temperature gradient or heat flow.
2. *Mechanical equilibrium* denotes uniformity of pressure or absence of unbalanced forces.
3. *Electrical equilibrium* denotes equality of electrical potential and absence of current flow.
4. *Chemical equilibrium* denotes absence of phase change or chemical reaction and there is no mass diffusion.

1.6.2 Point and Path Function

A property has a unique value at any state. The change in property between two equilibrium states is independent of the history of process. Mathematically speaking, a property is an exact differential.

$$\int_1^2 dp = p_2 - p_1$$

Since the change in the value of properties depends only on the initial and final states and is independent of the path of the process, they are called *state functions*. The change in the value of any property of the system for a cyclic process is zero.

$$\oint dp = 0$$

When two properties locate a point on the graph (coordinate axes) then these properties are called *point functions*. Pressure, temperature, volume, etc. are properties of the system and are point functions and also exact differentials.

$$\int_1^2 dV = V_2 - V_1$$

There are certain quantities which cannot be located on a graph by a point but are given by the area on the graph. The area for the graph for a process is a function of the path of the process. Such quantities are called *path functions.* Heat and work are path functions and also inexact differentials.

$$\int_1^2 \delta Q \neq Q_2 - Q_1 \text{ and is shown as } {}_1Q_2 \text{ or } Q_{1-2}$$

$$\int_1^2 \delta W \neq W_2 - W_1 \text{ and is shown as } {}_1W_2 \text{ or } W_{1-2}$$

The operator d is used to denote exact differential and δ as inexact differential.

1.6.3 Reversible and Irreversible Processes

(a) Reversible process

Reversible process is another concept for the ideal process. A process is said to be reversible if at any point during the course of the process, it can be reversed and both the system and surroundings can be restored to their original states. If a system passes through a continuous series of equilibrium states during a process, these states can be located on a property diagram.

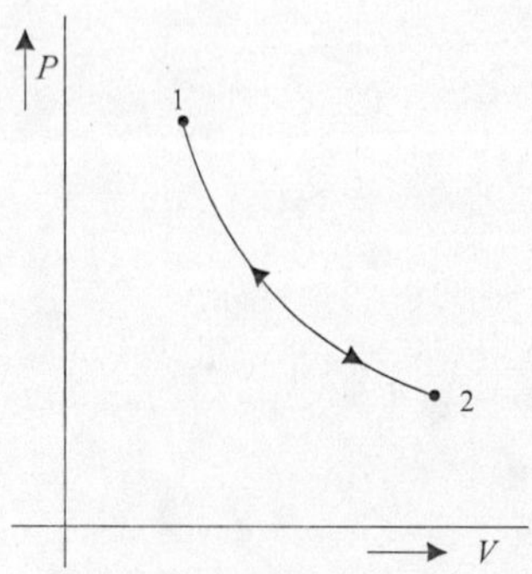

Fig. 1.7 Reversible process

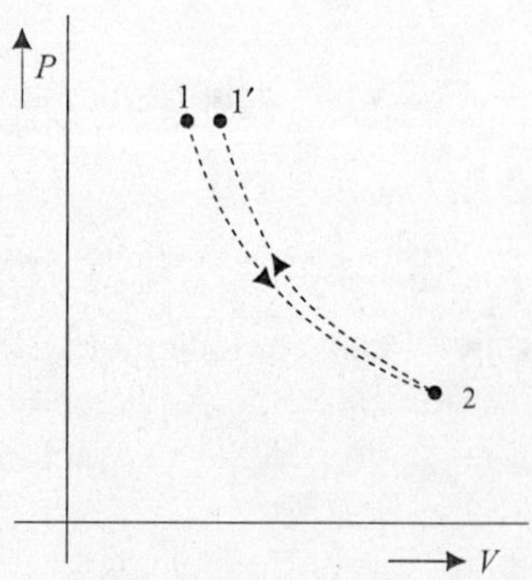

Fig. 1.8 Irreversible process

Line 1-2 has been drawn by joining points of equilibrium states of a process and represent a reversible process. If the process is repeated in the reverse direction, the initial state 1 is restored and the same line 2-1 is traced.

(b) Irreversible process

If a system passes through a sequence of non-equilibrium states during a process, these states cannot be located on any property diagram as each property does not have a unique value in the entire system. Such a process is called irreversible

Table 1.3 Comparison of reversible and irreversible process

Reversible Process	Irreversible Process
1. The system passes through a continuous series of equilibrium states.	1. The system passes through a sequence of non-equilibrium states.
2. Reversible process can be plotted by a line on any property diagram as the system has unique values of properties.	2. Irreversible process cannot be plotted on a property diagram as the system does not have unique values of properties.
3. The process can be reversed back to its initial state.	3. The process cannot be reversed back to its initial state.
4. It can be carried out in both the directions without violating second law of thermodynamics.	4. It can be carried out in one direction only.
5. It leaves no after effects or changes in the system or surrounding.	5. There are permanent changes in the surroundings.
6. There is no wastage of energy.	6. There is wastage of energy due to irreversibility.
7. It can be carried out at infinitesimal slow rate.	7. It occurs at finite rate.
8. It yields the maximum work in engines and all devices work with maximum efficiency.	8. There are losses in the process and operation does not take place at maximum efficiency.
9. It is an idealized process and cannot be achieved in practice.	9. All processes in the practice are irreversible processes.
10. There is no degradation of energy.	10. There is degradation of energy.
11. There is no change in entropy of a reversible cyclic process.	11. There is always increase in entropy of irreversible cyclic process.
12. The process takes place with the absence of fluid friction, mechanical friction and infinitely small difference of temperature and pressure between system and surrounding.	12. The process takes place with viscous flow, fluid friction, mechanical friction and energy transfer as heat with finite temperature difference.
13. Examples: Motion without friction, frictionless adiabatic process, controlled expansion and comression, Carnot cycle.	13. Examples: Free expansion and throttling process, fluid flow with friction, mixing of dissimilar gases, Rankine cycle.

process and is represented by a broken line on a property diagram joining initial and final state points which are in equilibrium. If irreversible process 1-2 is repeated in the reverse direction, the initial state 1 is not restored.

The real processes in engineering are irreversible processes due to the presence of the fluid friction, mechanical friction, heat transfer due to finite temperature difference, etc. These are the main causes of irreversibility.

The main features and examples of reversible and irreversible processes as can be differentiated by the comparison table (Table 1.3).

(c) Causes of irreversibility

(i) *Mechanical irreversibility:* It is associated with fluid friction between the molecules due to turbulence at high speeds as well as friction due to viscocity. Frictional resistance heats the fluid and there is loss of work. Heat cannot be completely recovered as work.

The mechanical irreversibility is also due to friction in bearings, piston and cylinder walls and other contact surfaces with relative motion. There is also windage losses due to friction between rotating parts and atmospheric air.

(ii) *Thermal irreversibility:* It is due to energy transfer as heat with finite temperature difference between system and surrounding. The heat transfer takes place due to temperature gradients. The heat transfer from low temperature to higher temperature is not possible without the aid of external work.

(iii) *Free expansion:* No work is done in free expansion and the process cannot be reversed, i.e., restoring initial parameters without doing work of compression.

(iv) *Mixing of fluids:* No work is done during mixing of fluids. The fluids cannot be separated into original condition without the help of work.

1.6.4 Non-flow Processes

The processes occurring in a closed system where there is no transfer of mass across the boundary are called non-flow processes. In such processes, the energy in the form of heat and work crosses the boundary of the system. The heating (or cooling) and expansion (or compression) of a gas may be performed in the following ways.

(a) Reversible non-flow processes

(i) Constant volume process (or isochoric process)

The equation of state,

$$pV = mRT$$

$$\therefore \quad \frac{pV}{T} = mR = \text{constant}$$

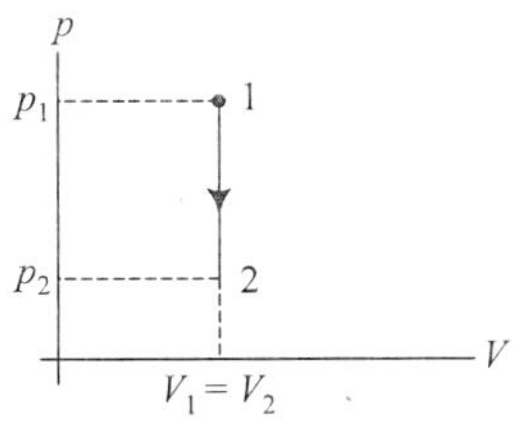

Fig. 1.9 Constant volume process

$$\frac{p_1 V_1}{T_1} = \frac{p_2 V_2}{T_2}$$

Now, $V_1 = V_2 = V = \text{constant}$

$\therefore$ $\dfrac{p_1}{T_1} = \dfrac{p_2}{T_2}$

or $\dfrac{p_1}{p_2} = \dfrac{T_1}{T_2}$

(ii) *Constant pressure process (or isobaric process)*

$$p_1 = p_2 = p = \text{constant}$$

$\therefore$ $\dfrac{V_1}{V_2} = \dfrac{T_1}{T_2}$

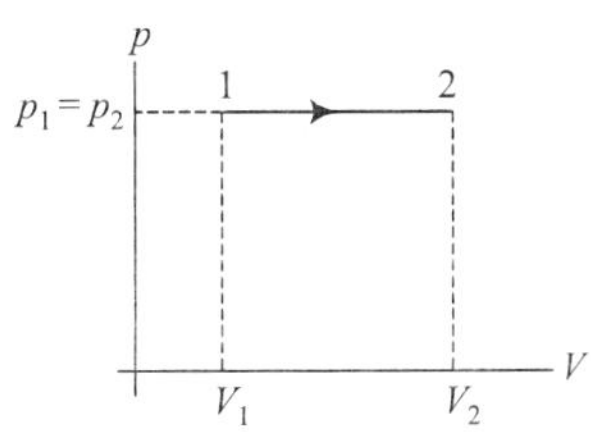

Fig. 1.10 Constant pressure process

(iii) *Constant temperature process (isothermal process)*

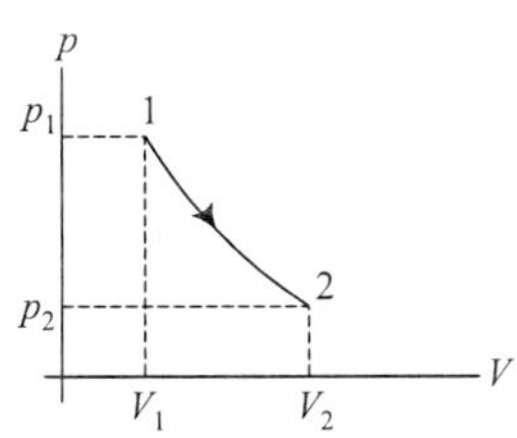

Fig. 1.11 Constant temperature process

$$T_1 = T_2 = T = \text{constant}$$

$$p_1V_1 = p_2V_2 = pV = \text{constant}$$

(iv) *Adiabatic process:* There is no heat transfer during the process between system and surrounding.

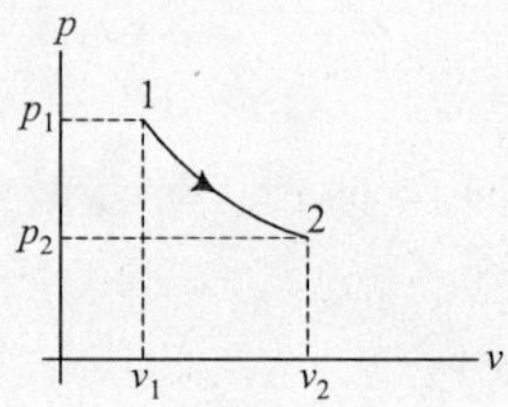

Fig. 1.12 Adiabatic process

$$pV = mRT$$

$$\therefore \quad pv = RT$$

Differentiating $\quad pdv + vdp = RdT$

$$\therefore \quad dT = \frac{pdv + vdp}{R}$$

$$= \frac{pdv + vdp}{(Cp - Cv)} \qquad [\because \quad R = Cp - Cv] \qquad \text{...(a)}$$

$$\delta q - \delta w = du \text{ (from 1st law)}$$

$$0 - \delta w = du$$

$$0 - pdv = CvdT$$

$$\therefore \quad dT = \frac{-pdv}{Cv} \qquad \text{...(b)}$$

Equating (a) and (b) for dT

$$\frac{-pdv}{Cv} = \frac{pdv + vdp}{(Cp - Cv)}$$

$$\therefore \quad \frac{Cp - Cv}{Cv} = \frac{pdv + vdp}{-pdv}$$

$$\frac{Cp}{Cv} - 1 = -1 - \frac{vdp}{pdv} = -1 - \left[\frac{v}{dv} \times \frac{dp}{p}\right]$$

$$\therefore \quad \gamma = -\left[\frac{v}{dv} \times \frac{dp}{p}\right] \qquad \left[\because \quad \frac{Cp}{Cv} = \gamma\right]$$

$$\therefore \qquad \gamma\left[\frac{dv}{v}\right] = -\frac{dp}{p}$$

or $$\gamma\frac{dv}{v} + \frac{dp}{p} = 0$$

Integrating,

$$\gamma \ln v + \ln p = \text{constant}$$

$$\ln pv^{\gamma} = \ln(\text{constant})$$

$$pv^{\gamma} = \text{constant}$$

$$p_1 v_1^{\gamma} = p_2 v_2^{\gamma} = pv^{\gamma} = \text{constant}$$

$$\frac{p_1}{p_2} = \left(\frac{v_2}{v_1}\right)^{\gamma}$$

or $$\frac{v_1}{v_2} = \left(\frac{p_2}{p_1}\right)^{\frac{1}{\gamma}}$$

But, $$\frac{p_1 v_1}{T_1} = \frac{p_2 v_2}{T_2}$$

or $$\frac{v_1}{v_2} = \frac{T_1}{T_2} \times \frac{p_2}{p_1}$$

$$\therefore \qquad \left(\frac{p_2}{p_1}\right)^{\frac{1}{\gamma}} = \frac{T_1}{T_2} \times \frac{p_2}{p_1}$$

or $$\frac{T_1}{T_2} = \left(\frac{p_2}{p_1}\right)^{\frac{1}{\gamma}} \times \frac{p_1}{p_2} = \left(\frac{p_1}{p_2}\right)^{-\frac{1}{\gamma}+1}$$

$$= \left(\frac{p_1}{p_2}\right)^{\frac{\gamma-1}{\gamma}}$$

(v) *Polytropic Process:* The general equation for a polytropic process

$$pV^n = C$$

when $n = 0, p = C$ (Isobaric process)

when $n = \infty, pV^{\infty} = C$

$$p^{\frac{1}{\infty}} V = C^{\frac{1}{\infty}}, V = C \text{ (Isochoric process)}$$

when $n = 1, pV = C$ (Isothermal process)

when $n = \gamma, pV^{\gamma} = C$ (Adiabatic process)

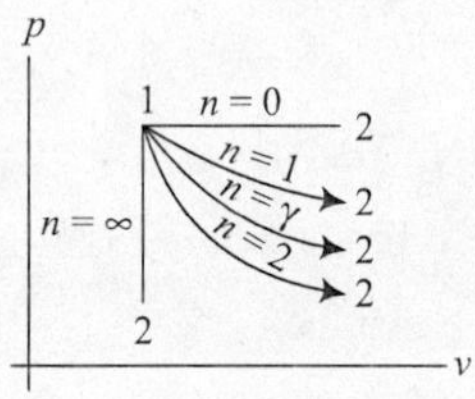

Fig. 1.13 Polytropic process

(b) *Irreversible non-flow process:* Free expansion process or unrestricted process is an irreversible non-flow process. The process will be discussed in a subsequent section.

1.6.5 Flow Processes

The processes in open systems permit the transfer of mass to and from the system. Such processes are called flow processes. The mass enters the system and leaves after exchange of energy.

The flow processes may be:-

(a) Steady flow processes

The mass flow rate, heat flow rate, work flow rate through the system remain constant and there is no change in any property or chemical composition of the working fluid at any given point within the system. Nozzles, turbines, compressors, pumps and other thermal machines operate as steady flow process.

The following conditions must hold good in a steady flow process:

1. The mass flow rate through the system remains constant.

$$\dot{m}_1 = \dot{m}_2 = \dot{m} = \text{constant}$$

2. The rate of heat transfer is constant.

$$\dot{Q}_{1-2} = \text{constant}$$

3. The rate of work transfer remains constant.

$$\dot{W}_{1-2} = \text{constant}$$

4. The state of working substance at any point in the system is same at all times.
5. There is no change in the chemical composition of the system.

If any one condition is not satisfied, the process is called unsteady flow process. The reversible steady flow processes include isochoric process, isobaric process, isothermal process, adiabatic process, polytropic process. Important example of irreversible flow process is the throttling process which is the expansion process of a gas through an obstruction in the form of a partly opened valve or orifice. This process is discussed in subsequent section.

(b) Unsteady flow process

The filling and evacuation of tanks and vessels are unsteady flow processes.

1.7 FORMS OF ENERGY

Man arrived on the earth and started exploiting different sources of energy for raising his standard of living. Energy has been an important component to meet the day-to-day need of human beings. Energy is a fundamental concept such as God, and is very difficult to define. However, energy has been defined as the capacity to produce an effect or capacity to do work. Man needs energy as motive power for industry and transport sector, thermal energy for heating and lighting.

1.7.1 Energy Characteristics

Energy has certain characteristics:

1. Energy is available in different forms like electrical energy, mechanical energy, chemical energy, heat energy, nuclear energy, etc.
2. Energy can neither be created nor destroyed. Energy can be transformed from one form to another form.
3. Energy can be transported from one system to another system or from one place to another. Energy can also be stored.
4. An energy is associated with a potential. For example, thermal potential (temperature difference) for heat energy or pressure difference for mechanical energy or electrical potential for electrical energy. Free flow of energy takes place only from a higher potential to a lower potential.
5. The energy is measured as N-m or in joules. The energy density is expressed as J/kg. Different sources and forms of energy have different energy densities.
6. Energy does not have absolute value. It is measured relative to a datum. The common data are ground level, atmospheric pressure and a temperature at OK. Heat energy is assumed to have zero value at OK. Similarly, potential energy is zero at ground level.
7. The transfer of heat energy takes place due to temperature difference. All other forms of energy exchange may take place due to difference in other intensive properties. These later forms of energy can also be called work.
8. The forms of energy are graded as per their availability or exergy content. A high grade energy like electrical energy can be completely converted into

mechanical work without significant loss. On the other hand, whole of low grade energy like heat cannot be converted into mechanical work. Some part of heat is further degraded (into a low temperature heat) and rejected to environment.

1.7.2 Classification of Energy Forms

There are many forms of energy, however, thermodynamics deals with two forms of energy, i.e., thermal energy and mechanical energy. Each form has two types:

Stored energy: The energy possessed by a system within its boundaries is called stored energy.

Transient energy: The energy which is not possessed by the system but can cross its boundaries is called transient energy.

The stored energy is a thermodynamic property whereas transient energy is not a thermodynamic property as it depends upon the path.

1. *Mechanical Energy:* It includes kinetic energy and potential energy as stored energy and shaft energy, flow energy as work or transient energy.
2. *Thermal Energy:* It includes internal energy as stored energy and sensible heat as transient energy.

1.8 MECHANICAL ENERGY

1.8.1 Kinetic Energy

The energy possessed by a body or a system by virtue of its mass and velocity of motion is called kinetic energy.

$$\text{K.E.} = \frac{mV^2}{2} \text{ [N-m]}$$

where, m = mass of the body [kg]

V = velocity of the body [m/s]

The work done, $$W_{1-2} = \frac{1}{2} m (V_2^2 - V_1^2)$$

The work done by a body is the change of its kinetic energy. Kinetic energy can be computed by knowing the mass of the body and velocity relative to a specific coordinate frame at any instant without the consideration how this velocity was attained. Therefore, kinetic energy is a state function and hence it is the property of the system. It depends upon the mass of the system, hence, it is an extensive property.

The energy is the capacity to do work and power is the rate of doing work. The total power of a fluid stream (air or water) is equal to the rate of the incoming energy of that stream.

Total power of a stream

$$P = \dot{m}\, ke = \dot{m}\, \frac{V^2}{2} \; [W]$$

where, $\dot{m}$ = mass flow rate [kg/s]

V = stream velocity [m/s]

ke = kinetic energy of a unit mass [J/kg]

The mass flow rate of stream is given by the continuity equation

$$\dot{m} = \rho A V$$

where, ρ = density of incoming stream [kg/m^3]

A = cross-sectional area of stream [m^2]

$$\therefore \quad P = \frac{\rho A V^3}{2}$$

The kinetic energy of air is utilized in a windmill and that of water in a hydro-turbine.

1.8.2 Potential Energy

The energy possessed by a body or system by virtue of its position or elevation above the datum (ground) level is called potential energy. The work is done due to its falling on earth's surface.

Potential energy,

$$PE = WZ = mgZ \; [\text{Nm}]$$

where, W = weight of body [N]

m = mass of body [kg]

Z = distance of fall of body [m]

g = acceleration due to gravity = 9.81 m/s^2

The total work done by the body in falling from initial position Z_1 to final position Z_2,

$$W_{1-2} = \int_{Z_1}^{Z_2} mgdZ = mg(Z_2 - Z_1)$$

Therefore, work done by the body represents the change in its potential energy. The change in potential energy in a gravitational field can be determined from the values of Z_1 and Z_2 and does not depend how the change has occurred. Therefore, potential energy is a property of the system as it does not depend upon the path of the fall. The magnitude of potential energy depends upon the mass of the system and hence it is an extensive property.

The total power of a stream

$$P = \dot{m}\,pe = \dot{m}\,gZ\,[W]$$

where,

$\dot{m}$ = mass flow rate [kg/s]

Z = distance of fall of stream [m]

pe = potential energy of unit mass [J/kg]

$\dot{m} = \rho\dot{Q}$

ρ = density of incoming stream [kg/m^3]

$\dot{Q}$ = volumetric flow rate [m^3/s]

$$\therefore \quad P = \rho\dot{Q}gZ\,[W]$$

The potential energy of water stored at a height is utilized in a hydroelectric power plant.

1.8.3 Flow Energy

Flow energy or pressure energy is possessed by a flowing fluid by virtue of its pressure. It can be expressed as the work done by the flowing fluid to push the quantity represented by mass m into and out of a system.

If a working fluid with pressure p_1(N/m^2) flows through an area $A_1(m^2)$ and moves through a distance $x(m)$ the energy or work required to move the fluid.

$$FE = \text{Force} \times \text{distance}$$

$$= (p_1A_1) \times x_1$$

$$= p_1V_1\,(J) \qquad (\because \quad A_1x_1 = V_1)$$

Flow work per unit mass,

$$fe = pv\,(\text{J/kg})$$

The rate of work done on the fluid is the product of normal force and fluid velocity.

$$\dot{W}_i = (pA)_i \times V_i\,[W],$$

where,

p_i = Fluid pressure at inlet (N/m^2)

A_i = Inlet area normal to flow (m^2)

V_i = Normal velocity of fluid at inlet (m/s)

Similarly, time rate of work done at exit of control volume,

$$\dot{W}_e = (pA)_i \times V_e\,[W]$$

Net flow work rate,

$$\dot{W}_{\text{net}} = \dot{W}_e - \dot{W}_i$$

$$\dot{W}_{net} = (pA)_e \times V_e - (pA)_e \times V_i$$

From equation of continuity,

$$AV = \dot{m}v$$

$\therefore$ $$\dot{W}_{net} = (\dot{m}pv)_e - (\dot{m}pv)_i$$

The flow work is supplied to the system when fluid enters and it is supplied to the surrounding when it leaves the system.

1.8.4 Shaft Energy

A rotating shaft has rotational energy. This can be expressed by the shaft work. Shaft energy,

$$W - \int_1^2 Td\theta$$

where, T = torque applied to the shaft (N-m)

$d\theta$ = angular displacement of shaft (rad.)

Shaft power, $$P = \int_1^2 T\frac{d\theta}{d\tau} = T\omega$$

where, ω = angular velocity of the shaft (rad/s)

$\therefore$ $P = 2\pi NT$ [W]

where, N = shaft speed (rps)

The rotating shaft possesses kinetic energy. Kinetic energy of rotation,

$$KE = \frac{1}{2}I\omega^2$$

where, I = mass moment of inertia

$= mk^2$

m = mass of rotating shaft (kg)

k – radius of gyration (m)

The work done by a rotating shaft is equal to change in kinetic energy.

$$W_{1-2} = \text{Initial KE} - \text{Final KE}$$

The rotational energy is supplied to a shaft by a prime mover and received by a driven machine.

1.9 WORK

Work is done when the point of application of a force moves in the direction of force.

$$W = \text{Force} \times \text{displacement}$$
$$= F \times x$$

If the force changes with displacement, then

$$W = \int_1^2 F(x)\, dx$$

The work is the scalar product of force and displacement. The thermodynamic definition of work is related with the concept of system and surrounding. The work may be defined as follows:

(i) *Obert:* "Work is defined as the energy transferred (without transfer of mass) across the boundary of a system because of an intensive property difference other than temperature that exists between the system and surrounding".

Pressure difference results in mechanical work and electrical potential difference results in electrical work.

(ii) *Keenan:* "Work is said to be done by a system during a given operation if the sole effect of the system on things external to the system (surroundings) can be reduced to the raising of a weight".

Example 1 If a piston is pushed by a gas pressure in a cylinder, the expansion of gas can be reduced to the raising of a weight W against gravity by manipulating surroundings external to system as shown in Fig. 1.14.

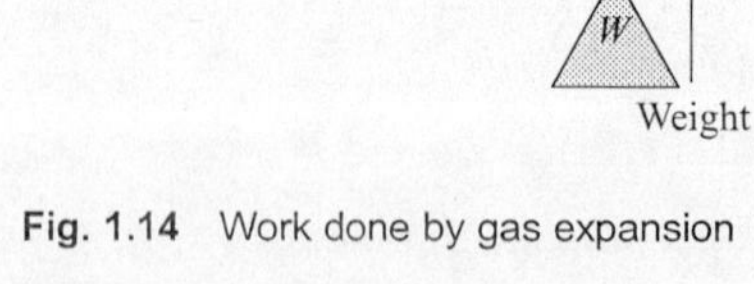

Fig. 1.14 Work done by gas expansion

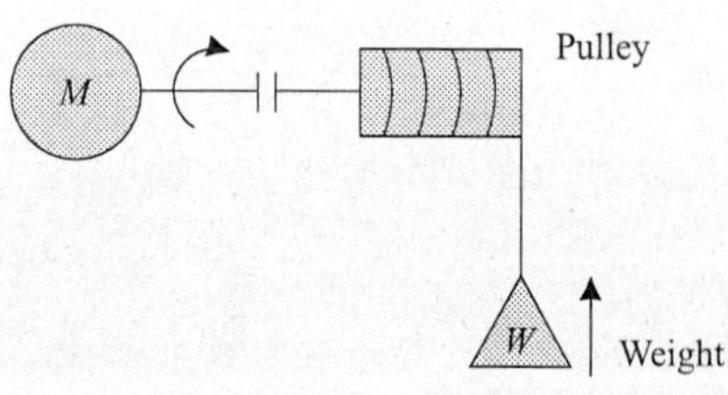

Fig. 1.15 Work done by electric motor

Example 2 When a motor rotates, it can be reduced to raising of a weight W through a rope and pulley arrangement as shown in Fig. 1.15.

The gas engine and motor have been manipulated to raise a weight. Hence, these devices are performing work as per the definition of Keenan.

The main *characteristics* of work are:

1. Work is a means of transferring energy and exists only so long as this transfer takes place.

2. Work is energy in transit.
3. *Unit of work:* In SI system, unit of work is joule (J)

$$1 \text{ J} = 1 \text{ N-m}$$

Power is the time rate of doing work and is designated as ρ. The unit of power is watt (W) or (kW)

$$1 \text{ W} = 1 \text{ J/s}$$

4. *Sign convection:* The work is positive when done by the system and negative when done on the system.

$$W_{\text{system}} + W_{\text{surroundings}} = 0$$

When a system does positive work, its surroundings do an equal amount of negative work.

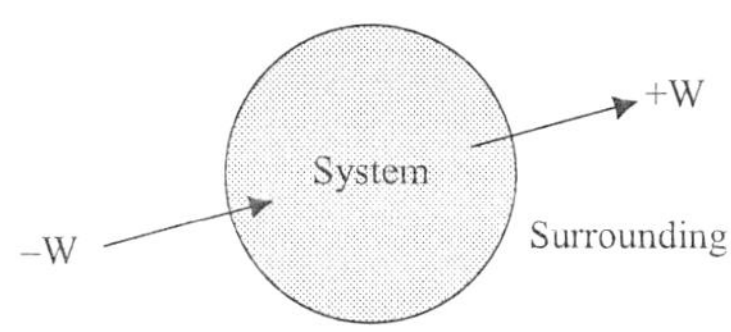

Fig. 1.16 Sign convection of work

5. *Expansion or compression work:* In a cylinder-piston arrangement, when the gas expands, it pushes the piston out. Positive work is done by the system during the expansion of gas:

When the piston is pushed inside the cylinder due to the pressure of the surrounding or any external force, the gas is compressed. Negative work is done by the system (as it receives work from the surrounding) in compression of gas.

6. *Work—a path function:* Work is evaluated in terms of macroscopic force and displacement

$$W_{1\text{-}2} = \int_1^2 F dX$$

where, F = Force, N

X = Displacement, m

W depends upon the details of interaction taking place between the system and surroundings and not just the initial and final state of the system. Therefore, work is a path function and, therefore, not a property of the system. The differential of work is not an exact differential.

$$\therefore \qquad \int_1^2 \delta W \neq W_2 - W_1$$

$$\int_1^2 \delta W = {}_1W_2 \text{ or } W_{1\text{-}2}$$

1.9.1 Piston Energy or Displacement Work

When a piston moves in a cylinder from position 1 to position 2 with volume changing from V_1 to V_2, the amount of work done by the system is given by

$$W_{1-2} = \int_{V_1}^{V_2} p dV$$

The value of work done is given by the area under the process 1-2 on p-V diagram (Fig. 1.17). The magnitude of work done will depend upon the thermodynamic process.

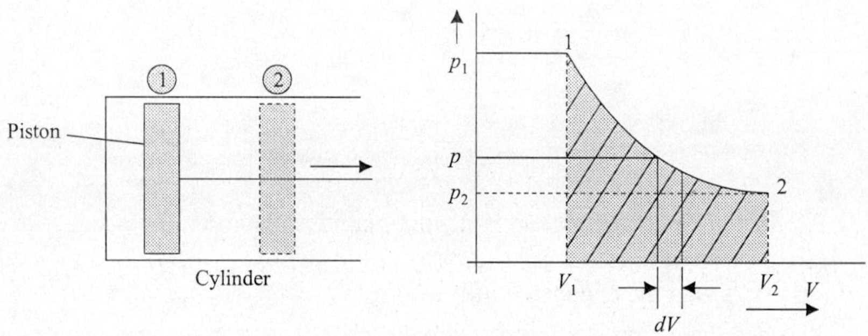

Fig. 1.17 Displacement work

1. *Isobaric process* (p = constant)

$$W_{1\text{-}2} = \int_{V_1}^{V_2} p dV$$

$$= p(V_2 - V_1)$$

2. *Isochoric process* (V = constant)

$$W_{1\text{-}2} = \int_{V_1}^{V_2} p dV$$

$$= p(V_2 - V_1) = 0$$

$$\because \quad V_1 = V_2$$

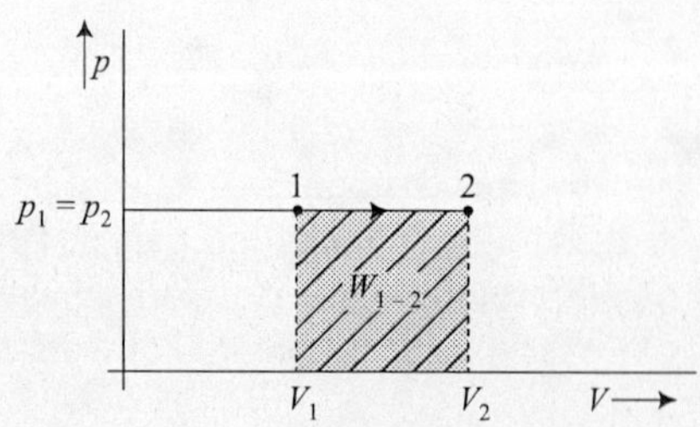

Fig. 1.18 Isobaric process

3. *Isothermal process* (pV = constant)

$$W_{1-2} = \int_{V_1}^{V_2} p\,dV$$

$$p_1V_1 = p_2V_2 = pV = C$$

$$\therefore \quad \frac{V_1}{V_2} = \frac{p_2}{p_1}$$

$$\therefore \quad p = \frac{p_1V_1}{V}$$

$$\therefore \quad W_{1-2} = \int_{V_1}^{V_2} \frac{p_1V_1}{V}\,dV$$

$$= p_1V_1 \int_{V_1}^{V_2} \frac{dV}{V}$$

$$= p_1V_1 \ln \frac{V_2}{V_1}$$

$$\text{or} \quad W_{1-2} = p_1V_1 \ln \frac{p_1}{p_2}$$

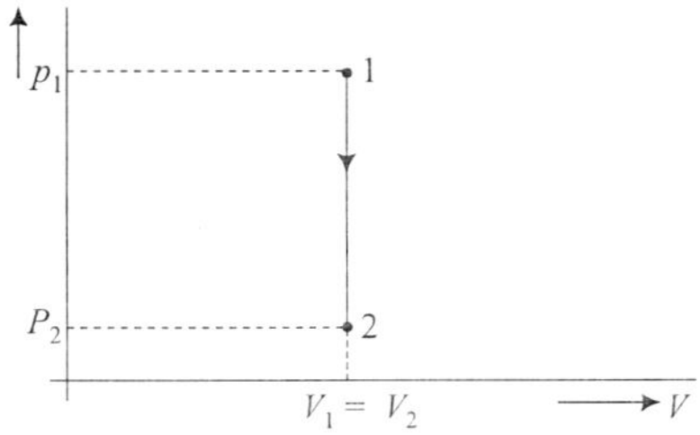

Fig. 1.19 Isochoric process

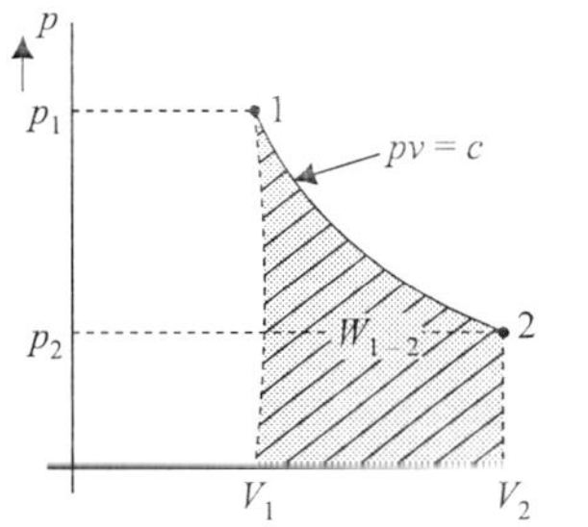

Fig. 1.20 Isothermal process

4. *Polytropic process* ($pV^n = C$)

$$p_1V_1^n = p_2V_2^n = pV^n = C$$

$$\therefore \quad p = \frac{p_1V_1^n}{V^n}$$

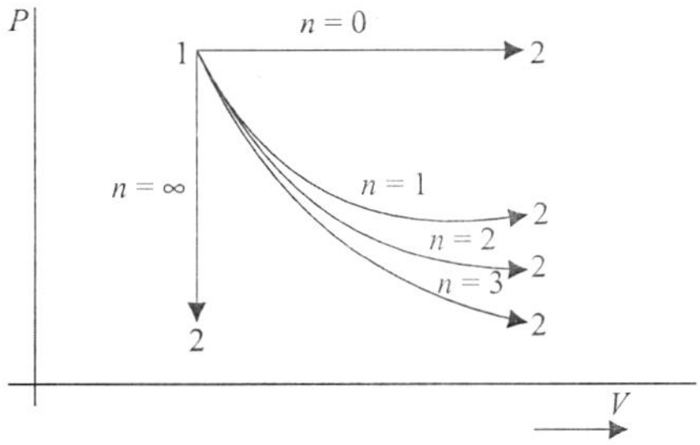

Fig. 1.21 Polytropic process

$$W_{1-2} = \int_{V_1}^{V_2} p\,dV$$

$$= \int_{V_1}^{V_2} \frac{p_1V_1^n}{V^n}\,dV = p_1V_1^n \left| \frac{V^{-n+1}}{-n+1} \right|_{V_1}^{V_2}$$

$$= \frac{p_1V_1^n}{1-n}\left(V_2^{1-n} - V_1^{1-n}\right)$$

$$\because \quad p_1V_1^n = p_2V_2^n$$

$$\therefore \quad W_{1\text{-}2} = \frac{p_2V_2^n \times V_2^{1-n} - p_1V_1^n \times V_1^{1-n}}{1-n}$$

$$= \frac{p_1V_1 - p_2V_2}{n-1} = \frac{p_1V_1}{n-1}\left[1-\left(\frac{p_2}{p_1}\right)^{\frac{n-1}{n}}\right]$$

5. *Adiabatic process:* In adiabatic process, $n = \gamma$, where $\gamma = \frac{C_p}{C_v}$.

$$\therefore \quad W_{1\text{-}2} = \frac{p_1V_1 - p_2V_2}{\gamma - 1}$$

Example 1.4 Calculate the work done in a piston cylinder arrangement during the expansion process, where the pressure is given by the equation:

$$p = (V^2 + 6V) \text{ bar}$$

The volume changes from 1 m^3 to 4 m^3 during expansion.

Solution

$$V_1 = 1 \text{ m}^3$$
$$V_2 = 4 \text{ m}^3$$

$$W_{1-2} = \int_1^2 pdV$$

$$= \int_{V_1}^{V_2} (V^2 + 6V) \times 10^5 \, dV$$

$$= 10^5 \left|\frac{V^3}{3} + \frac{6V^2}{2}\right|_1^4$$

$$= 10^5 \left[\frac{(4^3 - 1^3)}{3} + 3(4^2 - 1^2)\right]$$

$$= 66 \times 10^5 \text{ J} = \mathbf{6.6 \text{ MJ Ans.}}$$

Example 1.5 During a reversible process, air is compressed from initial pressure of 12 kN/m^2 to 6 times the initial pressure. The volume of air decreases from 4 m^3 to 1.8 m^3. Calculate:

(i) Law of the process

(ii) Work done in compressing the air

Solution Initial pressure of air, $p_1 = 12 \text{ kN/m}^2$

Final pressure of air, $p_2 = 72 \text{ kN/m}^2$

Initial volume of air, $V_1 = 4 \text{ m}^3$

Final volume of air, $V_1 = 1.8 \text{ m}^3$

$$p_1 V_1^n = p_2 V_2^n$$

$$\left(\frac{V_1}{V_2}\right)^n = \frac{p_2}{p_1}$$

$$n \ln \frac{V_1}{V_2} = \ln \frac{p_2}{p_1}$$

$$n \ln \frac{4}{1.8} = \ln 6$$

$\therefore$ $$n = 2.26$$

(i) The process is polytropic with $n = 2.26$. **Ans.**

(ii) Work done,

$$W_{1-2} = -\frac{p_1 V_1 - p_2 V_2}{n-1}$$

$$= \frac{p_2 V_2 - p_1 V_1}{n-1} = \frac{72 \times 1.8 - 12 \times 4}{2.26 - 1}$$

$$= \mathbf{64.76 \text{ kJ}} \quad \textbf{Ans.}$$

1.9.2 Mechanical Work in a Steady Flow Process

The mechanical work in a steady flow process is shown on p-V diagram (Fig. 1.22)

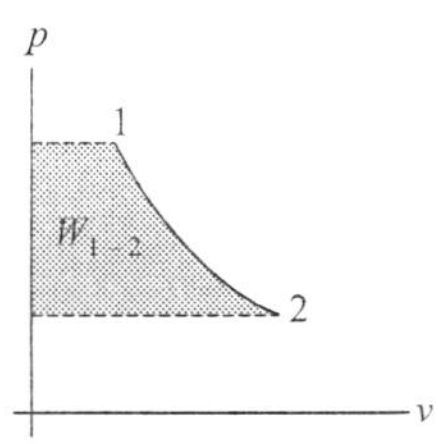

Fig. 1.22 Steady flow work

$$W_{1-2} = -\int_1^2 V dp$$

The polytropic process,

$$p_1 V_1^n = p_2 V_2^n = p V^n = \text{constant}$$

$$\left(\frac{V}{V_1}\right)^n = \frac{p_1}{p}$$

$$V = V_1\left(\frac{p_1}{p}\right)^{\frac{1}{n}}$$

$$W_{1-2} = -\int_{p_1}^{p_2} V dp$$

$$W_{1-2} = -\int_{p_1}^{p_2} V_1\left(\frac{p}{p_1}\right)^{n} dp = -V_1 p_1^{\frac{1}{n}}\int_1^2 p^{-\frac{1}{n}} dp$$

$$= -V_1\, p_1^{\frac{1}{n}}\left|\frac{p^{-\frac{1}{n}+1}}{-\frac{1}{n}+1}\right|_{p_1}^{p_2}$$

$$= -V_1\, p_1^{\frac{1}{n}}\left|\frac{p^{\frac{n-1}{n}}}{\frac{n-1}{n}}\right|_{p_1}^{p_2}$$

$$= \frac{n}{n-1}\left(-V_1 p_1^{\frac{1}{n}}\right)\left[p_2^{\frac{n-1}{n}} - p_1^{\frac{n-1}{n}}\right]$$

$$\therefore \qquad W_{1-2} = \frac{n}{n-1}(p_1V_1 - p_2V_2)$$

The mechanical work for different steady flow processes is compared with that in non-flow processes in Table 1.4.

Table 1.4 Mechanical work (Unit mass)

S.No.	Process	n	Non-flow work $\int p\,dv$	Steady flow work $-\int v\,dp$
1.	Isochoric ($v = c$)	∞	0	$v(p_1 - p_2)$
2.	Isobaric ($p = c$)	0	$p(v_2 - v_1)$	0
3.	Isothermal ($t = c$)	1	$p_1v_1 \ln\frac{v_2}{v_1}$	$p_1v_1 \ln\frac{v_2}{v_1}$
4.	Adiabatic ($pv^{\gamma} = c$)	γ	$\frac{p_1v_1 - p_2v_2}{\gamma - 1}$	$\frac{\gamma}{\gamma-1}(p_1v_1 - p_2v_2)$
5.	Polytropic ($pv^{n} = c$)	n	$\frac{p_1v_1 - p_2v_2}{n - 1}$	$\frac{n}{n-1}(p_1v_1 - p_2v_2)$
6.	Free expansion ($u = c$)	–	0	–
7.	Throttling ($h = c$)	–	–	0

Example 1.6 Three kg of air kept at 100 kPa and 300 K is compressed polytropically to 1500 kPa and 500 K in a steady flow process. Calculate:

(i) Index n

(ii) Final volume

(iii) Work done

Solution

$$p_1 = 100 \text{ kPa}$$
$$T_1 = 300 \text{ K}$$
$$p_2 = 1500 \text{ kPa}$$
$$T_2 = 500 \text{ K}$$

(i) $$p_1 V_1^n = p_2 V_2^n$$

and $$\frac{p_1 V_1}{T_1} = \frac{p_2 V_2}{T_2}$$

$\therefore$ $$\frac{T_2}{T_1} = \left(\frac{p_2}{p_1}\right)^{\frac{n-1}{n}}$$

$\therefore$ $$\frac{n-1}{n} = \frac{\ln \dfrac{T_2}{T_1}}{\ln \dfrac{p_2}{p_1}} = \frac{\ln \dfrac{500}{300}}{\ln \dfrac{1500}{100}} = 0.1886$$

$\therefore$ $n = 1.23$ **Ans.**

(ii) $$p_1 V_1 = mRT_1$$

$$V_1 = \frac{mRT_1}{p_1} = \frac{3 \times 0.287 \times 300}{100} = 2.583 \text{ m}^3$$

$$p_2 V_2 = mRT_2$$

$$V_2 = \frac{mRT_2}{p_2} = \frac{3 \times 0.287 \times 500}{1500} = \mathbf{0.287 \text{ m}^3}$$ **Ans.**

(iii) Work done,

$$W_{1-2} = \frac{n}{n-1}(p_1 V_1 - p_2 V_2) = \frac{n}{n-1} mR(T_1 - T_2)$$

$$= \frac{1.23}{0.23} \times 3 \times 0.287\,(300 - 500)$$

$$= \mathbf{-920.9 \text{ kJ}}$$ **Ans.**

Work is done on the air during compression.

1.10 THERMAL ENERGY

Thermal energy or heat is transferred across the boundary of a system due to temperature difference between the system and surrounding. The heat can be transferred by conduction, convection, radiation or combination.

1.10.1 Main Characteristics

The main characteristics of heat are:

1. Heat flows from a system at a higher temperature to a system at a lower temperature.
2. Heat is energy in transit and exists only during transfer into or out of a system. Once energy transfer is over there will be no heat. It is energy and not heat that is stored in the system.
3. *Sign convection.* Heat is positive when it flows into the system and negative when it flows out of the system. This sign convection is adopted as the main scope of thermodynamics is to generate work by the system when heat is supplied to it.
4. *Adiabatic process.* A process in which there is no heat transfer is called an adiabatic process.
5. Heat is a path function. The amount of heat transferred as a system undergoes a change of state depends on the path of the process. Its differential is not exact,

$$\int_1^2 \delta Q \neq Q_2 - Q_1$$

$$\int_1^2 \delta Q = {}_1Q_2 \text{ or } Q_{1-2}$$

6. The rate of heat transfer is denoted as:

$$\frac{\delta Q}{dt} = \dot{Q}$$

7. *Units.* In SI system, the unit of heat is joule (J) or kilojoule (kJ). The rate of heat transfer is kJ/s or kW.
8. Heat required to raise the temperature of a body or a system is:

$$Q_{1-2} = mC(T_2 - T_1) \quad \text{[kJ]}$$

where, m = mass, kg

C = specific heat, kJ/kg-K

T_1 and T_2 = Temperatures in °C or K.

Specific heat for gases can be specific heat at constant pressure (C_p) and constant volume (C_v).

The product mC = thermal or heat capacity, kJ

mC = water equivalent, kg

1.10.2 Sources of Thermal Energy

The various sources of thermal energy are:

1. Chemical energy of fuels

Fossil fuels in the form of solids, liquids, and gases are hydrocarbons or mixture of hydrocarbons which combine with oxygen of air during combustion to produce CO_2 and H_2O and exothermic heat. The combustion heat produced per kg of solid or liquid fuel or per m^3 of gaseous fuel is called its calorific value. The calorific value or heating value of a fuel depends upon its composition.

The heating values of some important fuels are given in Table 1.5.

Table 1.5 Heating value of fuels

Fuel	*Heating value*
Bituminous Coal	25 to 30 MJ/kg
Anthracite Coal	35 MJ/kg
Lignite	15 to 20 MJ/kg
Diesel Oil	50 MJ/kg
Petrol	45 MJ/kg
Ethane gas (C_2H_6)	52 MJ/kg

2. Nuclear energy

(i) *Nuclear Fusion:* Energy is produced in the sun and stars by continuous fusion reactions in which four nuclei of hydrogen fuse resulting in one nucleus of helium and two positrons.

$$4\ {}_1H^1 \rightarrow {}_2He^4 + 2\ {}_{+1}e^0$$

This results in a decrease in mass of about 0.0276 amu corresponding to 25.7 MeV. The heat produced in this reaction maintains temperatures of the order of several million degrees.

Many problems have to be solved before an artificially made fusion reactor becomes a reality. The most important difficulties are generating and maintaining high temperatures in the plasma and conversion of fusion energy into mechanical energy or electrical energy.

(ii) *Nuclear Fission:* There are many fission reactions that release different energy values. On an average, the fusion of a U^{235} nucleus yields about 193 MeV.

$$_{92}U^{235} + {}_0n^1 \rightarrow {}_{56}Ba^{137} + {}_{36}Kr^{97} + 2{}_0n^1$$

The mass balance gives

$$235.0439 + 1.00867 \rightarrow 136.9061 + 96.9212 + 2 \times 1.00867$$

$$236.052 \rightarrow 235.8446$$

$$\therefore \quad \Delta m = 235.8446 - 236.0526 = -0.2082 \text{ amu}$$

$$\Delta E = 931 \times (-0.52082) = -193.6 \text{ MeV}$$

$$= -3.1 \times 10^{-11} \text{ J}$$

This is called prompt energy. The total energy produced per fission reaction, therefore, is greater than prompt energy and is about 200 Mev. Burning of 1 g of fissionable material generates 1 MW-day of energy.

(iii) *Nuclear Breeding:* Nuclear fuel is defined as all uranium, plutonium and thorium isotopes. Reactor fuel contains fissionable nuclei as well as other nonfissionable isotopes of uranium, plutonium or thorium. Nuclear breeding is to convert U^{238} to Pu^{239} or thorium U^{234} to U^{235} with the help of fast neutrons.

$$_{92}U^{238} + {}_0n^1 \rightarrow {}_{94}Pu^{239} + {}_{-1}e^0$$

Pu^{239} is fissionable material and can be used in a fission reactor.

3. Solar energy

The sun is a large sphere of hot gases. As viewed from the earth, the sun is a solar disc of uniform brightness with a surface temperature of 5762 K. The solar radiations reach the earth as heat energy and can be calculated by Stefen-Boltzmann equation. The rate at which solar energy arrives at the top of the atmosphere is called solar constant. This is the amount of heat energy received in unit time on a unit area perpendicular to the sun's direction at the mean distance of the earth from the sun. The value of solar constant is 1353 watt per square metre. The value of solar radiations reaching the earth's surface is reduced due to diffusive and scattering effect of the atmospheric gases and dust particles.

4. Geothermal energy

Geothermal energy is primarily heat energy from the earth's own interior. It is almost as inexhaustible as solar energy. The centre of the earth is estimated at temperatures upto 10,000 K due to decay process of radioactive isotopes. The total steady geothermal energy flow towards earth's surface is 4.2×10^{10} kW. But the average energy flow density is only 0.063 W/m^2 and average temperature gradient is 30 K per kilometre of depth. Locations with very high temperature gradients are suitable for geothermal energy utilization. Geothermal energy can be used for generating heat and power.

1.10.3 Forms of Thermal Energy

In the study of engineering thermodynamics, thermal energy is expressed in different forms.

1. Internal energy

The energy possessed by a body or a system by virtue of its molecular arrangement and motion of molecules is called internal energy or intrinsic energy or microscopic energy. This energy is associated with thermodynamic state of the system. For perfect gases, internal energy depends only on the temperature of the system as it is assumed that the molecular forces of attraction and repulsion are absent. For non-permanent gas, vapour or liquid, the internal energy strongly depends on temperature and weakly on pressure.

For all gases,

$$u = u\,(T, p)$$

For ideal gases,

$$u = u(T)$$

The main characteristics of internal energy are:

(i) Internal energy is associated with the thermodynamic state of the system. Therefore, it is a point function and a thermodynamic property.

(ii) Internal energy depends upon the mass of the system, hence it is extensive property.

(iii) As per Joule's law

$$dU = mC_V\, dT$$

$$\therefore \quad U_2 - U_1 = m\int_{T_1}^{T_2} C_V\, dT$$

$$= m\, C_V\,(T_2 - T_1)$$

For ideal gases,

$$C_V = \left(\frac{du}{dT}\right)$$

(iv) When a system undergoes a thermodynamic process or change of state, both heat and work transfers take place. The net energy transfer is stored within the system and is called stored energy.

$$\delta Q - \delta W = dE$$

or $$Q_{1-2} - W_{1-2} = E_2 - E$$

The total energy of the system,

$$E = PE + KE + U$$

For a closed system or non-flow process, when there is no change of PE and also there is no flow of mass into or out of the system, there is no kinetic energy present.

$$\therefore \qquad Q_{1-2} - W_{1-2} = U_2 - U_1$$

(v) When a closed system undergoes a thermodynamic cycle, the net heat transfer is equal to net work transfer.

$$\sum_i Q_i = \sum_i W_i$$

or,

$$\oint \delta Q = \oint \delta W$$

The cyclic integral of heat transfer is equal to cyclic integral of work transfer.

$$\therefore \qquad \oint dU = 0$$

(vi) For an isolated system,

$$Q_{1-2} = 0$$

$$W_{1-2} = 0$$

$$\therefore \qquad U_2 = U_1$$

2. Enthalpy

The sums $U + pV$ and $u + pv$ appear together very frequently in thermodynamics. The combination, therefore, has been given the name *enthalpy* and the symbols H and h. The sum of internal energy (U) and product of pressure and volume (pV) or flow energy is called enthalpy (H).

$$H = U + pV$$

For unit mass, specific enthalpy

$$h = u + pv$$

For an ideal gas

$$p\bar{v} = \bar{R}T$$

and

$$pv = RT$$

$$h = u + (RT)$$

But,

$$u = u(T)$$

$$\therefore \qquad h = h(T)$$

The enthalpy of an ideal gas is a function of temperature only.

As the enthalpy is a sum of different variables which are thermodynamic properties, therefore, enthalpy is a thermodynamic property. Its units are kJ.

For all processes,

$$dh = CpdT$$

$$\therefore \qquad h_2 - h_1 = Cp(T_2 - T_1)$$

The enthalpy of steam or hot water from well is utilized in a steam turbine to produce power. This enthalpy is called geothermal energy.

3. Gibb's free energy

It is the maximum useful work that can be obtained from a steady flow stream when it undergoes a process at the beginning and end of which the pressure and temperature are in equilibrium with the atmosphere.

$$\Delta G = \Delta W_{max}$$

For a reversible isobaric and isothermal process,

$$\Delta G = 0$$

or $$G = \text{constant}$$

In a fuel cell, a chemical reaction takes place where reactants and products can be converted with a steady flow process and both are in equilibrium with the atmosphere.

4. Sensible heat energy

Heat is supplied to raise the temperature of a body or a system is called sensible heat.

$$Q_{1-2} = mC_n (T_2 - T_1) \quad \text{[kJ]}$$

where, m = mass, [kg]

C_n = specific heat (kJ/kg-k)

T_1 and T_2 = Temperature [K]

C_n depends upon the process taking place during heat transfer. The important specific heats for gases are specific heat at constant pressure (C_p) and constant volume (C_V).

The product $m\ C_n$ = thermal or heat capacity [kJ].

1.11 RELATION BETWEEN ENTHALPY AND SPECIFIC HEATS

The total heat supplied to a gas at constant volume,

$$Q_{1-2} = mC_V (T_2 - T_1)$$

There is no work done during heating of a gas at constant volume. The heat supplied is stored as internal energy or as increase in internal energy of gas.

$$C_V = \left(\frac{du}{dT}\right)$$

The total heat supplied at constant pressure,

$$Q_{1-2} = mC_p (T_2 - T_1)$$

The heat supplied is used to increase the temperature of gas and to do some external work of expansion.

$$
\begin{aligned}
Q_{1-2} &= dU + W_{1-2} \\
&= mC_V(T_2 - T_1) + p(V_2 - V_1) \\
&= mC_V(T_2 - T_1) + mR(T_2 - T_1) \qquad [\because \quad pV = mRT]
\end{aligned}
$$

For an ideal gas

$$C_p = \left(\frac{dh}{dT}\right)$$

$$h = u + RT$$

$$\frac{dh}{dT} = \frac{du}{dT} + R$$

$$C_p = C_V + R$$

or $$(C_p - C_V) = R$$

For molar quantities

$$(\overline{C}_p - \overline{C}_v) = \overline{R}$$

where $\overline{C}_p$ and $\overline{C}_v$ are molar specific heats.

Internal energy

$$du = C_v \, dT$$

For ideal gas, $$\Delta u = \int_{T_1}^{T_2} C_v \, dT$$

Enthalpy $$dh = C_p \, dT$$

$\therefore$ $$\Delta h = \int_{T_1}^{T_2} C_p \, dT$$

$$(C_p - C_v) = R$$

$$C_v\left(\frac{C_p}{C_v} - 1\right) = R$$

$$C_v \, (\gamma - 1) = R$$

$$\frac{C_p}{C_v} = \gamma \text{ by definition}$$

$\therefore$ $$C_v = \frac{R}{\gamma - 1}$$

1.12 COMPARISON BETWEEN WORK AND HEAT TRANSFER

There are many similarites and dissimilarities between heat and work.

Table 1.6 Comparison between work and heat

Heat	*Work*
I. Similarities	
1. Heat is a transient phenomenon. The systems do not process heat. When a system undergoes a change, heat transfer takes place.	1. Work is a transient phenomenon. The systems do not process work. When a system undergoes a change, work done may occur.
2. Heat is a boundary phenomenon. It is observed at the boundary of the system.	2. Work is a boundary phenomenon. It is observed at the boundary of the system.
3. Heat is the energy crossing the boundary of the system.	3. Work is the energy crossing the boundary of the system.
4. Heat is a path function and hence it is an inexact differential.	4. Work is a path function and hence it is an inexact differential.
5. Heat is not a property of the system.	5. Work is not a property of the system.
6. The magnitude of heat transfer depends upon the path followed by the system during change of state.	6. The magnitude of work done depends upon the path followed by the system during change of state.
II. Dissimilarities	
7. Heat transfer can take place in a stable system.	7. Work transfer cannot take place in a stable system.
8. Heat transfer is the energy interaction due to temperature difference only.	8. Work transfer is the energy interaction due to other reasons.
9. Heat is a low grade energy.	9. Work is high grade energy.
10. The quantity of heat transfer can be found out by area under the process curve on a T-S diagram.	10. The quantity of work done can be found out by the area under the process curve on *p-V* diagram.

Example 1.7 Estimate the heat required to raise the temperature of a fluid from 300 K to 400 K if its specific heat is given by

$$C = (0.2 + 0.002\ \text{T}),\ \text{kJ/kg-K}$$

The mass of fluid is 2 kg. What will be the mean specific heat?

Solution

$$Q_{1-2} = m\int_{T_1}^{T_2} C dT$$

$$= 2\int_{300}^{400} (0.2 + 0.002\ \text{T})\ dT$$

$$= 2\left|0.2 + \frac{0.002\,T^2}{2}\right|_{300}^{400}$$

$$Q_{1-2} = 180 \text{ kJ}$$

Also, $$Q_{1-2} = m\,C_m (T_2 - T_1)$$

$$\therefore \quad C_m = \frac{Q_{1-2}}{m(T_2 - T_1)} = \frac{180}{2(400 - 300)} = 0.9 \text{ kJ/kg-K}$$

Example 1.8 Calculate the change of enthalpy as 1 kg of oxygen is heated from 500 K to 2000 K. The value of specific heat at constant pressure is given as:

$$C_p = 11.515 - \frac{172}{\sqrt{T}} - \frac{1530}{T} \text{ kJ/kmol K}$$

Solution $$dh = \int_1^2 C_p\, dT$$

$$= \int_1^2 \left(11.515 - \frac{172}{\sqrt{T}} - \frac{1530}{T}\right) \text{dT}$$

$$= \left| 11.515T + \frac{172\sqrt{T}}{0.5} + 1530 \right|_{500}^{2000}$$

$$= \mathbf{232559.55 \text{ kJ/kg Ans.}}$$

Example 1.9 A system undergoes a cyclic process through four states, 1–2, 2–3, 3–4, 4–1. Find the values of X_1, X_2, Y_1, Y_2, Y_3 in the following table.

Process	*Heat transfer*	*Work transfer*		*Change of internal energy*
		kW	kJ/min	*kJ/min*
1-2	800	5.0	300	Y_1
2-3	400	X_1		600
3-4	–400	X_2		Y_2
4-1	0	3.0	180	Y_3

Solution Process 1-2, $Q_{1-2} - W_{1-2} = U_2 - U_1$

$$Y_1 = 800 - 300 = 500 \text{ kJ/min}$$

Process 2-3, $$Q_{2-3} - W_{2-3} = U_3 - U_2$$

$$400 - X_1 = 600$$

$$X_1 = -200 \text{ kJ/min}$$

Process 4-1, $$Q_{4-1} - W_{4-1} = U_1 - U_4$$

$$0 - 180 = Y_3$$

$$Y_3 = -180 \text{ kJ/min}$$

For the cyclic process,

$$\oint \delta Q - \oint \delta W = \oint \delta U$$

$$Y_1 + 600 + Y_2 - Y_3 = 0$$

$$500 + 600 + Y_2 - 180 = 0$$

$$\therefore \quad \mathbf{Y_2 = -920\ kJ/min}$$

QUESTION BANK 1

1. Define engineering thermodynamics. Discuss its scope.
2. Explain special technique and assumptions made for the study of engineering thermodynamics.
3. Give various classifications of thermodynamics. What is the difference between classical and statistical thermodynamics?
4. Define continuum. How is the concept used to define density and pressure?
5. Differentiate between:
   (i) Characteristic gas constant and universal gas constant.
   (ii) Open and closed system.
   (ii) Absolute and gauge pressure.
6. What are the main differences between reversible and irreversible processes? What are the causes of irreversibility?
7. What are the different forms of energy? What are their applications?
8. What are different sources of thermal energy?
9. In what forms, thermal energy is used in engineering thermodynamics?
10. What is the relation between internal energy and enthalpy?

CHAPTER 2

The First Law and Its Applications

2.1 INTRODUCTION

The first law of thermodynamics was formulated on the basis of Paddle Wheel Experiment conducted by Joule. (Fig. 2.1) A number of experiments were conducted by him wherein a paddle wheel was rotated by different forms of inputs. His findings were that the work expended was proportional to increase in thermal energy.

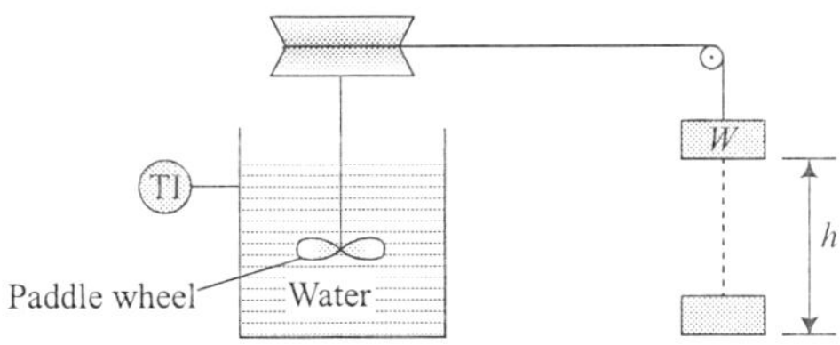

Fig. 2.1 Joule's paddle wheel experiment

$$Q \propto W$$

or
$$Q = \frac{W}{J}$$

where, J = proportionality constant called mechanical equivalent.

In SI system,

$$J = 1$$

The first law of thermodynamics states that work and heat are mutually convertible. The present tendency is to include all forms of energy.

The first law can be stated in many ways:

1. Energy can neither be created nor destroyed; it is always conserved. However, it can change from one form to another.
2. All energy that goes into a system comes out in some form or the other. Energy does not vanish and has the ability to be converted into any other form of energy.
3. If the system is carried through a cycle, the summation of work delivered to the surroundings is equal to summation of heat taken from the surrounding.
4. Total energy of an isolated system, in all its forms, remains constant.
5. No machine can produce energy without corresponding expenditure of energy. It is impossible to construct a perpetual motion machine of the first kind.

$$Q_1 \neq 0$$

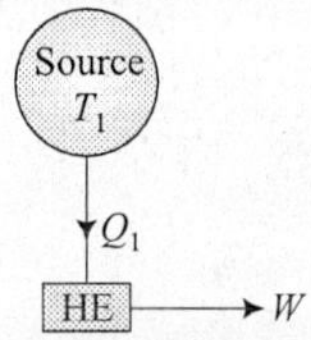

PM M–I, impossible

Fig. 2.2 PMM-1

2.2 LIMITATIONS OF FIRST LAW OF THERMODYNAMICS

There are some important limitations of First Law of Thermodynamics.

1. When a closed system undergoes a thermodynamic cycle, the net heat transfer is equal to the net work transfer.

$$\sum_i Q_i = \sum_i W_i$$

The cyclic integral of heat transfer is equal to the cyclic integral of work transfer.

$$\oint \delta Q = \oint \delta W$$

Where $\oint$ stands for cyclic integral (integral around complete cycle), δQ and δW are small elements of heat and work transfer and have the same units.

The law neither specifies the direction of flow of heat and work nor gives any conditions under which energy transfers can take place.

2. The heat energy and mechanical work are mutually convertible. The mechanical energy can be fully converted into heat energy but only a part

of heat energy can be converted into mechanical work. Therefore, there is a limitation on the amount of conversion of one form of energy into another form.

2.3 APPLICATIONS OF FIRST LAW TO THERMODYNAMIC PROCESSES

According to the first law, when a system undergoes a thermodynamic process (change of state) both heat and work transfer take place. The net (difference) energy transfer is stored within the system and is called stored energy or total energy of the system.

$$\delta Q - \delta W = dE.$$

where, E is the extensive property and represents the total energy of the system at a given state.

$$Q_{1-2} - W_{1-2} = E_2 - E_1$$

Q, W and E have same units.

For a unit mass,

$$q_{1-2} - w_{1-2} = e_2 - e_1$$

where, Q_{1-2} – Heat transferred to the system during the process 1-2. It is positive when supplied to the system and negative when rejected by the system.

W_{1-2} = Work done by the system on the surrounding during the process 1-2.

E_1 = Total energy of the system at state 1.

$$= PE_1 + KE_1 + U_1 + FE_1$$

$$= mgZ_1 + \frac{m_1 V_1^2}{2} + U_1 + p_1 V_1$$

E_2 = total energy of the system at state 2.

$$= PE_2 + KE_2 + U_2 + FE_2$$

$$= mgZ_2 + KE_2 + U_2 + p_2 V_2$$

$$\therefore \quad Q_{1-2} - W_{1-2} = E_2 - E_1$$

$$= (PE_2 + KE_2 + U_2 + FE_2) - (PE_1 + KE_1 + U_1 + FE_1)$$

$$= (PE_2 - PE_1) + (KE_2 - KE_1) + (H_2 - H_1) \quad [\because H = U + FE]$$

$$= mg\,(Z_2 - Z_1) + \frac{m}{2}\,(V_2^2 - V_1^2) + (H_2 - H_1)$$

For unit mass,

$$q_{1-2} - w_{1-2} = g(Z_2 - Z_1) + \frac{V_2^2 - V_1^2}{2} + (h_2 - h_1)$$

Case I For a closed system, there is no flow energy (FE)

$$\therefore \quad q_{1-2} - w_{1-2} = g(Z_2 - Z_1) + \frac{V_2^2 - V_1^2}{2} + (u_2 - u_1)$$

This is called non-flow energy equation.

Case II When there is no change in the potential energy of a closed system, i.e., $PE_2 = PE_1$

$$\therefore \quad q_{1-2} - w_{1-2} = (KE_2 - KE_1) + (u_2 - u_1) = \frac{V_2^2 - V_1^2}{2} + (u_2 - u_1)$$

Case III Closed or non-flow thermodynamic system, when there is no change of *PE* and also there is no flow of mass into or out of a system, i.e.,

$$PE_2 = PE_1 \text{ and } KE_2 = KE_1$$

$$\therefore \quad q_{1-2} - w_{1-2} = u_2 - u_1$$

Case IV Isolated system,

$$q_{1-2} = 0; \quad w_{1-2} = 0$$

$$e_2 = e_1 \quad \text{and} \quad u_2 = u_1$$

This shows that the first law of thermodynamics is the law of conservation of energy.

Case V Cyclic process

There is no change in the internal energy and the stored energy is zero.

$$\oint \delta q = \oint \delta w.$$

2.4 FIRST LAW ANALYSIS OF PROCESSES FOR IDEAL GAS

For an ideal gas,

$$du = Cv\, dT$$

$$dh = Cp\, dT$$

$$pv = RT.$$

These equations are valid for all processes.

2.4.1 Constant Volume Process

The first law equation for a non-flow process when kinetic energy and potential energy are negligible:

$$\delta q - \delta w = du.$$

For a reversible process,

$$\delta w = p\, dv \text{ and } du = Cv\, dT$$

$\therefore \qquad \delta q - p\,dv = Cv\,dT$

For constant volume process

$$\delta w = p\,dv = 0$$

$\therefore \qquad \delta q = Cv\,dT = du$

or $$q_{1-2} = u_2 - u_1 = \int_1^2 Cv\,dT$$

If the value of Cv as a function of T is known, integral can be evaluated.

2.4.2 Constant Pressure Process

The first law equation is:

$$\delta q - \delta w = du$$

or $$\delta q = du + p\,dv$$

$$\delta q = d(u + pv) \qquad [\because \; p = \text{constant}]$$

$$= dh = Cp\,dT$$

$\therefore$ $$q_{1-2} = \int_1^2 Cp\,dT = h_2 - h_1$$

If the value of Cp as a function of temperature is known, integral can be evaluated.

2.4.3 Constant Temperature Process

The first law equation is:

$$\delta q - \delta w = du$$

$$du = Cv\,dT = 0 \qquad (\because \; T \text{ is constant})$$

$\therefore \qquad \delta q = p\,dv$

$$q_{1-2} = \int_1^2 p\,dv$$

For an ideal gas,

$$pv = RT$$

$\therefore$ $$p = \frac{RT}{v}$$

$$\therefore \qquad q_{1-2} = w_{1-2} = \int_1^2 \frac{RT}{v} dv$$

$$= RT \ln \frac{v_2}{v_1}$$

$$= p_1 v_1 \ln \frac{v_2}{v_1}$$

Also, $$w_{1-2} = p_1 v_1 \ln \frac{p_1}{p_2} \qquad (\because \quad p_1 v_1 = p_2 v_2)$$

2.4.4 Adiabatic Process

The first law equation is,

$$\delta q - \delta q = du.$$

No heat leaves or enters the system.

$$\therefore \qquad 0 - \delta w = du$$

$$0 - p\, dv = Cv\, dT$$

$$\therefore \qquad dT = \frac{-pdv}{Cv} \qquad \text{...(a)}$$

But, $$pv = RT$$

Differentiating,

$$pdv + vdp = RdT$$

$$\therefore \qquad dT = \frac{pdv + vdp}{R} = \frac{pdv + vdp}{(Cp - Cv)} \qquad [\because R = Cp - Cv] \quad \text{...(b)}$$

Equating (a) and (b) for dT

$$\frac{-pdv}{Cv} = \frac{pdv + vdp}{(Cp - Cv)}$$

$$\therefore \qquad \frac{(C_p - Cv)}{Cv} = \frac{pdv + vdp}{-pdv} = -1 - \frac{vdp}{pdv}$$

$$\therefore \qquad \frac{Cp}{Cv} - 1 = -1 - \left[\frac{v}{dv} \times \frac{dp}{p}\right]$$

or $$\gamma = -\left[\frac{v}{dv} \times \frac{dp}{p}\right] \qquad \left[\because \frac{C_p}{Cv} = \gamma\right]$$

$\therefore$ $$\gamma\left[\frac{dv}{v}\right] = -\frac{dp}{p}$$

or $$\gamma\frac{dv}{v} + \frac{dp}{p} = 0$$

Integrating,

$$\gamma \ln v + \ln p = \text{constant}$$

$$\ln pv^{\gamma} = \ln (\text{constant})$$

$$p_1 v_1^{\gamma} = p_2 v_2^{\gamma} = pv^{\gamma} = \text{constant}$$

$$\frac{p_1}{p_2} = \left(\frac{v_2}{v_1}\right)^{\gamma}$$

$$\frac{v_1}{v_2} = \left(\frac{p_2}{p_1}\right)^{\frac{1}{\gamma}}$$

But $$\frac{p_1 v_1}{T_1} = \frac{p_2 v_2}{T_2}$$

or $$\frac{v_1}{v_2} = \frac{T_1}{T_2} \times \frac{p_2}{p_1}$$

$\therefore$ $$\left(\frac{p_2}{p_1}\right)^{\frac{1}{\gamma}} = \frac{T_1}{T_2} \times \frac{p_2}{p_1}$$

or $$\frac{T_1}{T_2} = \left(\frac{p_2}{p_1}\right)^{\frac{1}{\gamma}} \times \frac{p_2}{p_1}$$

$$= \left(\frac{p_1}{p_2}\right)^{-\frac{1}{\gamma}+1}$$

$$\therefore \qquad \frac{T_1}{T_2} = \left(\frac{p_1}{p_2}\right)^{\frac{\gamma-1}{\gamma}}$$

2.4.5 Polytropic Process

The first law equation is

$$\delta q - \delta w = du$$

$$q_{1-2} - w_{1-2} = (u_2 - u_1)$$

$$\delta q - pdv = du$$

For polytropic process,

$$w_{1-2} = \frac{p_1 v_1 - p_2 v_2}{n-1} = \frac{R(T_1 - T_2)}{n-1}$$

$$\therefore \qquad q_{1-2} = \frac{R(T_1 - T_2)}{n-1} + Cv(T_2 - T_1)$$

$$= \frac{R(T_1 - T_2)}{n-1} + \frac{R}{\gamma - 1}(T_2 - T_1) \qquad \left[\because \; Cv = \frac{R}{\gamma-1}\right]$$

$$= R(T_1 - T_2)\left[\frac{1}{n-1} - \frac{1}{\gamma-1}\right]$$

$$= R(T_1 - T_2)\left[\frac{(\gamma-1)-(n-1)}{(n-1)(\gamma-1)}\right]$$

$$= R(T_1 - T_2)\left[\frac{\gamma - n}{(n-1)(\gamma-1)}\right]$$

$$= \frac{\gamma - n}{\gamma - 1} \times \frac{R(T_1 - T_2)}{n-1}$$

Also,

$$q_{1-2} = \frac{\gamma - n}{\gamma - 1} \times \frac{p_1 v_1 - p_2 v_2}{n-1}$$

$$q_{1-2} = \frac{\gamma - n}{\gamma - 1} \times w_{1-2}$$

2.4.6 Free Expansion Process

Free expansion process is unrestricted expansion of a gas without any work output. As the gas cannot be compressed back to its initial state without the use of external work, the process of free expansion is highly irreversible process. The process is also called constant internal energy process as there is no exchange of heat and work between the system and the surrounding. The process can be explained by a simple example.

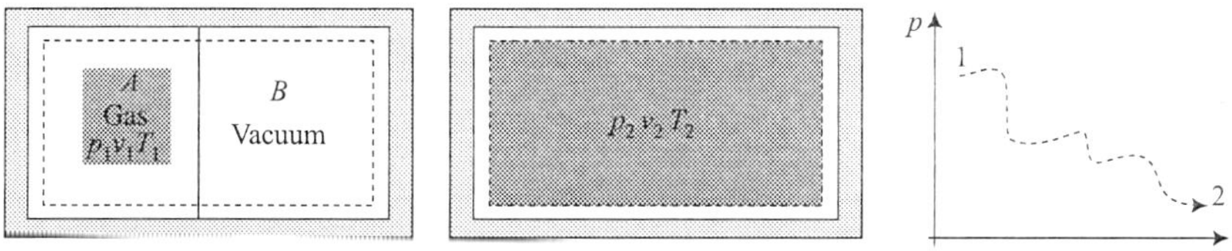

(a) Before expansion (b) After expansion p-v diagram

Fig. 2.3 Free expansion process

There are two chambers A and B completely insulated and separated by a membrane. Chamber A has gas at parameters $p_1 v_1 T_1$ and chamber B is completely evacuated. The state is plotted as point 1 on p-v diagram. The partition is removed and the gas occupies the total volume of chambers A and B. The final conditions of the system are $p_2 v_2 T_2$ are shown by point 2 on the p-v diagram. The free expansion is shown by a dotted line as the process is uncontrolled, irreversible and unspecified.

1. The work done is zero as there is no expansion of boundary

$$W_{1-2} = 0$$

2. The system is insulated and there is no exchange of heat with the surrounding.

$$Q_{1-2} = 0$$

3. Applying the first law of thermodynamics to the closed system,

$$Q_{1-2} - W_{1-2} = U_2 - U_1$$

$$0 - 0 = U_2 - U_1$$

$$\therefore \quad U_2 = U_1$$

The internal energy of the system during free expansion remains constant.

4. $$U_2 - U_1 = m\,Cv\,(T_2 - T_1) = 0$$

$$\therefore \quad T_2 = T_1.$$

The process of free expansion is isothermal process.

5. $$H_2 - H_1 = m\,Cp\,(T_2 - T_1) = 0$$

The enthalpy of the system remains constant.

2.4.7 Summary of Thermodynamic Relations for Non-flow Processes

The thermodynamic relations for an ideal gas in different non-flow processes are summarised in Table 2.1. The following relationships have been tabulated:

1. Governing equation or *p-v-T* relationship
2. Work done
3. Change of internal energy
4. Change of enthalpy
5. Heat exchanged.

The values given are for unit mass. The mechanical work in a steady flow process is compared with that in non-flow process in Table 2.2.

Example 2.1 Three kg of air kept at 100 kPa and 300 K is compressed polytropically to 1500 kPa and 500 K. Calculate:

(i) Index *n*
(ii) Final volume
(iii) Work done
(iv) Heat exchanged

Solution:

(i) *Index n*

$$p_1 = 100 \text{ kPa} = 100 \times 10^3 \text{ N/m}^3$$

$$T_1 = 300 \text{ K}$$

$$p_2 = 1500 \text{ kPa} = 1500 \times 10^3 \text{ N/m}^2$$

$$T_2 = 500 \text{ K}$$

$$p_1 V_1^n = p_2 V_2^n$$

and
$$\frac{p_1 V_1}{T_1} = \frac{p_2 V_2}{T_2}$$

$\therefore$
$$\frac{T_2}{T_1} = \left(\frac{p_2}{p_1}\right)^{\frac{n-1}{n}}$$

Table 2.1 Thermodynamic relations (Unit mass)

Process	*Equation*	*Index* n	*p-v-T relation*	*Specific heat C*	*Heat added*	*Work done* $\int_1^2 pdv$	*Change of enthalpy*	*Change of internal energy*	*Remark*
Units	–	–	–	kJ/kg-K	kJ/kg	kJ/kg	kJ/kg	kJ/kg	–
Constant volume (Isochoric)	$v = c$	$n = \infty$	$\frac{p_1}{p_2} = \frac{T_1}{T_2}$	Cv	$Cv(T_2 - T_1)$	0	$Cp(T_2 - T_1)$	$Cv(T_2 - T_1)$	–
Constant pressure (Isobaric)	$p = c$	$n = 0$	$\frac{v_1}{v_2} = \frac{T_1}{T_2}$	Cp	$Cp(T_2 - T_1)$	$p(v_2 - v_1)$	$Cp(T_2 - T_1)$	$Cv(T_2 - T_1)$	–
Constant temperature (Isothermal)	$t = c$	$n = 1$	$p_1 v_1 = p_2 v_2$	∞	$p_1 v_1 \ln \frac{v_2}{v_1}$	$p_1 v_1 \ln \frac{v_2}{v_1}$	0	0	–
Reversible Adiabatic (Isentropic)	$pv^\gamma = c$	$n = \infty$	$p_1 v_1^\gamma = p_2 v_2^\gamma$	0	0	$\frac{p_1 v_1 - p_2 v_2}{\gamma - 1}$	$Cp(T_2 - T_1)$	$\frac{R(T_1 - T_2)}{\gamma - 1}$	$\frac{T_2}{T_1} = \left(\frac{v_1}{v_2}\right)^{\gamma-1} = \left(\frac{p_2}{p_1}\right)^{\frac{\gamma-1}{\gamma}}$ $u_2 - u_1 = \frac{p_1 v_1 - p_2 v_2}{\gamma - 1}$
Polytropic	$pv^n = c$	$n = n$	$p_1 v_1^n = p_2 v_2^n$	Cn	$Cn(T_2 - T_1)$	$\frac{p_1 v_1 - p_2 v_2}{n - 1}$	$Cp(T_2 - T_1)$	$Cv(T_2 - T_1)$	$\frac{T_2}{T_1} = \left(\frac{v_1}{v_2}\right)^{n-1} = \left(\frac{p_2}{p_1}\right)^{\frac{n-1}{n}}$ $C_n = Cv\left(\frac{\gamma - n}{1 - n}\right)$; $Q_{1\text{-}2} = \frac{\gamma - n}{r - 1} W.D$ $W.D = \frac{R(T_2 - T_1)}{n - 1}$
Free expansion	$u_1 = u_2$ $T_1 = T_2$	–	$T_1 = T_2$	–	–	0	0	0	–

$$\therefore \quad \frac{n-1}{n} = \frac{\ln \frac{T_2}{T_1}}{\ln \frac{p_2}{p_1}} = \frac{\ln \frac{500}{300}}{\ln \frac{1500}{100}} = \frac{0.5108}{2.708} = 0.1886.$$

$$\therefore \quad n = 1.23$$

(ii) *Final volume*

$$p_1 V_1 = m\,RT_1$$

Take $R = 287$ J/kg-K for air

$$V_1 = \frac{m\,RT_1}{p_1} = \frac{3 \times 287 \times 300}{100 \times 10^3} = 2.583 \text{ m}^3$$

$$p_2 V_2 = m\,RT_2$$

$$V_2 = \frac{m\,RT_2}{p_2} = \frac{3 \times 287 \times 500}{1500 \times 10^3} = \mathbf{0.287\ m^3} \quad \textbf{Ans.}$$

(iii) *Work done*

$$W_{1-2} = \frac{p_1 V_1 - p_2 V_2}{n-1} = \frac{m\,R(T_1 - T_2)}{n-1} = \frac{3 \times 287\,(300 - 500)}{1.23 - 1}$$

$$= -748696 \text{ J} = -748.7 \text{ kJ}$$

Work is done on the air during compression.

(iv) *Heat exchanged*

$$Q_{1-2} = \frac{\gamma - n}{\gamma - 1}\, W_{1-2} = \frac{1.4 - 1.23}{1.4 - 1}\,(-748.7)$$

$$= -318.2 \text{ kJ}.$$

Heat is rejected by the system.

2.5 THROTTLING PROCESS

The expansion of gas through an obstruction in the form of a partly opened valve or orifice is called throttling.

There is reduction in pressure and increase in the volume of the fluid. The process is adiabatic as there is no exchange of heat but it is irreversible. No work output occurs during throttling. It is not possible to compress the fluid back to initial pressure without the aid of external work. Therefore, the process is irreversible.

Table 2.2 Mechanical work (Unit mass)

Sl. No.	*Process*	*Non-flow work* $\int pdv$	*Steady flow work* $-\int vdp$
1.	Isochoric ($v = c$)	0	$v(p_1 - p_2)$
2.	Isobaric ($p = c$)	$p(v_2 - v_1)$	0
3.	Isothermal ($t = c$)	$p_1 v_1 \ln \frac{v_2}{v_1}$	$p_1 v_1 \ln \frac{v_2}{v_1}$
4.	Adiabatic ($pv^\gamma = c$)	$\frac{p_1 v_1 - p_2 v_2}{\gamma - 1}$	$\frac{\gamma}{\gamma - 1}(p_1 v_1 - p_2 v_2)$
5.	Polytropic ($pv^n = c$)	$\frac{p_1 v_1 - p_2 v_2}{n - 1}$	$\frac{n}{n - 1}(p_1 v_1 - p_2 p_2)$
6.	Free expansion ($u = c$)	0	–
7.	Throttling ($h = c$)		0

The Joule-Thomson Porous Plug Experiment

This experiment was performed by Joule and Thomson in 1852. A gas was made to pass steadily through a porous plug placed in an insulated horizontal pipe. The inlet parameters p_i and T_i are kept constant. The resistance of porous plug is varied by compressing it. The pressure drop through the plug varies. The pressure (P_e) and temperature (T_e) at the exit are measured.

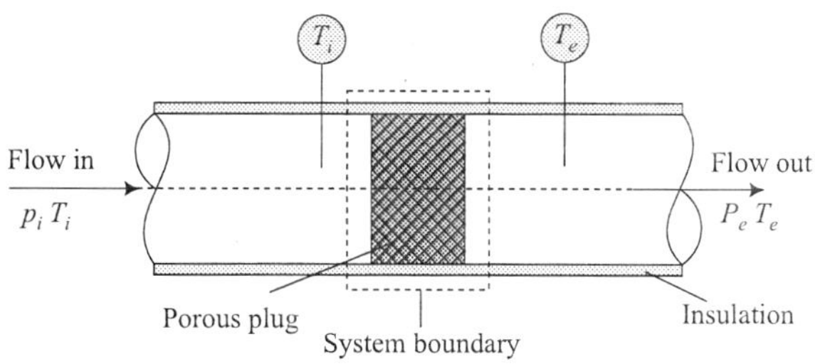

Fig. 2.4 Process plug apparatus

During this experiment:

1. There is no work output

$$W_{1-2} = 0$$

2. There is no heat exchange as the surface is insulated

$$Q_{1-2} = 0$$

3. The pipe is placed horizontally

$$\therefore \quad Z_1 = Z_2$$

$$PE_1 = PE_2$$

4. The change in velocity is negligible

$$V_1 \approx V_2$$

$$KE_1 = KE_2$$

Applying steady flow energy equation to unit mass flow

$$h_1 + \frac{v_1^2}{2} + gz_1 + q_{1-2} = h_2 + \frac{v_2^2}{2} + gz_2 + w_{1-2}$$

$$\therefore \quad h_1 = h_2$$

Therefore, throttling process is a constant enthalpy process

$$h_1 = h_2 = h_3 = h_4 = h_5 = \ldots$$

If the readings of pressure and temperature of the experiment are plotted on *T-p* diagram, a constant enthalpy line is obtained.

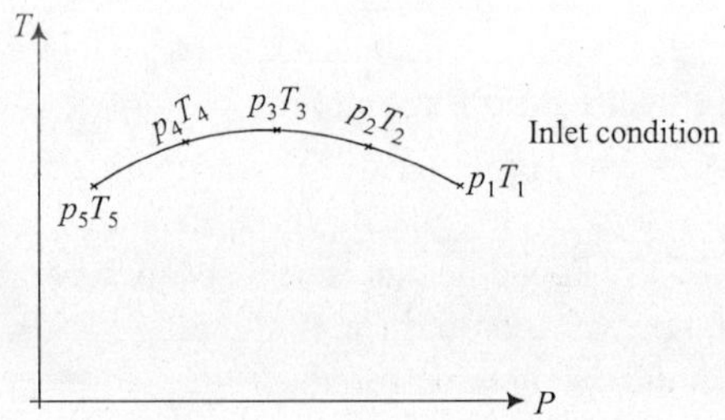

Fig. 2.5 Constant enthalpy process

The slope of the constant enthalpy curve is called Joule-Thomson coefficient.

$$\mu = \left(\frac{dT}{dp}\right)_h$$

For perfect gas, $\mu = 0$.

Applications of Throttling

Although throttling is a energy loss process, it is used for the following:

1. To find out the dryness fraction of steam in a throttling calorimeter.
2. The speed of steam turbine is controlled in throttle governing.
3. Refrigeration effect (cooling) is obtained by throttling the refrigerant in a valve or capillary tube at inlet to evaporator.

$$\because \quad h_1 = h_2$$

$$\therefore \quad h_1 - h_2 = 0$$

$$Cp(T_1 - T_2) = 0$$

$$\therefore \quad T_1 = T_2$$

$$Cv(T_1 - T_2) = 0$$

$$\therefore \qquad u_1 = u_2$$

For an ideal gas, throttling takes place at:

1. Constant enthalpy
2. Constant internal energy
3. Constant temperature.

However, in a real gas or fluid (refrigerant),

$$T_1 \neq T_2.$$

2.6 STEADY FLOW ENERGY EQUATION

According to the first law of thermodynamics, the total energy entering a system must be equal to total energy leaving the system. For unit mass,

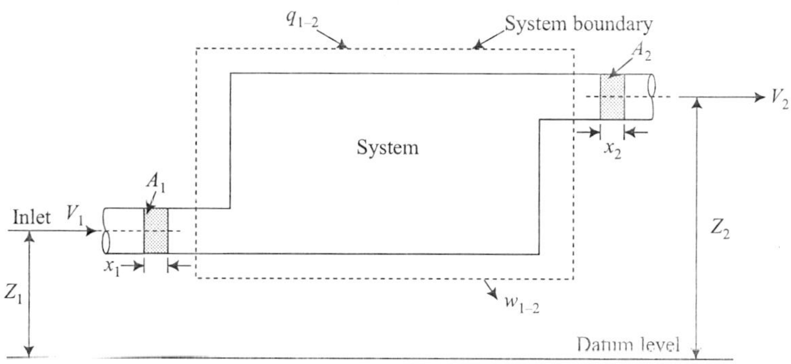

Fig. 2.6 Steady flow process

$$e_1 = e_2$$

$$u_1 + p_1v_1 + \frac{V_1^2}{2} + gz_1 + q_{1-2} = u_2 + p_2v_2 + \frac{V_2^2}{2} + gz_2 + w_{1-2}$$

where, suffix 1 is for inlet and 2 for outlet.

u = Specific internal energy

pv = Flow work

V = Fluid velocity

Z = Height

q_{1-2} = Heat exchange

w_{1-2} = Work exchange

Now, $\qquad h = u + pv$

$$\therefore \qquad h_1 + \frac{V_1^2}{2} + qz_1 + q_{1-2} = h_2 + \frac{V_2^2}{2} + qz_2 + w_{1-2}$$

This is called steady flow energy equation. This equation may also be written as follows:

$$q_{1-2} - w_{1-2} = (h_2 - h_1) + \frac{V_2^2 - V_1^2}{2} + q(z_2 - z_1)$$

In differential form,

$$dq - dw = dh + d(ke) + d(pe)$$

Case I. If the effect of gravity can be neglected, i.e., $z_2 \approx z_1$

$\therefore$ $$d(pe) = 0$$

$$dq - dw = dh + (dke)$$

or $$q_{1-2} - w_{1-2} = (h_2 - h_1) + \left(\frac{V_2^2 - V_1^2}{2}\right)$$

Case II. If gravity can be neglected and the change in velocity is negligible, i.e., $V_2 \approx V_1$

$$d(pe) = 0$$

$$d(ke) = 0$$

$\therefore$ $$q_{1-2} - w_{1-2} = (h_2 - h_1).$$

Case III. Applying the steady flow energy equation to a closed system (non-flow process)

$$d(pe) = 0$$

$$d(ke) = 0$$

$$p_1 v_1 = 0$$ (Flow energy or displacement energy at inlet and outlet is zero)

$$p_2 v_2 = 0$$

$\therefore$ $$h_1 = u_1$$

$$h_2 = u_2$$

$\therefore$ $$q_{1-2} - w_{1-2} = u_2 - u_1$$

This is called energy equation for a non-flow process.

2.6.1 Equation of Continuity

The mass flow rate ($\dot{m}$) of the working substance entering the system is same as leaving the system. The steady flow energy equation will be:

$$\dot{m}\left(h_1 + \frac{V_1^2}{2} + qZ_1 + q_{1-2}\right) = \dot{m}\left(h_2 + \frac{V_2^2}{2} + qZ_2 + W_{1-2}\right)$$

or $$\dot{Q}_{1-2} - \dot{W}_{1-2} = \dot{m}(h_2 - h_1) + \frac{\dot{m}}{2}(V_1^2 - V_1^2) + \dot{m}g(z_2 - z_1)$$

where, $\dot{m} = \frac{A_1 V_1}{v_1} = \frac{A_2 V_2}{v_2}$ (kg/s)

This is called equation of continuity.

Example 2.2 0.5 kg/s of a fluid flows in a steady state process. The properties of the fluid at the entrance are measured as $p_1 = 1.4$ bar, density = 2.5 kg/m^3, $u_1 = 920$ kJ/kg while at exit the properties are $p_2 = 5.6$ bar, density = 5 kg/m^3 and $u_2 = 720$ kJ/kg. The velocity at entrance is 200 m/s while at exit is 180 m/s. It rejects 60 kW of heat and rises through 60 m during the flow. Find the change of enthalpy and the rate of work done.

Solution: The data given,

$$\dot{m} = 0.5 \text{ kg/s}$$

$$p_1 = 1.4 \text{ bar} = 140 \text{ kPa}$$

$$\rho_1 = 2.5 \text{ kg/m}^3 \quad \therefore v_1 = \frac{1}{\rho_1} = 0.4 \text{ m}^3/\text{kg}$$

$$u_1 = 920 \text{ kJ/kg}$$

$$p_2 = 5.2 \text{ bar} = 560 \text{ kPa}$$

$$\rho_2 = 5 \text{ kg/m}^3 \quad \therefore v_2 = \frac{1}{\rho_2} = \frac{1}{5} = 0.25 \text{ m}^3/\text{kg}$$

$$u_2 = 720 \text{ kJ/kg}$$

$$V_1 = 200 \text{ m/s}$$

$$V_2 = 180 \text{ m/s}$$

$$Q_{1-2} = 60 \text{ kW} = 60 \text{ kJ/s}$$

$$Z_2 - Z_1 = 60 \text{ m}$$

(i) *Change of enthalpy*

$$\Delta H = \dot{m}\,(h_2 - h_1) = \dot{m}[(u_2 + p_2 v_2) - (u_1 + p_1 v_1)]$$

$$= 0.5\,[(720 + 560 \times 0.25) - (920 + 140 \times 0.4)]$$

$$= 0.5\,[860 - 976] = -58 \text{ kJ}$$

There is a decrease in the enthalpy.

(ii) *Rate of work done*

Apply steady flow energy equation to the system

$$\dot{Q}_{1-2} - \dot{W}_{1-2} = \dot{m}\left[(h_2 - h_1) + \frac{V_2^2 - V_1^2}{2 \times 10^3} + \frac{g(Z_2 - Z_1)}{10^3}\right]$$

$$-60 - \dot{W}_{1-2} = 0.5\left[(860 - 976) + \frac{180^2 - 200^2}{2 \times 10^3} + \frac{9.81 \times 60}{10^3}\right]$$

$$= 0.5\,[-116 - 3.8 + 0.5886] = -59.6$$

$$\therefore \quad \dot{W}_{1-2} = -60 + 59.6 = \mathbf{0.4\ kW} \quad \textbf{Ans.}$$

Work is done on the system.

2.7 APPLICATIONS OF STEADY FLOW ENERGY EQUATION

The steady flow energy equation can be applied to various energy systems and devices such as boilers, condensers, evaporators, nozzles, turbines, compressors, etc.

2.7.1 Heat Exchanger

A heat exchanger is a device in which heat is transferred from one fluid to another. There are two steady flow streams, one of heating fluid and other fluid to be heated. The flow through a heat exchanger is characterized by,

1. No work exchange,

$$W_{1-2} = 0$$

2. No change in potential energy

$$Z_2 = Z_1$$

3. No change in kinetic energy

$$V_2 \approx V_1$$

4. Normally, no external heat interaction, if heat exchanger is insulated.

   Heat gained by cold fluid = Heat lost by hot fluid

$$\dot{m}_c(h_1 - h_2) = \dot{m}_h(h_4 - h_3)$$

where, $\dot{m}_c$ and $\dot{m}_h$ are flow rate of cold fluid and hot fluid respectively.

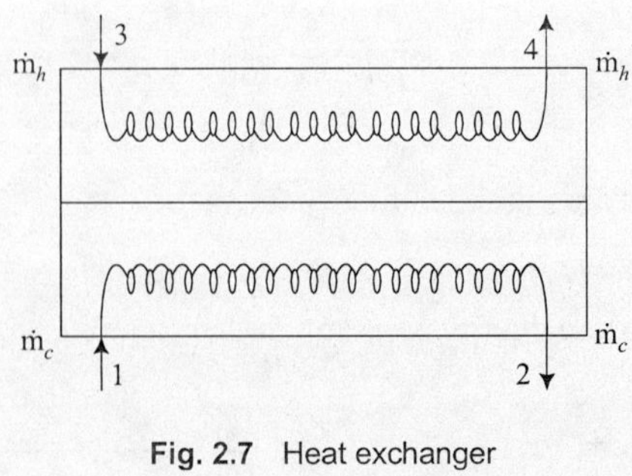

Fig. 2.7 Heat exchanger

The boiler, condenser, evaporator, etc. are designed in a similar manner by using steady flow energy equation. The heat supplied or removed is equal to change of enthalpy,

$$q_{1-2} = (h_2 - h_1)$$

Example 2.3 In a heat exchanger, 50 kg of water is heated per minute from 50°C to 110°C by hot gases which enter the heat exchanger at 250°C. If the flow rate of gases is 100 kg/min, estimate the net change of enthalpy. For water $C_p = 4.186$ kJ/kg-K and for air, $C_p = 1$ kJ/kg-K.

Solution:

$$\dot{m}_w = 50 \text{ kg/min}$$
$$t_{w1} = 50°C$$
$$t_{w2} = 110°C$$
$$\dot{m}_g - 100 \text{ kg/min}$$
$$t_{g1} = 250°C$$
$$Cp_w = 4.186 \text{ kJ/kg-K}$$
$$Cp_g = 1 \text{ kJ/kg-K}$$

Using energy balance,

Heat gained by water = Heat lost by hot gas

$$\dot{m}_w\, Cp_w\,(t_{w2} - t_{w1}) = \dot{m}_g\, Cp_g(t_{g1} - t_{g2})$$
$$50 \times 4.186\,(110 - 50) = 100 \times 1\,(250 - t_{g2})$$
$$\therefore \quad t_{g2} = 125°C$$

Net change of enthalpy of gas = $100 \times 1 \times (250 - 125)$

= 12500 kJ/min

2.7.2 Compressor

Rotary and reciprocating compressors are used to increase the pressure of air and other gases. These are rotated by electric motor or engines. The main characteristics of the system are:

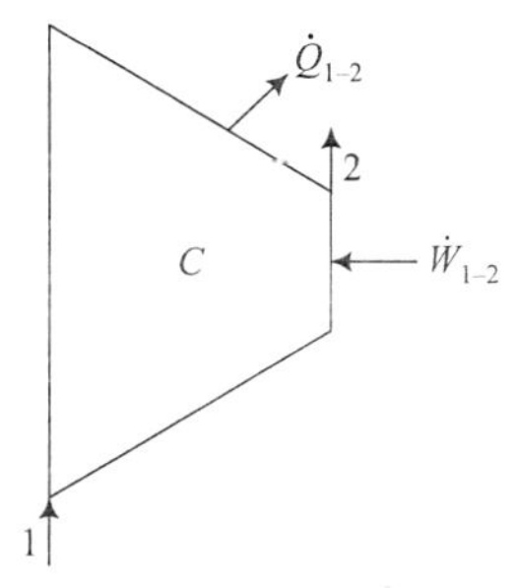

Fig. 2.8 Compressor

1. Work is done on the system and, hence, it is negative.

   $\dot{W}_{1-2}$ is –ve.

2. Potential energy and kinetic energy can normally be neglected.

$$Z_2 \approx Z_1$$
$$V_2 \approx V_1$$

3. Heat is lost from the compressor either by radiation or through a coolant (air or water).

   The heat exchange is –ve.

   $\dot{Q}_{1-2}$ is –ve or zero.

$$-\dot{Q}_{1-2} - (-\dot{W}_{1-2}) = \dot{m}(h_2 - h_1)$$

$$\therefore \quad \dot{W}_{1-2} = \dot{Q}_{1-2} + \dot{m}(h_2 - h_1)$$

Therefore, work is done on the system to increase the enthalpy of the fluid.

Example 2.4 0.8 kg of air flows through a compressor under steady state conditions. The properties of air at entry are: pressure 1 bar, velocity 10 m/s, specific volume 0.95 m^3/kg and internal energy 30 kJ/kg. The corresponding values at exit are: 8 bar, 6 m/s, 0.2 m^3/kg and 124 kJ/kg. Neglecting the change in potential energy, determine the power input and pipe diameter at entry and exit.

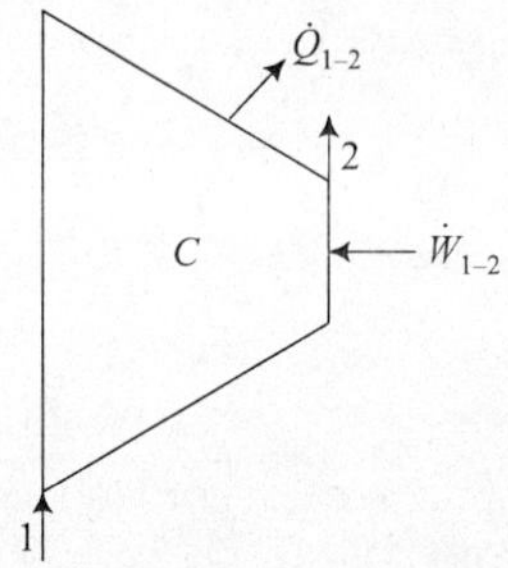

Solution: The data given,

$$\dot{m} = 0.8 \text{ kg/s}$$
$$p_1 = 1 \text{ bar} = 100 \text{ kPa}$$
$$V_1 = 10 \text{ m/s}$$
$$v_1 = 0.95 \text{ m}^3/\text{kg}$$
$$u_1 = 0.9 \text{ kJ/kg}$$
$$p_2 = 8 \text{ bar} = 800 \text{ kPa}$$

$$V_2 = 6 \text{ m/s}$$
$$v_2 = 0.2 \text{ m}^3\text{/kg}$$
$$u_2 = 124 \text{ kJ/kg}$$
$$\therefore \quad h_1 = u_1 + p_1v_1 = 30 + 100 \times 0.95 = 125 \text{ kJ/kg}$$
$$h_2 = u_2 + p_2v_2 = 124 + 800 \times 0.2 = 284 \text{ kJ/kg}$$

(i) Apply steady flow energy equation to the compressor,

$$Q_{1-2} - W_{1-2} = \dot{m}\left[(h_2 - h_1) + \frac{V_2^2 - V_1^2}{2 \times 10^3} + \frac{g}{10^3}(Z_2 - Z_1)\right]$$

$$Q_{1-2} = 0$$

W_{1-2} is –ve.

$$Z_2 = Z_1$$

$$\therefore \quad W_{1-2} = 0.8\left[(284 - 125) + \frac{6^2 - 10^2}{2 \times 10^3}\right] = \mathbf{127 \text{ kW}} \quad \textbf{Ans.}$$

(ii) Apply equation of continuity,

$$\dot{m} = \frac{A_1V_1}{v_1} = \frac{A_2V_2}{v_2}$$

$$\therefore \quad A_1 = \frac{\pi}{4}D_1^2 = \frac{\dot{m}v_1}{V_1}$$

$$\therefore \quad D_1 = \sqrt{\frac{4\dot{m}v_1}{\pi V_1}} = \sqrt{\frac{4 \times 0.8 \times 0.95}{\pi \times 10}} = 0.311 \text{ m}$$

$$= \mathbf{311 \text{ mm}} \quad \textbf{Ans.}$$

$$D_2 = \sqrt{\frac{4\dot{m}v_2}{\pi V_2}} = \sqrt{\frac{4 \times 0.8 \times 0.2}{\pi \times 6}} = 0.184 \text{ m}$$

$$= \mathbf{184 \text{ mm}} \quad \textbf{Ans.}$$

Example 2.5 An air compressor compresses atmospheric air at 0.1 MPa and 27°C by ten times of inlet pressure. During compression, the heat loss to the surrounding is estimated to be 5% of compression work. Air enters the compressor with a velocity of 40 m/s and leaves with 100 m/s. Inlet and exit cross- sectional areas are 100 cm^2 and 20 cm^2 respectively. Estimate the temperature of air at exit from compressor and power input to the compressor.

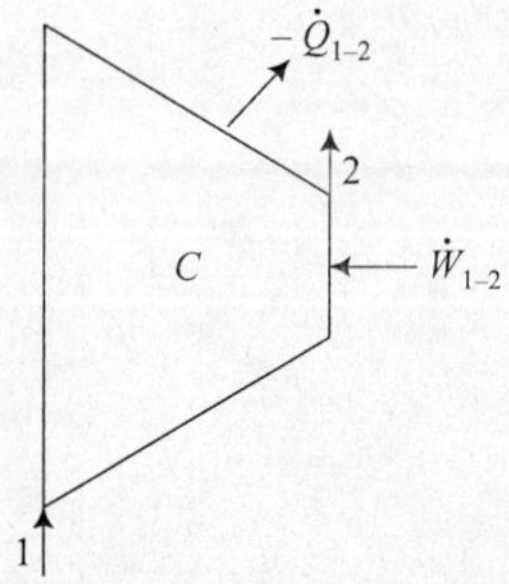

Solution:

1. *Inlet conditions*

$$p_1 = 0.1 \text{ MPa} = 0.1 \times 10^3 \text{ kN/m}^2$$
$$T_1 = 27°\text{C} + 273 = 300 \text{ K}$$
$$V_1 = 40 \text{ m/s}$$
$$A_1 = 100 \text{ cm}^2 = 100 \times 10^{-4} \text{ m}^2$$

2. *Outlet conditions*

$$p_2 = 10 \times 0.1 \times 10^3 \text{ kN/m}^2 = 10^3 \text{ kN/m}^2$$
$$V_2 = 100 \text{ m/s}$$
$$A_2 = 20 \text{ cm}^2 = 20 \times 10^{-4} \text{ m}^2$$

3. *Equation of continuity*

The equation of state,

$$p_1 V_1 = mRT_1$$

or $$p_1 v_1 = RT_1$$

$$\therefore \quad v_1 = \frac{RT_1}{p_1} = \frac{0.287 \times 300}{100}$$

$$= 0.861 \text{ m}^3\text{/kg} \quad [R = 0.287 \text{ kJ/kg-K for air}]$$

$$\dot{m} = \frac{A_1 V_1}{v_1} = \frac{(100 \times 10^{-4}) \times 40}{0.861} = 0.4646 \text{ kg/s}$$

$$\dot{m} = \frac{A_2 V_2}{v_2}$$

$$\therefore \quad v_2 = \frac{A_2 V_2}{\dot{m}} = \frac{(20 \times 10^{-4}) \times 100}{0.4646} = 0.43 \text{ m}^3\text{/kg}$$

$$p_2 v_2 = RT_2$$

$$T_2 = \frac{p_2 v_2}{R} = \frac{10^3 \times 0.43}{0.287} = 1498 \text{ K}$$

4. *Steady flow energy equation*

$$Q_{1-2} - W_{1-2} = \dot{m}\left[(h_2 - h_1) + \frac{V_2^2 - V_1^2}{2 \times 10^3} + \frac{g}{10^3}(Z_2 - Z_1)\right]$$

$$-0.05\, W_{1-2} - W_{1-2} =$$

$$0.43\left[1.005(1498 - 300) + \frac{(100)^2 - (40)^2}{2 \times 10^3} + 0\right]$$

W_{1-2} = **546.86 kW Ans.**

2.7.3 Gas Turbine

A gas turbine converts the heat energy of hot gases into mechanical work. A compressor driven by the gas turbine compresses air or gas to a higher pressure. The high pressure air or gas is heated by combustion of fuel. The high pressure and high temperature air or gas is admitted to a gas turbine. Power is produced in the turbine at the expense of enthalpy drop of the gas. The main characteristics of the system are:

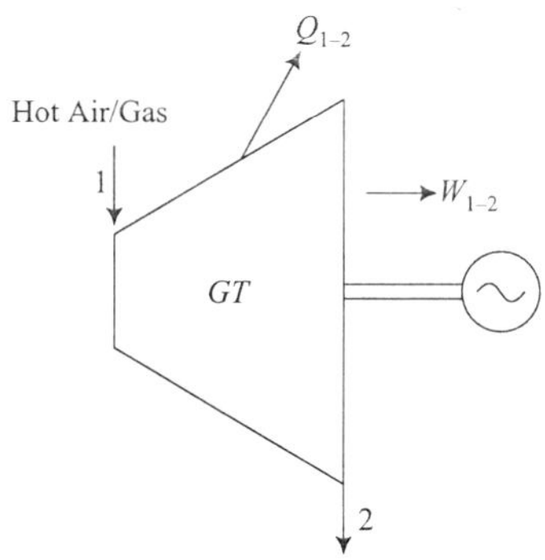

Fig. 2.9 Gas turbine

1. The heat loss by radiation to the surrounding is –ve.
2. Normally, *ke* and *pe* are neglected.

$$V_2 \approx V_1$$

$$Z_2 = Z_1$$

$$\dot{W}_{1-2} = Q_{1-2} + \dot{m}(h_1 - h_2)$$

Example 2.6 Air passes through a gas turbine system at the rate of 4.5 kg/s. It enters the turbine system with a velocity of 90 m/s and a specific volume of 0.85 m^3/kg. It leaves the turbine system with a specific volume of 1.45 m^3/kg. The exit area of the turbine system is 0.38 m^2. In its passage through the turbine, the specific enthalpy of air is reduced by 200 kJ/kg and there is a heat loss of 40 kJ/kg. Determine:

(i) The inlet area of turbine.
(ii) The exit velocity of air in m/s.
(iii) Power developed by the turbine system in kW.

Solution:

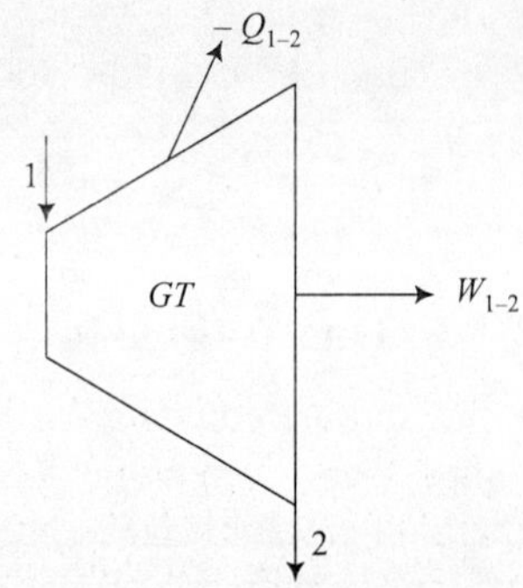

1. *Inlet conditions*

$$\dot{m} = 4.5 \text{ kg/s}$$
$$V_1 = 90 \text{ m/s}$$
$$v_1 = 0.85 \text{ m}^3\text{/kg}$$

2. *Outlet conditions*

$$v_2 = 1.45 \text{ m}^3\text{/kg}$$
$$A_2 = 0.38 \text{ m}^2$$
$$Q_{1-2} = -40 \text{ kJ/kg}$$
$$\Delta h = 200 \text{ kJ/kg}$$

3. *Equation of continuity*

$$\dot{m} = \frac{A_1 V_1}{v_1} = \frac{A_2 V_2}{v_2}$$

$$\therefore \quad A_1 = \frac{\dot{m} V_1}{v_1} = \frac{4.5 \times 0.85}{90} = \mathbf{0.0425 \text{ m}^2} \quad \textbf{Ans.}$$

$$V_2 = \frac{\dot{m} V_2}{A_2} = \frac{4.5 \times 1.45}{0.38} = \mathbf{17 \text{ m/s}} \quad \textbf{Ans.}$$

4. *Steady flow energy equation*

$$Q_{1-2} - W_{1-2} = \dot{m}\left[(h_2 - h_1) + \frac{V_2^2 - V_1^2}{2 \times 10^3} + \frac{g}{10^3}(Z_2 - Z_1)\right]$$

$$-40 \times 4.5 - W_{1-2} = 4.5\left[200 + \frac{(17)^2 - (90)^2}{2 \times 10^3} + 0\right]$$

$$W_{1-2} = \mathbf{702.43 \text{ kW}} \quad \textbf{Ans.}$$

2.7.4 Steam Turbine

High pressure and high temperature steam from a steam generator or a boiler is admitted through a steam turbine. During expansion of steam through the turbine, there is enthalpy drop. The turbine gives out positive mechanical work. Although it is insulated, there can be some heat loss due to radiation. The main characteristics of the system are:

1. There is enthalpy drop through the turbine.
2. Change of potential energy is usually neglected

$$Z_2 = Z_1$$

3. If $V_2 = V_1$, change of kinetic energy is neglected.
4. The heat loss due to radiation is –ve.

$$W_{1-2} = \dot{m}(h_1 - h_2) + Q_{1-2}$$

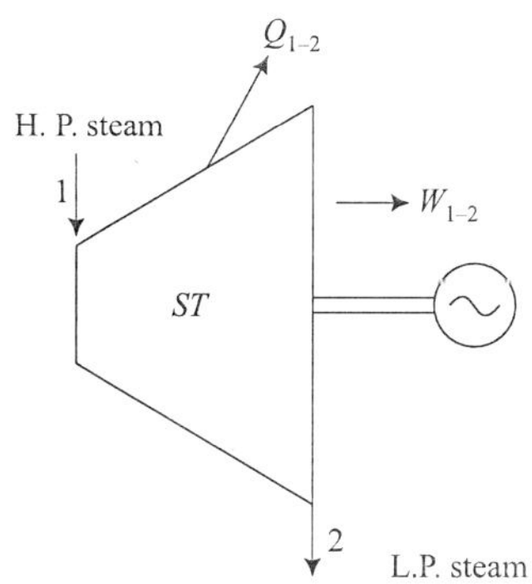

Fig. 2.10 Steam turbine

Example 2.11: The following details refer to a steam turbine:

Steam flow rate = 1 kg/sec

Inlet velocity of steam = 100 m/sec

Exit velocity of steam = 150 m/sec

Enthalpy at inlet = 2900 kJ/kg

Enthalpy at outlet = 1600 kJ/kg

Write the steady flow energy equation. Assuming that the change in potential energy is negligible, determine the power available from the turbine.

Solution:

1. *Inlet Data*

$\dot{m}$ = 1 kg/sec

V_1 = 100 m/s

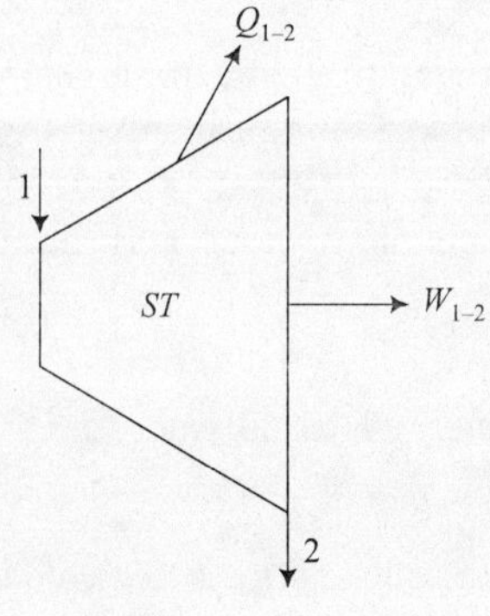

$$h_1 = 2900 \text{ kJ/sec}$$

2. *Outlet data,*

$$V_2 = 150 \text{ m/sec}$$
$$h_2 = 1600 \text{ kJ/sec}$$

3. *Steady flow energy equation*

$$Q_{1-2} - W_{1-2} = \dot{m}\left[(h_2 - h_1) + \frac{V_2^2 - V_1^2}{2 \times 10^3} + \frac{g}{10^3}(Z_2 - Z_1)\right]$$

Here $\quad Q_{1-2} = 0$

$$Z_2 - Z_1 = 0$$

$$\therefore \quad -W_{1-2} = \dot{m}\left[(h_2 - h_1) + \frac{V_2^2 - V_1^2}{2 \times 10^3}\right]$$

$$= 1\left[(1600 - 2900) + \frac{(150)^2 - (100)^2}{2 \times 10^3}\right]$$

$$= -1300 + 6.25$$

$$\therefore \quad W_{1-2} = 1293.75 \text{ kW}$$

Example 2.8 A steam turbine operating under steady flow conditions receives 3600 kg of steam per hour. The steam enters the turbine at a velocity of 80 m/s an elevation of 10 m and specific enthalpy of 3276 kJ/kg. It leaves the turbine at a velocity of 150 m/sec, an elevation of 3 m and a specific enthalpy of 2465 kJ/kg. Heat losses from the turbine to the surrounding amount to 36 MJ/hr. Estimate the power output of turbine.

Solution:

1. *Inlet data,*

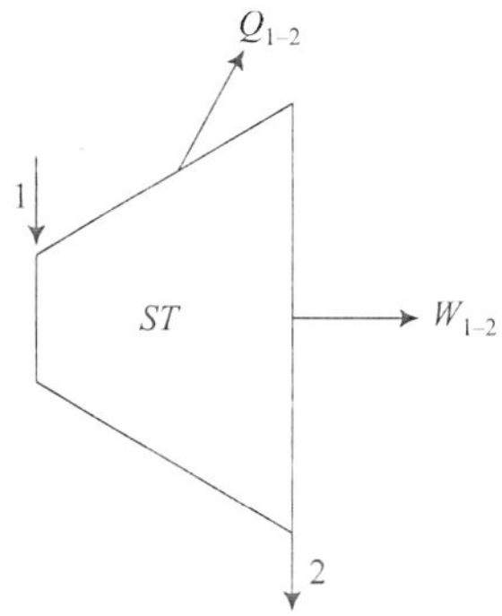

$$\dot{m} = 3600 \text{ kg/hour} = 1 \text{ kg/sec}$$
$$V_1 = 80 \text{ m/s}$$
$$Z_1 = 10 \text{ m}$$
$$h_1 = 3276 \text{ kJ/kg}$$
$$Q_{1-2} = -36 \text{ MJ/hour} = -\frac{36 \times 10^3}{3600} \text{ kJ/sec}$$
$$= -10 \text{ kJ/sec}$$

2. *Outlet conditions,*

$$V_2 = 150 \text{ m/sec}$$
$$Z_2 = 3 \text{ m}$$
$$h_2 = 2465 \text{ kJ/kg}$$

3. *Steady flow energy equation,*

$$Q_{1-2} - W_{1-2} = \dot{m}\left[(h_2 - h_1) + \frac{V_2^2 - V_1^2}{2 \times 10^3} + \frac{g}{10^3}(Z_2 - Z_1)\right]$$
$$-10 - W_{1-2} = 1\left[(2465 - 3276) + \frac{(150)^2 - (80)^2}{2 \times 10^3} + \frac{9.81}{10^3}(3 - 10)\right]$$
$$-10 - W_{1-2} = -811 + 8.05 - 0.06867$$
$$W_{1-2} = \mathbf{793 \text{ kW}} \quad \textbf{Ans.}$$

Example 2.9 The steam supply to an engine comprises two streams which mix before entering the engine. One stream is supplied at the rate of 0.01 kg/s with an enthalpy of 2950 kJ/kg and a velocity of 20 m/s. The other stream is supplied at the rate of 0.1 kg/s with an enthalpy of 2665 kJ/kg and a velocity of 120 m/s. At the exit from the engine the fluid leaves as two streams, one of water at the rate of 0.001 kg/s with an enthalpy of 421 kJ/kg and the other of steam. The fluid velocities at the exit are negligible. The engine develops a shaft power of 25 kW. The heat transfer is negligible. Evaluate the enthalpy of the second exit stream.

Solution:

1. *Inlet data,*

$$m_1 = 0.01 \text{ kg/s}$$
$$h_1 = 2950 \text{ kJ/kg}$$

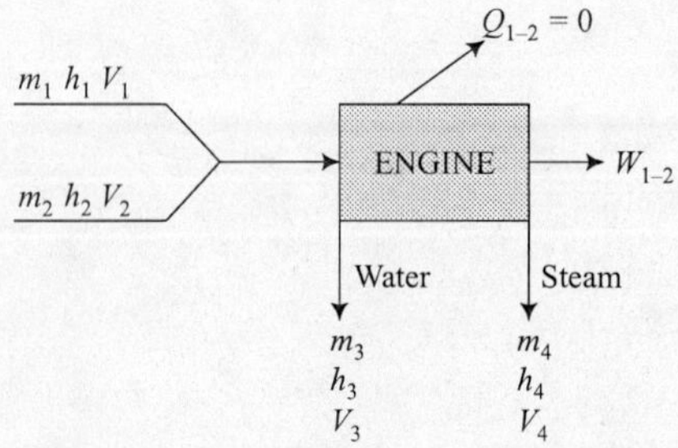

$$V_1 = 20 \text{ m/s}$$
$$m_2 = 0.1 \text{ kg/s}$$
$$h_2 = 2665 \text{ kJ/kg}$$
$$V_2 = 120 \text{ m/s}$$

Total inlet energy

$$E_i = m_1 h_1 + \frac{m_1 V_1^2}{2 \times 10^3} + \frac{g}{10^3} Z_1 + Q_{1-2} + m_2 h_2 + \frac{m_1 V_2^2}{2 \times 10^3} + \frac{g}{10^3} Z_2$$

$$= 0.01 \times 2950 + \frac{0.01 \times (20)^2}{2 \times 10^3} + 0 + 0 + 0.1 \times 2665 + \frac{0.1 \times 120^2}{2 \times 10^3} + 0$$

$$E_i = 29.5 + 0.002 + 266.5 + 0.72 = 296.722 \text{ kJ}$$

$$m_i =1\ 0.01 + 0.1 = 0.11 \text{ kg/s}$$

2. *Outlet data*

$$m_3 = 0.001 \text{ kg/s}$$
$$h_3 = 421 \text{ kJ/kg}$$
$$V_3 = 0$$
$$Z_3 = 0$$
$$m_4 = m_i - m_3 = 0.11 - 0.001 = 0.109 \text{ kg/sec}$$

Total energy,

$$E_e = m_3 h_3 + m_4 h_4 + W_{1-2}$$
$$= 0.001 \times 421 + 0.109 h_4 + 25$$
$$= 0.109 h_4 + 25.421$$

For steady flow conditions,

$$E_i = E_e$$

$$\therefore \quad 296.722 = 0.109 h_4 + 25.421$$

$$\therefore \quad h_4 = \frac{296.722 - 25.421}{0.109} = \mathbf{2489\ kJ/kg} \quad \textbf{Ans.}$$

2.7.5 Nozzle

The nozzle converts the pressure energy of a stream into its kinetic energy and as a result the velocity of the stream increases. The enthalpy drop of the fluid is used to accelerate the flow. Nozzles are used in steam turbines, gas turbines, pumps, etc. The operating characteristics of a nozzle are:

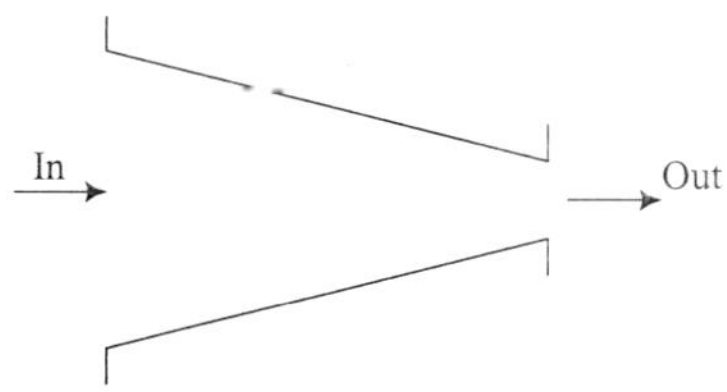

Fig. 2.11 Nozzle

1. There is no work output,

   $$W_{1-2} = 0$$

2. The heat loss is normally absent,

   $$Q_{1-2} = 0$$

3. The change of potential energy is negligible,

   $$Z_2 = Z_1$$

4. The steady flow energy equation,

   $$\frac{V_2^2 - V_1^2}{2} = (h_1 - h_2)$$

   $$V_2 = \sqrt{2(h_1 - h_2) + V_1^2}$$

   If $V_2 >> V_1$ and V_1 can be neglected,

   $$V_2 = \sqrt{2(h_1 - h_2)}$$

   The mass flow rate,

   $$\dot{m} = \frac{A_1 V_1}{v_1} = \frac{A_2 V_2}{v_2}$$

The area of nozzle at the inlet and outlet can be estimated.

Example 2.10 In an isentropic flow through a nozzle, air flows at the rate of 600 kg/hr. At inlet to nozzle, pressure is 2 MPa and temperature is 127°C. The exit pressure is 0.5 MPa. If the initial air velocity is 300 m/s, determine:

(i) Exit velocity of air, and
(ii) Inlet and exit area of nozzle.

Solution:

1. *Inlet conditions,*

$$p_1 = 2 \text{ MPa} = 2 \times 10^3 \text{ kPa}$$

$$T_1 = 127°\text{C} + 273 = 400 \text{ K}$$

$$\dot{m} = 600 \text{ kg/hr} = \frac{600}{3600} = \frac{1}{6} \text{ kg/s}$$

$$V_1 = 300 \text{ m/s}$$

$$Q_{1-2} = 0$$

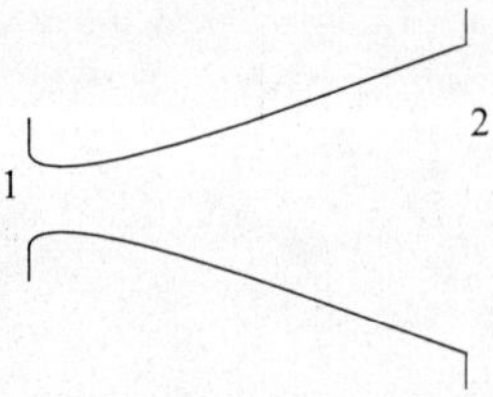

2. *Outlet conditions,*

$$p_2 = 0.5 \text{ MPa} = 0.5 \times 10^3 \text{ kPa}$$

The flow through nozzle is isentropic,

$$\therefore \quad \frac{T_2}{T_1} = \left(\frac{p_2}{p_1}\right)^{\frac{\gamma = 1}{\gamma}}$$

For air $\gamma = 1.4$

$$\therefore \quad T_2 = 400\left(\frac{0.5}{2}\right)^{\frac{1.4-1}{1.4}} = 269.18 \text{ K}$$

$$W_{1-2} = 0$$

3. *Steady flow energy equation,*

$$Q_{1-2} - W_{1-2} = \dot{m}\left[(h_2 - h_1) + \frac{V_2^2 - V_1^2}{2 \times 10^3} + \frac{g}{10^3}(Z_2 - Z_1)\right]$$

$$h_2 - h_1 = C_p(T_2 - T_1)$$

$$\therefore \quad 0 - 0 = \frac{1}{6}\left[1.005(400 - 269.18) + \frac{V_2^2 - (300)^2}{2 \times 10^3} + 0\right]$$

$$\therefore \quad 0 = 131.474 + \frac{V_2^2 - (300)^2}{2 \times 10^3}$$

$$\therefore \quad V_2 = \sqrt{131.474 \times 2 \times 10^3 + 300^2}$$

$$= \mathbf{594\ m/s} \quad \textbf{Ans.}$$

4. *Equation of continuity*

$$\dot{m} = \frac{A_1 V_1}{v_1} = \frac{A_2 V_2}{v_2}$$

From equation of state,

$$p_1 v_1 = RT_1$$

$$v_1 = \frac{RT_1}{p_1} = \frac{0.287 \times 400}{2 \times 10^3} = 0.0574 \text{ m}^3/\text{kg}$$

For air, $R = 0.287$ kJ/kg-K

$$\therefore \quad A_1 = \frac{\dot{m} V_1}{v_1} = \frac{\frac{1}{6} \times 0.0574}{300} = 3.1888 \times 10^{-5} \text{ m}^2$$

$$= 31.88 \text{ mm}^2$$

$$v_2 = \frac{RT_2}{p_2} = \frac{0.287 \times 269.18}{0.5 \times 10^3} = 0.1545 \text{ m}^3/\text{kg}$$

$$\therefore \quad A_2 = \frac{\dot{m} v_2}{V_2} = \frac{\frac{1}{6} \times 0.1545}{594} = 4.335 \times 10^{-5} \text{ m}^2$$

$$= \mathbf{43.35\ mm^2} \quad \textbf{Ans.}$$

The exit area of the nozzle is more than inlet area. Therefore, the nozzle is a convergent-divergent nozzle.

2.7.6 Diffuser

A diffuser has varying cross-section and reduces the velocity of the flowing fluid. There are two types of diffusers:

1. *Subsonic Diffuser*

   The velocity of the fluid is less then sonic speed and the area of cross-section of diffuser increases from inlet to exit.

2. *Supersonic Diffuser*

   The velocity of fluid is more than sonic velocity and the area of diffuser decreases.

The operating characteristics of a diffuser are similar to that for a nozzle.

Example 2.11 Water vapour at 90 kPa and 150°C enters a subsonic diffuser with a velocity of 150 m/s and leaves the diffuser at 190 kPa with a velocity of 55 m/s and during the process 1.5 kJ/kg of heat is lost to the surrounding. Determine:

(i) The final temperature
(ii) The mass flow rate
(iii) The exit diameter, assuming the inlet diameter as 10 cm and steady flow.

Solution:

1. *Inlet condition,*

$$p_1 = 90 \text{ kN/m}^2$$
$$T_1 = 150°\text{C} + 273 = 423 \text{ K}$$
$$V_1 = 150 \text{ m/s}$$

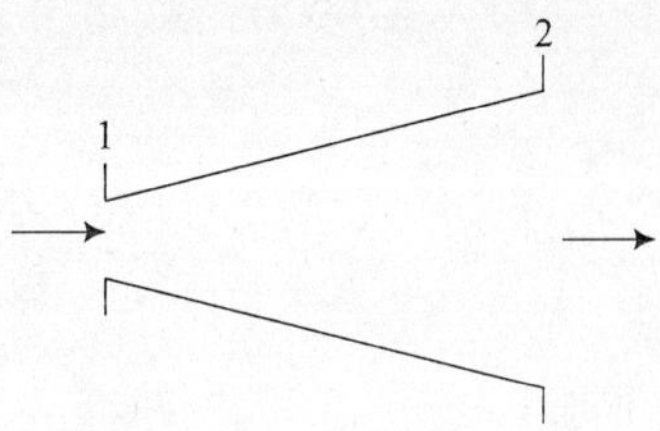

2. *Outlet condition,*

$$p_2 = 190 \text{ kN/m}^2$$
$$V_2 = 55 \text{ m/s}$$

For water vapours take Cp = 2.1 kJ/kg-K

3. *Steady flow energy equation,*

$$Q_{1-2} - W_{1-2} = (h_2 - h_1) + \frac{V_2^1 - V_1^2}{2 \times 10^3} + \frac{g}{10^3}(Z_2 - Z_1)$$

For the given diffuser,

$$W_{1-2} = 0$$
$$Z_2 = Z_1$$

$$\therefore \quad q_{1-2} = (h_2 - h_1) + \frac{V_2^1 - V_1^2}{2 \times 10^3}$$

$$-1.5 = Cp(T_2 - T_1) + \frac{(55)^2 - (150)^2}{2 \times 10^3}$$

$$Cp(T_2 - T_1) = -1.5 + 9.7375 = 8.2375$$

$$T_2 = \frac{8.2375}{2.1} + 423 = 427 \text{ K}$$

or $\quad t_2 = 427 - 273 = $ **154°C Ans.**

4. *Mass flow rate,*

From equation of state,

$$p_1 v_1 = RT_1$$

Take, $$R = \frac{\bar{R}}{M} = \frac{8.314}{18} = 0.4619 \text{ kJ/kg-K}$$

$\therefore$ $$v_1 = \frac{0.4619 \times 423}{90} = 2.17 \text{ m}^3\text{/kg}$$

$$v_2 = \frac{0.4619 \times 427}{190} = 1.038 \text{ m}^3\text{/kg}$$

$$\dot{m} = \frac{A_1 V_1}{v_1} = \frac{\frac{\pi}{4} \times (0.10)^2 \times 150}{2.17} = \mathbf{0.543 \text{ kg/s}} \quad \textbf{Ans.}$$

5. *Exit diameter*,

$$A_2 = \frac{\dot{m} v_2}{V_2} = \frac{0.543 \times 1.038}{55} = 0.010248 \text{ m}^2$$

$\therefore$ $$D_2 = \sqrt{\frac{A_2 \times 4}{\pi}} = \sqrt{\frac{0.010248 \times 4}{\pi}} = 0.1142 \text{ m}$$

$$= \mathbf{11.42 \text{ cm}} \quad \textbf{Ans.}$$

Example 2.12 An air compressor compresses atmospheric air at 0.1 MPa and 27°C by ten times of inlet pressure. During compression, the heat loss to the surrounding is estimated to be 5% of compression work. Air enters the compressor with a velocity of 40 m/s and leaves with 100 m/s. Inlet and exit cross-sectional areas are 100 cm^2 and 20 cm^2 respectively. Estimate the temperature of air at the exit from compressor and power input to the compressor.

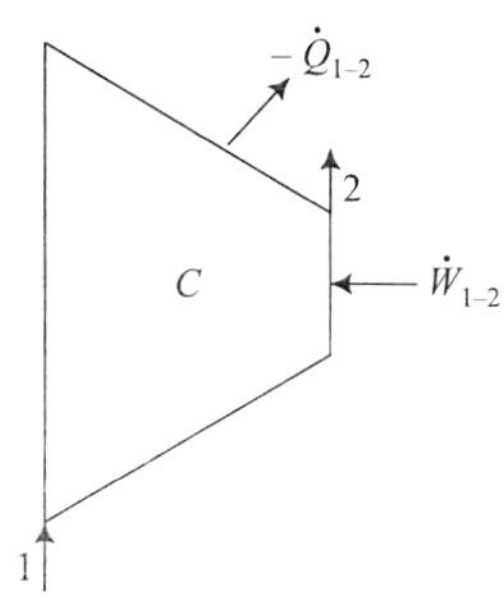

Solution

1. *Inlet conditions*

$$p_1 = 0.1 \text{ MPa} = 0.1 \times 10^3 \text{ kN/m}^2$$
$$T_1 = 27°\text{C} + 273 = 300 \text{ K}$$
$$V_1 = 40 \text{ m/s}$$
$$A_1 = 100 \text{ cm}^2 = 100 \times 10^{-4} \text{ m}^2$$

2. *Outlet conditions*

$$p_2 = 10 \times 0.1 \times 10^3 \text{ kN/m}^2 = 10^3 \text{ kN/m}^2$$
$$V_2 = 100 \text{ m/s}$$
$$A_2 = 20 \text{ cm}^2 = 20 \times 10^{-4} \text{ m}^2$$

3. *Equation of continuity*

The equation of state,

$$p_1 V_1 = mRT_1$$

or $$p_1 v_1 = RT_1$$

$$\therefore \quad v_1 = \frac{RT_1}{p_1} = \frac{0.287 \times 300}{100}$$

$$= 0.861 \text{ m}^3/\text{kg} \qquad [R = 0.287 \text{ kJ/kg-K for air}]$$

$$\dot{m} = \frac{A_1 V_1}{v_1} = \frac{(100 \times 10^{-4}) \times 40}{0.861} = 0.4646 \text{ kg/s}$$

$$\dot{m} = \frac{A_2 V_2}{v_2}$$

$$\therefore \quad v_2 = \frac{A_2 V_2}{\dot{m}} = \frac{(20 \times 10^{-4}) \times 100}{0.4646} = 0.43 \text{ m}^3/\text{kg}$$

$$p_2 v_2 = RT_2$$

$$T_2 = \frac{p_2 v_2}{R} = \frac{10^3 \times 0.43}{0.287} = 1498 \text{ K}$$

4. *Steady flow energy equation*

$$Q_{1-2} - W_{1-2} = \dot{m}\left[(h_2 - h_1) + \frac{V_2^2 - V_1^2}{2 \times 10^3} + \frac{g}{10^3}(Z_2 - Z_1)\right]$$

$$-0.05\,W_{1-2} - W_{1-2} = 0.43\left[1.005(1498-300) + \frac{(100)^2-(40)^2}{2\times 10^3} + 0\right]$$

W_{1-2} = **546.86 kW Ans.**

Example 2.13 A steam turbine operating under steady flow conditions receives 3600 kg of steam per hour. The steam enters the turbine at a velocity of 80 m/s, an elevation of 10 m and specific enthalpy of 3276 kJ/kg. It leaves the turbine at a velocity of 150 m/sec, an elevation of 3 m and a specific enthalpy of 2465 kJ/kg. Heat losses from the turbine to the surrounding amount to 36 MJ/hr. Estimate the power output of turbine.

Solution

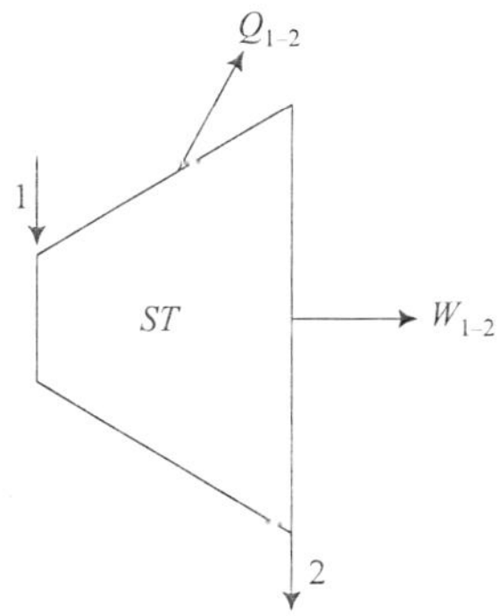

1. *Inlet data,*

$\dot{m}$ = 3600 kg/hour = 1 kg/sec

V_1 = 80 m/s

Z_1 = 10 m

h_1 = 3276 kJ/kg

$$Q_{1-2} = -36 \text{ MJ/hour} = -\frac{36\times 10^3}{3600} \text{ kJ/sec}$$

$$= -10 \text{ kJ/sec}$$

2. *Outlet conditions,*

V_2 = 150 m/sec

Z_2 = 3 m

h_2 = 2465 kJ/kg

3. *Steady flow energy equation,*

$$Q_{1-2} - W_{1-2} = \dot{m}\left[(h_2 - h_1) + \frac{V_2^2 - V_1^2}{2\times 10^3} + \frac{g}{10^3}(Z_2 - Z_1)\right]$$

$$-10 - W_{1-2} = 1\left[(2465-3276) + \frac{(150)^2-(80)^2}{2\times 10^3} + \frac{9.81}{10^3}(3-10)\right]$$

$$-10 - W_{1-2} = -811 + 8.05 - 0.06867$$

W_{1-2} = **793 kW Ans.**

Example 2.14 In an isentropic flow through a nozzle, air flows at the rate of 600 kg/hr. At inlet to nozzle, pressure is 2 MPa and temperature is 127°C. The exit pressure is 0.5 MPa. If the initial air velocity is 300 m/s, determine:

(i) Exit velocity of air, and
(ii) Inlet and exit area of nozzle.

Solution

1. *Inlet conditions,*

$$p_1 = 2 \text{ MPa} = 2 \times 10^3 \text{ kPa}$$
$$T_1 = 127°\text{C} + 273 = 400 \text{ K}$$
$$\dot{m} = 600 \text{ kg/hr} = \frac{600}{3600} = \frac{1}{6} \text{ kg/s}$$
$$V_1 = 300 \text{ m/s}$$
$$Q_{1-2} = 0$$

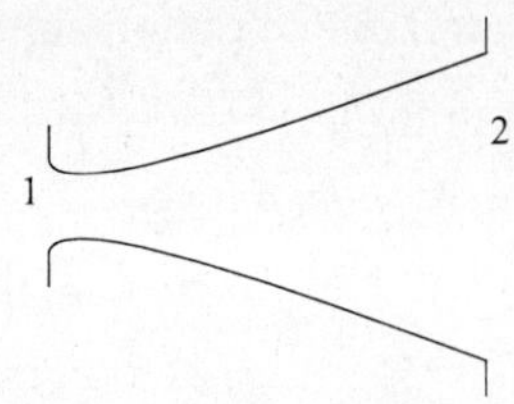

2. *Outlet conditions,*

$$p_2 = 0.5 \text{ MPa} = 0.5 \times 10^3 \text{ kPa}$$

The flow through nozzle is isentropic,

$$\therefore \quad \frac{T_2}{T_1} = \left(\frac{p_2}{p_1}\right)^{\frac{\gamma - 1}{\gamma}}$$

For air $\gamma = 1.4$

$$\therefore \quad T_2 = 400 \left(\frac{0.5}{2}\right)^{\frac{1.4-1}{1.4}} = 269.18 \text{ K}$$

$$W_{1-2} = 0$$

3. *Steady flow energy equation,*

$$Q_{1-2} - W_{1-2} = \dot{m}\left[(h_2 - h_1) + \frac{V_2^2 - V_1^2}{2 \times 10^3} + \frac{g}{10^3}(Z_2 - Z_1)\right]$$

$$h_2 - h_1 = C_p(T_2 - T_1)$$

$$\therefore \quad 0 - 0 = \frac{1}{6}\left[1.005(400 - 269.18) + \frac{V_2^2 - (300)^2}{2 \times 10^3} + 0\right]$$

$$\therefore \quad 0 = 131.474 + \frac{V_2^2 - (300)^2}{2 \times 10^3}$$

$$\therefore \quad V_2 = \sqrt{131.474 \times 2 \times 10^3 + 300^2}$$

$$= \mathbf{594 \text{ m/s}} \quad \textbf{Ans.}$$

4. *Equation of continuity*

$$\dot{m} = \frac{A_1 V_1}{v_1} = \frac{A_2 V_2}{v_2}$$

From equation of state,

$$p_1 v_1 = RT_1$$

$$v_1 = \frac{RT_1}{p_1} = \frac{0.287 \times 400}{2 \times 10^3} = 0.0574 \text{ m}^3/\text{kg}$$

For air, $R = 0.287$ kJ/kg-K

$$\therefore \quad A_1 = \frac{\dot{m} V_1}{v_1} = \frac{\frac{1}{6} \times 0.0574}{300} = 3.1888 \times 10^{-5} \text{ m}^2$$

$$= 31.88 \text{ mm}^2$$

$$v_2 = \frac{RT_2}{p_2} = \frac{0.287 \times 269.18}{0.5 \times 10^3} = 0.1545 \text{ m}^3/\text{kg}$$

$$\therefore \quad A_2 = \frac{\dot{m} v_2}{V_2} = \frac{\frac{1}{6} \times 0.1545}{594} = 4.335 \times 10^{-5} \text{ m}^2$$

$$= \mathbf{43.35 \text{ mm}^2} \quad \textbf{Ans.}$$

The exit area of the nozzle is more than inlet area. Therefore, the nozzle is a convergent-divergent nozzle.

QUESTION BANK NO. 2

1. State the 1st Law of Thermodynamics for a closed system undergoing a change of state. Also show that the total energy is a property of the system.
2. Derive general energy equation for steady flow.
3. Classify different types of thermodynamic processes. Compare the work done in non-flow and flow processes.
4. Show that the the work in done, W in non-flow process for polytropic process, i.e., pV^n = constant is given by

$$W = (p_1 V_1 - p_2 V_2)/n - 1$$

5. Prove that the work in a steady flow adiabatic process is given by

$$W_{1-2} = \frac{\gamma}{\gamma - 1}(p_1 V_1 - p_2 V_2)$$

6. Compare the following processes:
   (i) Free expansion process
   (ii) Throttling process
7. Briefly describe the following processes:
   (i) Joule's paddle wheel experiment
   (ii) Joule's-Thomson Porous Plug experiment

CHAPTER 3

The Second Law of Thermodynamics

3.1 SECOND LAW STATEMENTS

It is the directional law of energy and also the law of degradation of energy. It is based on the following statements.

(a) Kelvin-Planck statement

"It is impossible to construct an engine working on a cyclic process, whose sole purpose is to convert heat energy from a single reservoir into an equivalent amount of work." No cyclic engine can convert whole of heat into equivalent work. Second law dictates limits on the conversion of heat into work.

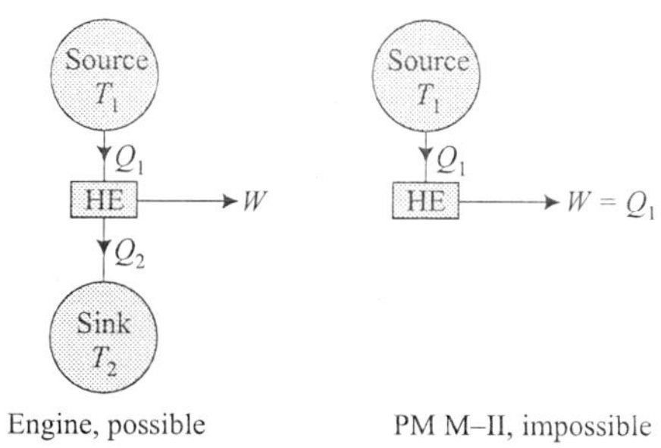

Fig. 3.1 Kelvin-Planck statement

(b) Clausius statement

"It is impossible for a self-acting machine, working in a cyclic process, to transfer heat from a body at a lower temperature to a body at a higher temperature without the aid of an external energy".

Heat cannot flow itself from a cold body to a hot body without expenditure of mechanical work. Second law deals with the direction of flow of heat energy.

(c) Carnot statement

"No heat engine operating in a cycle between two given thermal reservoirs, with fixed temperatures, can be more efficient than a reversible engine operating between the same thermal reservoirs".

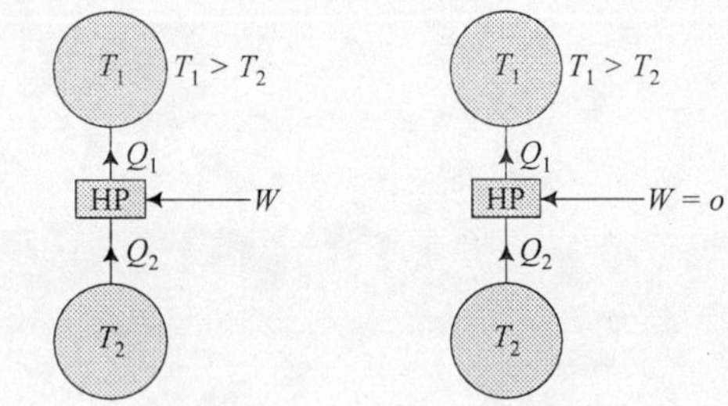

Fig. 3.2 Clausius statement

Second law provides a yardstick to evaluate the performance of an engine or a device. The portion of heat energy which is not available for conversion into work is measured by entropy.

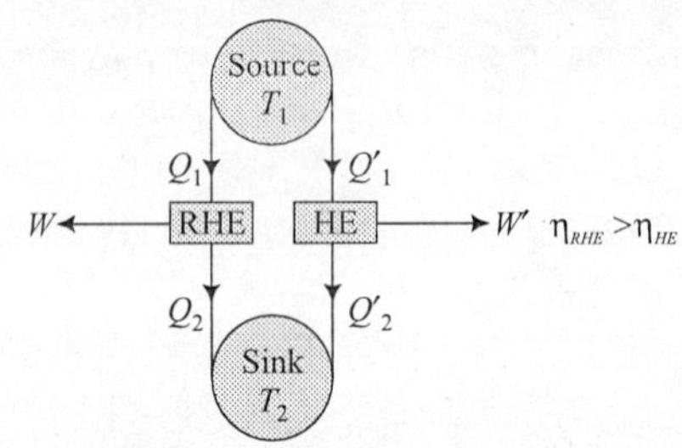

Fig. 3.3 Carnot statement

3.2 EQUIVALENCE OF SECOND LAW STATEMENTS

The Kelvin-Planck statement and Clausius statements are two statements of second law of thermodynamics (see para 3.1). These statements look different but these are two parallel statements and are equivalent in all respects. The equivalence of these statements will be proved by the logic that violation of one statement leads to violation of second statement and vice versa.

3.2.1 Violation of Clausius Statement

A cyclic heat engine (HE) operates between two reservoirs drawing heat Q_1 and producing work, W as shown in Fig. 3.4(a).

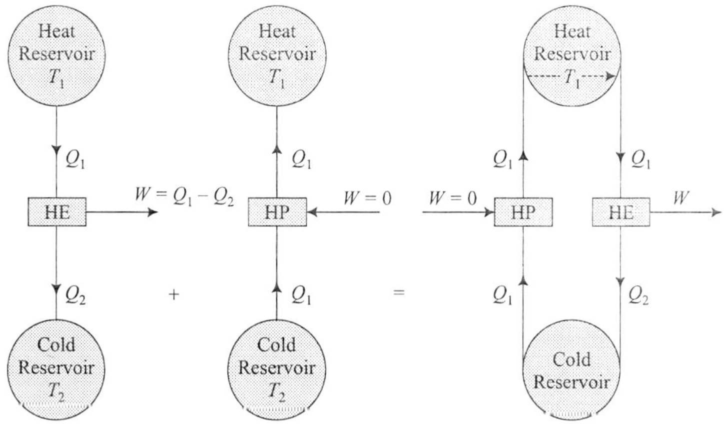

(a) Heat Engine (b) Heat Pump Violating Clausius Statement (c) Heat Engine system violating Kelvin-Planck statement

Fig. 3.4 Violation of Clausius statement

A cyclic heat pump (HP) is shown in Fig. 3.4(b) which is transferring heat from cold reservoir (T_2) to a hot reservoir (T_1) with no work input ($W = 0$). This violates Clausius statement as heat cannot flow from a cold temperature to a higher temperature without any external work input.

Figure 3.4(c) shows a heat engine (HE) and heat pump (HP) combined to form a heat engine system. The heat pump is supplying heat Q_1 to hot reservoir which is the requirement of heat engine. Therefore, hot reservoir can be by-passed. The heat pump and heat engine T form a system operating in cycles and producing work W continuously while exchanging heat with one reservoir (cold) only. Thus, the Kelvin-Planck statement is violated. Hence, when Clausius statement is violated, simultaneously, Kelvin-Plank statement is also violated.

3.2.2 Violation of Kelvin-Planck Statement

A heat pump (HP) is extracting heat Q_1 from low temperature reservoir (T_2) and discharging heat to high temperature reservoir (T_1) getting work from outside.

A heat engine (HE) is producing work W by exchanging heat with one reservoir at temperature T_1 only. The Kelvin-Planck statement is violated which requires two heat reservoirs for an engine to produce work continuously.

The heat pump and heat engine are combined in Fig. 3.5(c) to create a heat pump system. The heat pump is getting work W from heat engine. Therefore, heat pump system is transferring heat from a lower temperature (T_2) to a higher temperature (T_1) without any external work. Therefore, the Clausius statement is also violated.

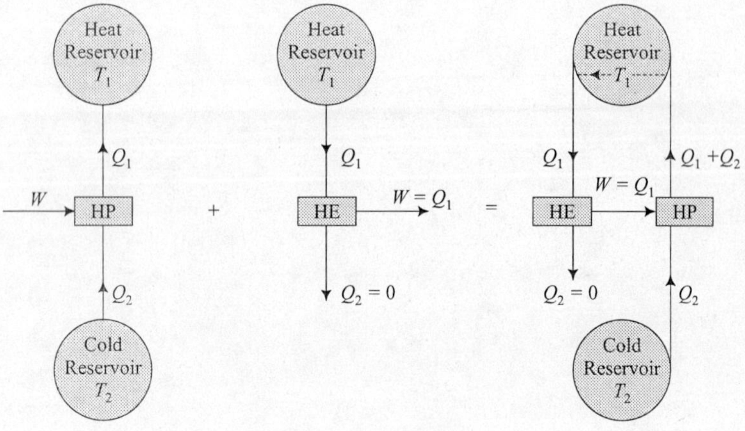

(a) Heat Pump (b) Heat Engine violating Kelvin-Planck statement (c) Heat Pump system violating Clausius statement

Fig. 3.5 Violation of Kelvin-Planck statement

From the above two cases, it is clear that when one of the statements of second law is violated, the other statement is also violated. Therefore, the two statements are equivalent.

3.3 PERFORMANCE OF HEAT ENGINE

A heat engine is a thermodynamic device for conversion of heat energy into mechanical work. It requires a source for continuous supply of heat. It also requires a sink to reject that part of heat which cannot be converted into work.

The heat engine can also be operated in reverse direction. The refrigerator and heat pump are reversed heat engines. These reversed engines transport heat energy from low temperature to higher temperature. For operation of these devices, power is required to operate them.

3.3.1 Thermal Reservoirs

A thermal reservoir exchanges heat with the system. It has the following characteristics:

1. It is a part of the environment.
2. It has sufficiently large capacity for supply or receipt of heat from the system.
3. The temperature of heat reservoir remains constant.
4. The temperature of heat reservoir is not affected by the quantity of heat transferred to or from it.

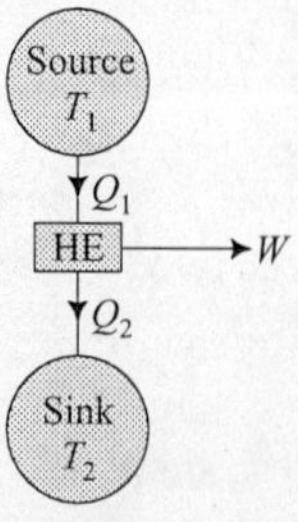

Fig. 3.6 Thermal reservoirs

5. The changes that take place in the thermal reservoir are very slow as heat enters or leaves it.
6. The processes within the thermal reservoirs are considered quasi-static.
7. There are two types of heat reservoirs; a source and a sink.
8. *Heat source* is the reservoir at high temperature and supplies heat to the system.

   Examples: boiler furnace, combustion chamber, nuclear reactor, sun, etc.
9. *Heat sink* is the reservoir at a low temperature and receives heat from the system.

   Examples: river, lake, ocean, atmospheric air, etc.

3.3.2 Heat Engine

It is a thermodynamic device which can produce mechanical work from heat continuously. It works on a cyclic process. Examples are steam engine, steam turbine, gas turbine, petrol engine, diesel engine, gas engine.

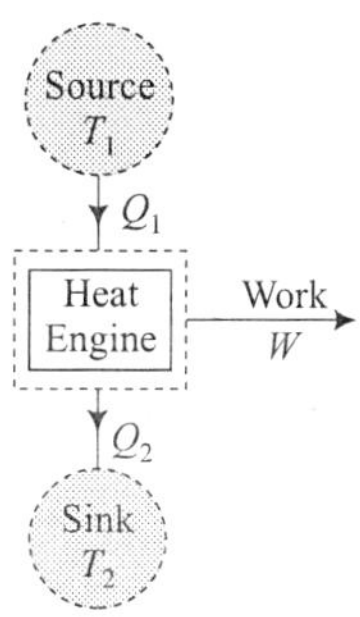

Fig. 3.7 Heat engine

The schematic diagram of a heat engine is shown in Fig. 3.7. The heat and work interactions take place across the boundary of the engine. There is a working substance which continuously flows through the engine to ensure continuous/ cyclic operation. It operates as follows:

1. It receives heat Q_1 from a high temperature thermal reservoir (source) at temperature T_1.
2. It converts part of the heat into mechanical work, W.
3. It rejects remaining heat Q_2 to a low temperature heat reservoir (sink) at temperature T_2.

   From 1st law of thermodynamics,

$$W = Q_1 - Q_2.$$

Thermal efficiency

The performance of a heat engine is measured by its thermal efficiency which is the degree of useful conversion of heat received into work.

The thermal efficiency,

$$\eta_{th} = \frac{\text{Network output}}{\text{Total heat supplied}} = \frac{\dot{W}}{\dot{Q}_1} = \frac{\dot{Q}_1 - \dot{Q}_2}{\dot{Q}_1}$$

$$\eta_{th} = 1 - \frac{\dot{Q}_2}{\dot{Q}_1}$$

For a reversible heat engine,

$$\frac{\dot{Q}_1}{\dot{Q}_2} = \frac{T_1}{T_2}$$

$$\therefore \qquad \eta_{th} = 1 - \frac{T_2}{T_1}$$

Observations

1. Thermal efficiency could be equal to unity, if $Q_1 \to \infty$ or $Q_2 \to 0$, which is practically not possible.
2. Thermal efficiency of a heat engine working between two thermal reservoirs is always less than unity.
3. If $\dot{Q}_2$ can be reduced by keeping $\dot{Q}_1$ constant, thermal efficiency can be increased.

3.3.3 Heat Pump

Heat pump is a reversed heat engine which is used to remove heat from a body at a low temperature and to transfer heat to a body at higher temperature. It is used to keep the room warm in winter.

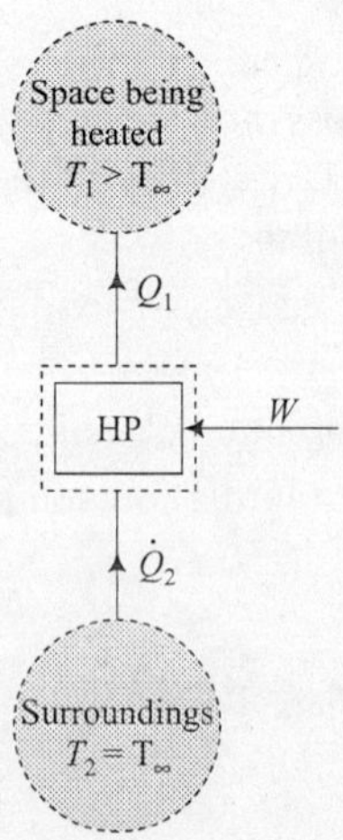

Fig. 3.8 Heat pump

If T_∞ is the atmospheric temperature and T_1 is the temperature of space to be heated, then

$$T_\infty = T_2$$

$$T_\infty < T_1$$

From 1st law of thermodynamics,

$$\dot{W} = \dot{Q}_1 - \dot{Q}_2$$

The schematic diagram of a heat pump is shown in Fig. 3.8. It operates as follows:

1. Heat pump receives heat $\dot{Q}_2$ for atmosphere which is at a temperature T_2 equal to atmospheric temperature, T_∞.
2. It receives external work, $\dot{W}$ and transfers heat from low temperature to a higher temperature.
3. It supplies heat $\dot{Q}_1$ to the space to be heated which is maintained at a temperature T_1 which is higher than atmospheric temperature.

Coefficient of performance (COP)

The performance of a heat pump is measured by its coefficient of performance. *COP* is the ratio of amount of heat supplied to the space and mechanical work received.

$$(COP)_{HP} = \frac{\dot{Q}_1}{\dot{W}} = \frac{\dot{Q}_1}{\dot{Q}_1 - \dot{Q}_2} = 1 + \frac{\dot{Q}_2}{\dot{Q}_1 - \dot{Q}_2}$$

For a heat pump working as reversible reversed heat engine.

$$COP = \frac{T_1}{T_1 - T_2}$$

3.3.4 Refrigerator

A refrigerator is a reversed heat engine which removes heat from a body at low temperature and transfers heat to a body at higher temperature. The objective of the system is to produce refrigeration effect and is used to preserve food and drugs by storing at low temperatures.

The schematic diagram of a refrigerator is shown in Fig. 3.9.

$$T_1 = T_\infty$$

$$T_\infty > T_2$$

$W = Q_1 - Q_2$ as per 1st law of thermodynamics.

The refrigerator works as follows:

1. It removes heat $\dot{Q}_2$ from space being cooled at low temperature T_2 which is lower than atmospheric temperature, T_∞.

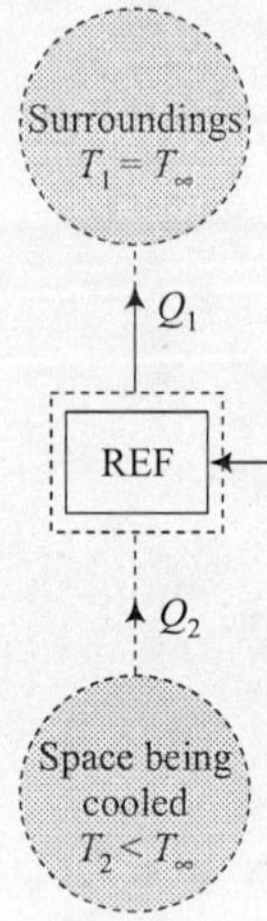

Fig. 3.9 Refrigerator

2. It rejects heat $\dot{Q}_1$ to the surrounding at temperature T_1 which is equal to atmospheric temperature, T_∞.
3. It receives power in the form of work, $\dot{W}$ to transfer heat from a lower temperature to higher temperature.

Coefficient of performance (COP)

The performance of a refrigerator is measured by its coefficient of performance. COP is the ratio of amount of heat extracted called *refrigerated effect* and work input

$$(CPO)_{REF} = \frac{\dot{Q}_2}{\dot{W}} = \frac{\dot{Q}_2}{\dot{Q}_1 - \dot{Q}_2}$$

Observation,

$$(COP)_{HP} = \frac{Q_1}{Q_1 - Q_2} = 1 + \frac{\dot{Q}_2}{\dot{Q}_1 - \dot{Q}_2}$$

$$(COP)_{REF} = \frac{\dot{Q}_2}{\dot{Q}_1 - \dot{Q}_2}$$

$$\therefore \quad (COP)_{HP} = 1 + (COP)_{REF}.$$

If refrigerator works as reversible reversed heat engine, $COP = \dfrac{T_2}{T_1 - T_2}$.

3.3.5 Carnot Refrigerator and Heat Pump

The coefficient of performance of a refrigerator working as reversed engine can be expressed as:

$$(\text{COP})_{\text{REF}} = \frac{Q_2}{Q_1 - Q_2}$$

From Carnot principle,

$$\frac{Q_1}{T_1} = \frac{Q_2}{T_2}$$

The COP of a Cannot refrigerator will be

$$\text{COP} = \frac{T_2}{T_1 - T_2}$$

The coefficient of performance of a heat pump is expressed as

$$(\text{COP})_{\text{HP}} = \frac{Q_2}{Q_1 - Q_2}$$

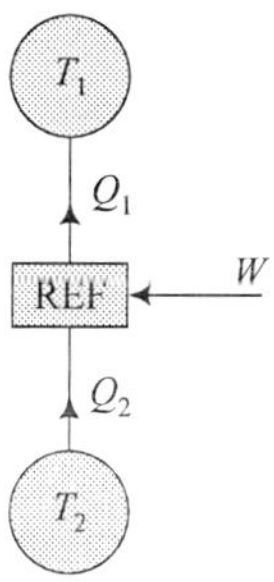

Fig. 3.10 Carnot refrigerator

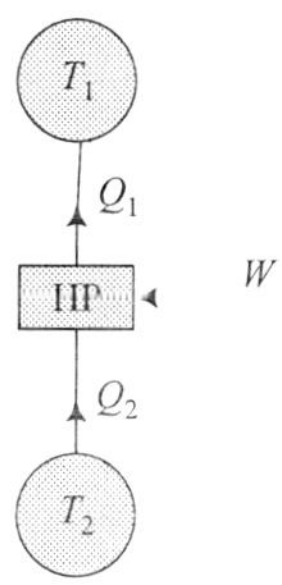

Fig. 3.11 Carnot heat pump

From Carnot principle,

$$\frac{Q_1}{T_1} = \frac{Q_2}{T_2}$$

The COP of a Carnot heat pump will be:

$$\text{COP} = \frac{T_1}{T_1 - T_2}$$

Example 3.1 Obtain the COP of the composite refrigerator system in which two reversible refrigerators A and B are arranged in series in terms of the COP of refrigerator A and COP of refrigerator B only.

Solution

1. Draw the schematic diagram of composite refrigerator system.

   Assume COP of refrigerator $1 = R_1$

COP of refrigerator 2 $= R_2$

COP of composite refrigerator $= R_{12}$

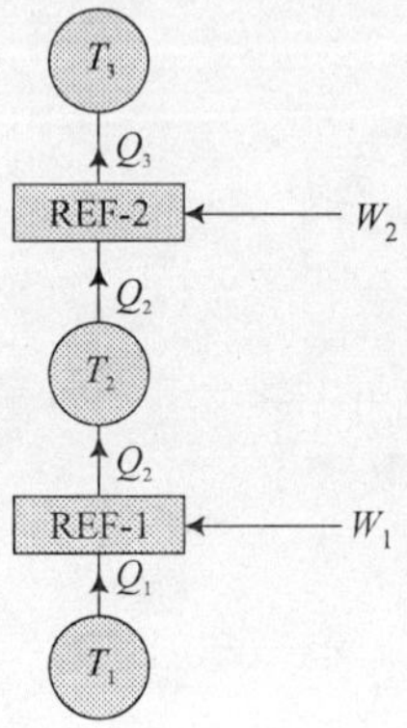

2. The refrigerators are reversible,

$$\therefore \quad R_1 = \frac{T_1}{T_2 - T_1} \qquad \text{...(1)}$$

$$R_2 = \frac{T_2}{T_3 - T_1} \qquad \text{...(2)}$$

$$R_{12} = \frac{T_1}{T_3 - T_1} \qquad \text{...(3)}$$

From equation (1),

$$T_2R_1 - T_1R_1 = T_1$$

$$T_2R_1 = T_1 + T_1R_1 = T_1(1 + R_1)$$

$$\therefore \quad T_2 = \frac{1 + R_1}{R_1} . T_1$$

From equation (2)

$$R_2 = \frac{\left(\frac{1 + R_1}{R_1}\right)T_1}{T_3 - \left(\frac{1 + R_1}{R_1}\right)T_1} = \frac{(1 + R_1)T_1}{T_3R_1 - (1 + R_1)T_1}$$

$$\therefore \quad T_3R_1R_2 - (1 + R_1)\,T_1R_2 = (1 + R_1)\,T_1$$

$$T_3 R_1 R_2 = (1 + R_1)T_1 + (1 + R_1)R_2 T_1$$

$$\therefore \quad T_3 = \frac{T_1[(1+R_1)+(1+R_1)R_2]}{R_1 R_2}$$

$$= \frac{T_1(1+R_1)(1+R_2)}{R_1 R_2}$$

Substituting T_3 into equation (3),

$$R_{12} = \frac{T_1}{\dfrac{T_1(1+R_1)(1+R_2)}{R_1 R_2} - T_1}$$

$$R_{12} = \frac{R_1 R_2}{(1+R_1)(1+R_2) - R_1 R_2}$$

Example 3.2 A Carnot engine E_1 operates between temperatures T_1 and T_2 and engine E_2 operates between temperatures T_2 and T_3 receiving heat rejected from engine E_1. What is the relationship of η_1 and η_2 with η_3 of engine E_3 working between T_1 and T_3?

Solution

1. Draw schematic diagram. The engines are reversible.

$$\therefore \quad \eta_1 = \frac{T_1 - T_2}{T_1} = 1 - \frac{T_2}{T_1}$$

$$1 - \eta_1 = \frac{T_2}{T_1} \qquad \text{...(1)}$$

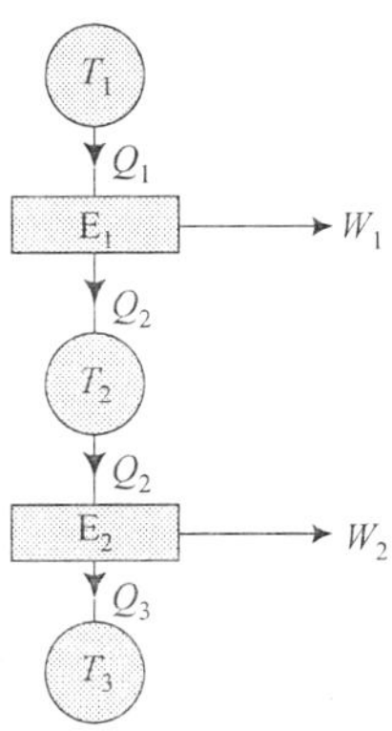

Composite engine

Similarly, $$1 - \eta_2 = \frac{T_3}{T_2} \qquad \text{...(2)}$$

Multiply equations (1) and (2),

$$\therefore \quad (1-\eta_1)(1-\eta_2) = \frac{T_2}{T_1} \cdot \frac{T_3}{T_2} = \frac{T_3}{T_1} \qquad \text{...(3)}$$

But, $$\eta_3 = \frac{T_1 - T_3}{T_1} = 1 - \frac{T_3}{T_1}$$

$$\therefore \quad 1 - \eta_3 = \frac{T_3}{T_1} \qquad \text{...(4)}$$

From equations (3) and (4),

$$1 - \eta_3 = (1 - \eta_1)(1 - \eta_2) = \frac{T_3}{T_1}$$

$$= 1 - \eta_2 - \eta_1 + \eta_1\eta_2$$

$$\therefore \quad \eta_3 = \eta_1 + \eta_2 - \eta_1\eta_2.$$

Example 3.3 Heat pump is used for heating the premises in winter and cooling the same in summer such that temperature inside remains 25°C. Heat transfer across the walls and roof is found 2 MJ per hour per degree temperature difference between interior and exterior. Determine the minimum power required for operating the pump in winter when outside temperature is 1°C and also give the maximum temperature in summer for which the device shall be capable of maintaining the premises at desired temperature for same power input.

Solution

(i) In winter, the device works as heat pump.

$$T_1 = 25°\text{C} + 273 = 298 \text{ K}$$

$$T_2 = 1°\text{C} + 273 = 274 \text{ K}$$

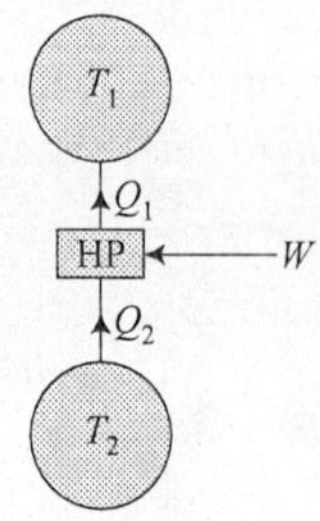

Heat pump

For a reversible heat pump,

$$(\text{COP})_{HP} = \frac{T_1}{T_1 - T_2} = \frac{298}{298 - 274} = 12.4$$

$$\text{Also, COP} = \frac{Q_1}{W}$$

$$Q_1 = \frac{2 \times 10^6}{3600}(298 - 274) = 13333.33 \text{ J} = 13.33 \text{ kJ}$$

$$W = \frac{13.33}{12.42} = \mathbf{1.074 \text{ kW}} \quad \textbf{Ans.}$$

(ii) In summer, the device works as air-conditioner,

$$(\text{COP})_{REF} = \frac{T_2}{T_1 - T_2} = \frac{Q_2}{W}$$

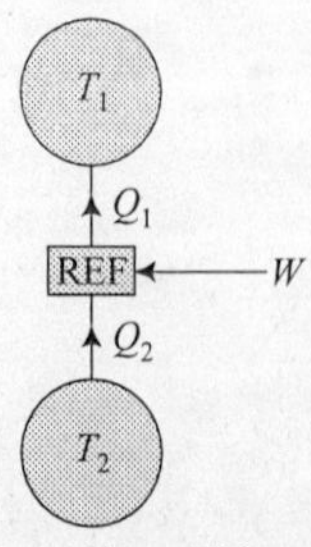

Refrigerator (Air-conditioner)

$$\therefore \quad \frac{298}{(298 - T_2)} = \frac{\frac{2 \times 10^6}{3600}(298 - T_2)}{1.074}$$

$$(298 - T_2)^2 = \frac{298 \times 1.074 \times 3600}{2 \times 10^6}$$

$$= 576$$

$$298 - T_2 = \sqrt{576} = 24$$

$$\therefore \quad T_2 = 298 + 24 = 322 \text{ K}$$

$$= 49°\text{C} \quad \textbf{Ans.}$$

3.4 CARNOT THEOREM

The Carnot theorem states:

1. "No heat engine operating in a cycle between two given thermal reservoirs, with fixed temperatures, can be more efficient than a reversible engine operating between the same thermal reservoirs".
2. "The efficiency of all reversible heat engines operating between the same temperature levels is the same".
3. "The efficiency of a reversible engine is independent of the nature and amount of the working substance undergoing the cycle".

E_A is any engine and E_B is a reversible engine working between the same thermal reservoirs at temperatures T_1 and T_2.

We have to prove that $\eta_B > \eta_A$. If it is not true, then $\eta_A > \eta_B$.

1. Assume

$$Q_{1A} = Q_{1B} = Q_1$$

and

$$\eta_A > \eta_B$$

$$\therefore \quad \frac{W_A}{Q_{1A}} > \frac{W_B}{Q_{1B}}$$

$$\therefore \quad \frac{W_A}{Q_1} > \frac{W_B}{Q_1}$$

or

$$W_A > W_B$$

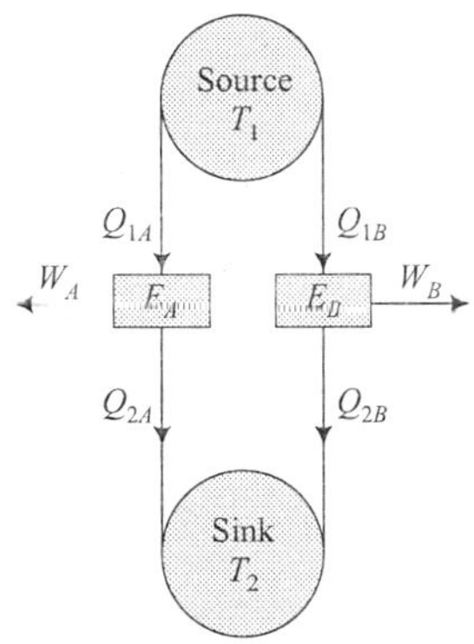

Fig. 3.12 Two engines working between same reservoir

2. *Reverse Engine E_B*

   The engine E_B is reversible. Reverse the engine B so that all input and output energy values are same but directions are reversed. Since W_A is more than W_B, therefore, part of W_A (equal to W_B) may be used to drive the reversed heat engine $\exists_B$.

   The heat Q_{1B} discharged by $\exists_B$ may be supplied to E_A. The source can be eliminated.

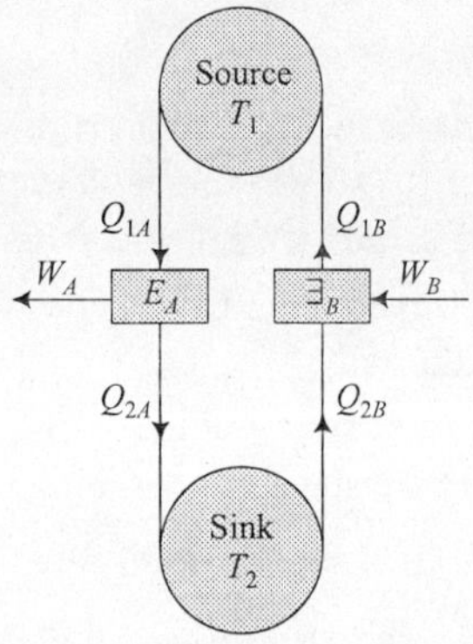

Fig. 3.13 Engine E_B is reversed

3. *Combined Engines ($E_A + \exists_B$)*

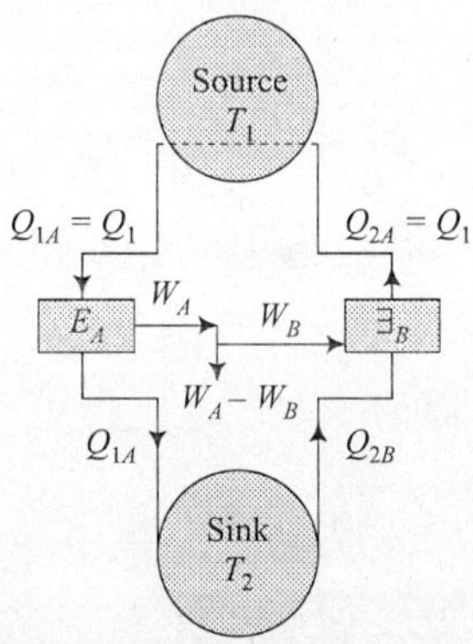

Fig. 3.14 Combined operation of $E_A + \exists_B$

The combined engines $E_A + \exists_B$ constitute a heat engine system operating on a single reservoir (T_2) and still produces work equal to $W_A - W_B$. This system violates the Kelvin-Planck statement of second law of thermodynamics.

Therefore, our assumption that $\eta_A > \eta_B$ is wrong.

$$\therefore \qquad \eta_B \geq \eta_A$$

The efficiency of an irreversible engine cannot be greater than that of the reversible engine if both operate between same thermal reservoirs with fixed temperatures.

Example 3.4 Identify the following machines and discuss the feasibility of each.

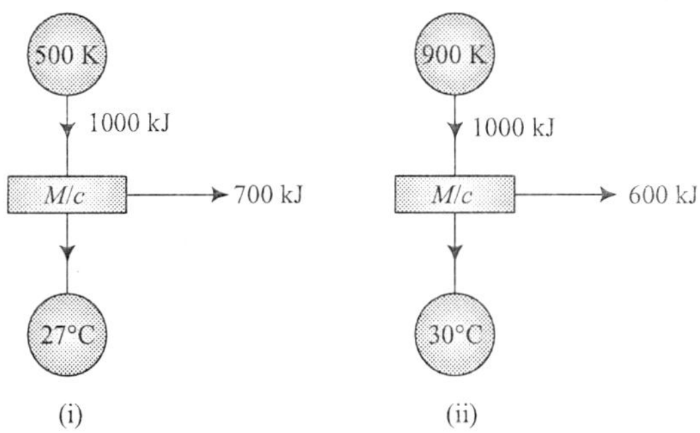

Fig. 3.15

(i) Machine is a heat engine producing 700 kJ of work using a heat input of 1000 kJ

$$T_1 = 500 \text{ K}$$
$$T_2 = 27°\text{C} + 273 = 300 \text{ K}$$

Carnot efficiency, $\eta_{carnot} = \dfrac{T_1 - T_2}{T_1} = \dfrac{500 - 300}{500} = 0.40$

Actual efficiency, $\eta = \dfrac{W}{Q_1} = \dfrac{700}{1000} = 0.70$

The machine is not possible as actual efficiency cannot be more than Carnot efficiency.

(ii) Machine is a heat engine producing 600 kJ of work using a heat input of 1000 kJ

$$T_1 = 900 \text{ K}$$
$$T_2 = 30°\text{C} + 273 = 303 \text{ K}$$

Carnot efficiency, $\eta_{carnot} = \dfrac{T_1 - T_2}{T_1} = \dfrac{900 - 303}{900} = 0.66$

Actual efficiency, $\eta = \dfrac{W}{Q_1} = \dfrac{600}{100} = 0.60$

The machine is possible as well as irreversible because actual efficiency is less than Carnot efficiency.

3.5 CLAUSIUS INEQUALITY

The clausius theorem states: "Whenever a closed system undergoes a cyclic process, the cyclic integral $\oint \frac{dQ}{T}$ is less than zero (i.e., negative) for an irreversible cyclic process and equal to zero for a reversible cyclic process".

Mathematically,

$$\oint \frac{dQ}{T} < 0 \text{ for an irreversible process}$$

$$\oint \frac{dQ}{T} = 0 \text{ for reversible process}$$

Combining the equations for reversible and irreversible processes

$$\oint \frac{dQ}{T} \geq 0$$

The equation for irreversible cyclic process may be written as,

$$\oint \frac{dQ}{T} + I = 0$$

where, I = Amount of irreversibility of a cyclic process

$I = 0$ for a reversible cyclic process.

1. *Reversible Heat Engine* (RHE)

The efficiency of a reversible heat engine operating between temperature limits T_1 and T_2 is given by,

$$\eta = \frac{Q_1 - Q_2}{Q_1} = \frac{T_1 - T_2}{T_1} = 1 - \frac{T_2}{T_1}$$

$$\therefore \quad \frac{T_2}{T_1} = \frac{Q_2}{Q_1}$$

or $$\frac{Q_1}{T_1} = \frac{Q_2}{T_2}$$

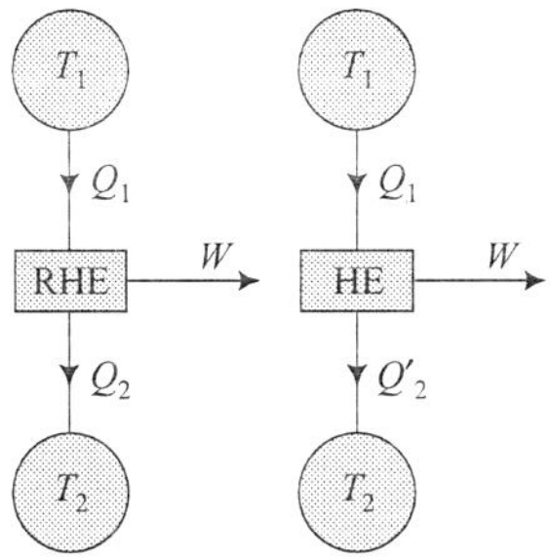

Fig. 3.16 Clausius inequality

The heat rejected by the system is negative

$$\therefore \qquad \left(\frac{Q_1}{T_1}\right) - \left(-\frac{Q_2}{T_2}\right) = 0$$

$$\text{or} \qquad \frac{Q_1}{T_1} + \frac{Q_2}{T_2} = 0$$

Extending the equation to any number of heat reservoirs,

$$\oint \frac{dQ}{T} = 0 \text{ for a reversible cycle.}$$

2. *Irreversible Heat Engine* (HE)

The efficiency of an irreversible heat engine operating between the same temperature limits T_1 and T_2.

$$\eta < \eta_c$$

$$\therefore \qquad \frac{Q_1 - Q_2'}{Q_1} < \frac{T_1 - T_2}{T_1}$$

$$\left(1 - \frac{Q_2'}{Q_1}\right) < \left(1 - \frac{T_2}{T_1}\right)$$

$$\frac{Q_2'}{Q_1} > \frac{T_2}{T_1}$$

$$\text{or} \qquad \frac{Q_2'}{T_2} > \frac{Q_1}{T_1}$$

$$\text{or} \qquad \frac{Q_1}{T_1} < \frac{Q_2'}{T_2}$$

If Q'_2 is –ve, being heat rejected by the system,

$$\frac{Q_1}{T_1} - \left(\frac{-Q'_2}{T_2}\right) < 0$$

$$\frac{Q_1}{T_1} + \frac{Q'_2}{T_2} < 0$$

Extending the case to arbitrary number of heat reservoirs

$$\oint \frac{dQ}{T} < 0 \text{ for an irreversible cycle.}$$

Statement for Clausius Inequality:

> "When a system operates on a cycle (reversible or irreversible) having heat interaction with a number of heat reservoirs, then the algebraic sum of the quotients of heat transfers from heat reservoirs to their respective absolute temperature is always equal to or less than zero (respectively)".

Example 3.5 A heat engine is supplied with 2512 kJ/min of heat at 650°C. Heat rejection takes place at 100°C. Specify the following results of heat rejection:

(i) 867 kJ/min
(ii) 1015 kJ/min
(iii) 1494 kJ/min

Solution The schematic diagram of the engine is shown,

$$T_1 = 650°\text{C} + 273 = 923 \text{ K}$$
$$T_2 = 100°\text{C} + 273 = 373 \text{ K}$$
$$Q_1 = 2512 \text{ kJ/min}$$

(i) $Q_2 = 867$ kJ/min

Applying Clausius inequality,

$$\frac{Q_1}{T_1} - \frac{Q_2}{T_2} = \frac{2512}{923} - \frac{867}{373} = 0.34770 > 0$$

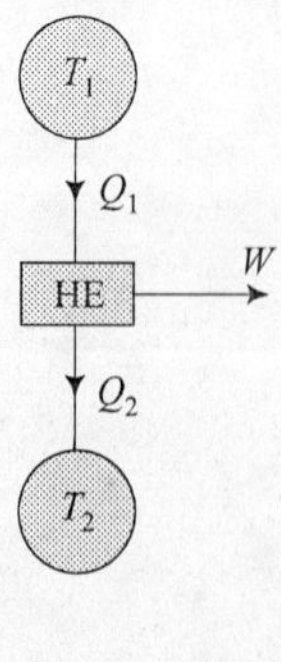

The cycle is not possible.

(ii) $Q_2 = 1015$ kJ/min

$$\frac{2512}{923} - \frac{1015}{373} = 0$$

The cycle is reversible.

(iii) $Q_2 = 1494$ kJ/min

$$\frac{2512}{923} - \frac{1494}{373} = -1.284 < 0$$

The cycle is irreversible and possible.

Example 3.6 A heat engine working on Carnot cycle absorbs heat from three reservoirs at 1000 K, 800 K and 600 K. The engine does 10 kW of net work and rejects 400 kJ/min of heat to a heat sink at 300 K. If heat supplied by the reservoir at 1000 K is 60% of the heat supplied by reservoir at 600 K. Calculate the quantity of heat absorbed from each reservoir.

Solution The schematic diagram of engine is shown,

$$T_1 = 600 \text{ K}$$

$$T_2 = 800 \text{ K}$$

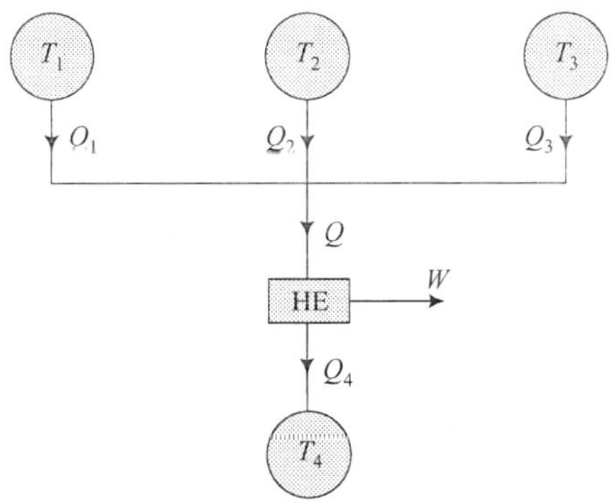

$$T_3 = 1000 \text{ K}$$

$$T_4 = 300 \text{ K}$$

$$Q_4 = 400 \text{ kJ/min}$$

$$W = 10 \text{ kW} = 10 \times 60 = 600 \text{ kJ/min}$$

$$Q = Q_1 + Q_2 + Q_3 = W + Q_4$$

$$= 600 + 400$$

$$= 1000 \text{ kJ/min}$$

$$Q_3 = 0.6Q_1$$

$$Q_2 = Q - (Q_1 + Q_3) = 1000 - Q_1 - 0.6Q_1$$

$$= 1000 - 1.6Q_1$$

Apply Clausius theorem,

$$\frac{Q_1}{T_1} + \frac{Q_2}{T_2} + \frac{Q_3}{T_3} - \frac{Q_4}{T_4} = 0$$

$$\frac{Q_1}{600} + \frac{(1000 - 1.6Q_1)}{800} + \frac{0.6Q_1}{1000} - \frac{400}{300} = 0$$

$$Q_1 = \mathbf{312.5\ kJ/min}$$

$$Q_2 = 1000 - 1.6(312.5) = \mathbf{500\ kJ/min}$$

$$Q_3 = 0.6\ (312.5) = \mathbf{187.5\ kJ/min} \qquad \textbf{Ans.}$$

QUESTION BANK NO. 3

1. State and explain the second law of thermodynamics.
2. Prove the equivalence of Kelvin-Planck and Clausius statements of second law of thermodynamics.
3. Define engine efficiency and COP of a heat pump and a refrigerator. Prove that COP of heat pump is more than COP of refrigerator.
4. State and prove Carnot theorem.
5. Explain Clausius inequality and prove it.

CHAPTER 4

Entropy and Thermodynamic Relations

4.1 CONCEPT OF ENTROPY

The concept of entropy was introduced by Clausius and it means transformation. It is a thermodynamic property of a working substance, which increases with the addition of heat, and decreases with its removal. In a reversible process, over a small range of temperature, the change in entropy multiplied by absolute temperature is equal to heat exchanged.

$$\delta Q = TdS$$

where, T = absolute temperature

dS = change of entropy

The main *characteristics* of entropy are:

1. Entropy is a thermodynamic property of the system.
2. It is a point (state) function and not a path function.
3. It increases with the addition of heat and decreases with its removal.
4. The area under T-s diagram of any thermodynamic process represents the heat added or heat rejected.
5. Entropy remains constant in a reversible adiabatic process, but increases in an irreversible adiabatic process.
6. The change in entropy represents the maximum amount of work obtainable per degree drop in temperature.
7. The change in entropy is a measure of rate of availability (or unavailability) of heat for transformation into work.
8. Entropy is an extensive property of the system.
9. The specific entropy is expressed as kJ/kg-K.
10. Increase in entropy is obtained from a given quantity of heat at low temperature.
11. The datum for zero entropy is OK.

4.2 ENTROPY: A THERMODYNAMIC PROPERTY

Entropy is a point function and not a path function. Therefore, it is a property of the system.

In Fig. 4.1, $1A2B1$ is a reversible cyclic process consisting of two reversible processes $1A2$ and $2B1$. As per Clausius theorem,

$$\int_{1A}^{2A} \frac{dQ}{T} + \int_{2B}^{1B} \frac{dQ}{T} = 0 \quad ...(1)$$

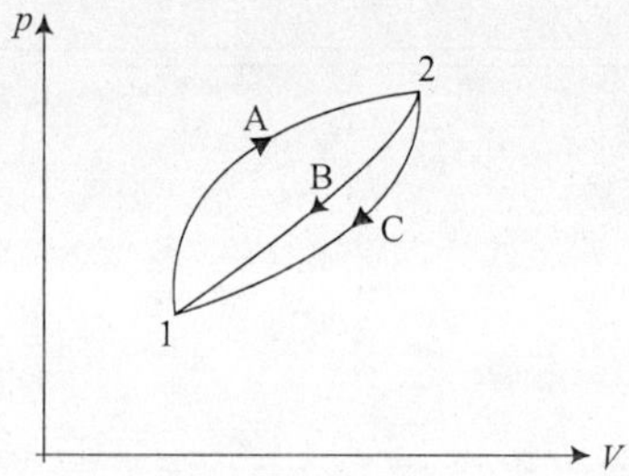

Fig. 4.1 Entropy

For another reversible cyclic process $1A\,2\,C1$ consisting of processes $1A2$ and $2C1$, Clausius theorem gives,

$$\int_{1A}^{2B} \frac{dQ}{T} + \int_{2C}^{1C} \frac{dQ}{T} = 0 \quad ...(2)$$

Subtracting equation (2) from equation (1),

$$\int_{2B}^{1B} \frac{dQ}{T} - \int_{2C}^{1C} \frac{dQ}{T} = 0$$

Changing the limits of integral

$$\int_{1B}^{2B} \frac{dQ}{T} - \int_{1C}^{2C} \frac{dQ}{T} = 0$$

or

$$\int_{1B}^{2B} \frac{dQ}{T} = \int_{1C}^{2C} \frac{dQ}{T}$$

Therefore, $\int \frac{dQ}{T}$ is independent of the path of the process and is a state function. The change of entropy,

$$dS = \frac{dQ}{T}$$

For a process 1–2,

$$\int_1^2 dS = \int_1^2 \frac{dQ}{T}$$

$$\therefore \qquad S_2 - S_1 = \int_1^2 \frac{dQ}{T}$$

Therefore, entropy is a property of the system.

4.3 TEMPERATURE ENTROPY DIAGRAM

Process 1–2 is a reversible heating process shown on *T-s* diagram.

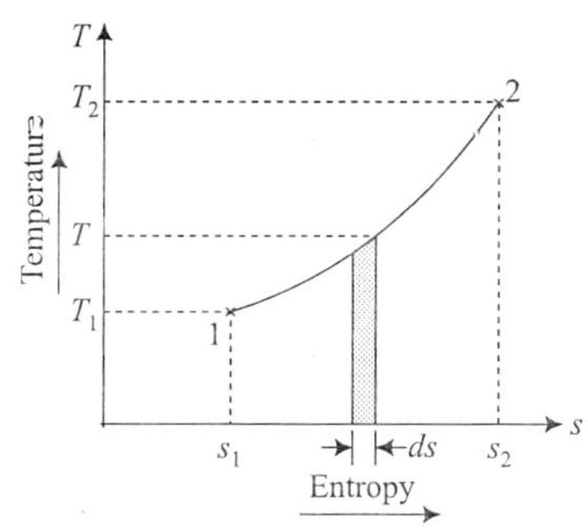

Fig. 4.2 *T-s* diagram

When a small quantity of heat dQ is supplied at temperature T, then

$$dQ = TdS$$

$$dS = \frac{dQ}{T}$$

$$\int_1^2 dS = \int_1^2 \frac{dQ}{T}$$

The area under any thermodynamic process on *T-s* diagram represents the heat added or rejected.

1. *Constant volume process*

A constant volume process 1–2 is shown on *T-s* diagram. For a constant volume process,

$$dQ = mC_v dT$$

$$\therefore \quad \frac{dQ}{T} = mCv\frac{dT}{T}$$

$$dS = mCv\frac{dT}{T}$$

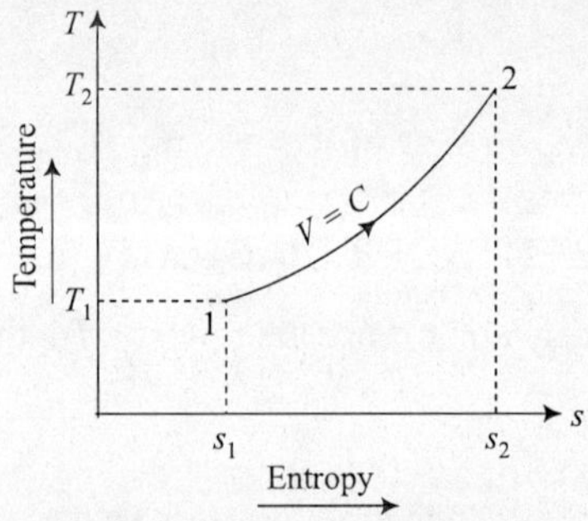

Fig. 4.3 Constant volume process

Integrating between limits 1 and 2,

$$\int_{S_1}^{S_2} ds = mCv \int_{T_1}^{T_2} \frac{dT}{T}$$

$$\therefore \quad S_2 - S_1 = mCv \ln\frac{T_2}{T_1}$$

For $V = C$,

$$\frac{T_2}{T_1} = \frac{p_2}{p_1}$$

$$\therefore \quad S_2 - S_1 = mCv \ln\left(\frac{p_2}{p_1}\right)$$

Again,

$$dS = mCv\frac{dT}{T}$$

$$\therefore \quad \frac{dT}{dS} = \frac{T}{mCv}$$

This equation gives the shape (or slope) of process 1–2 on *T-s* diagram. Equations are valid for both reversible and irreversible processes.

2. *Constant pressure process*

A constant pressure process is plotted on *T-s* diagram. For a constant pressure process,

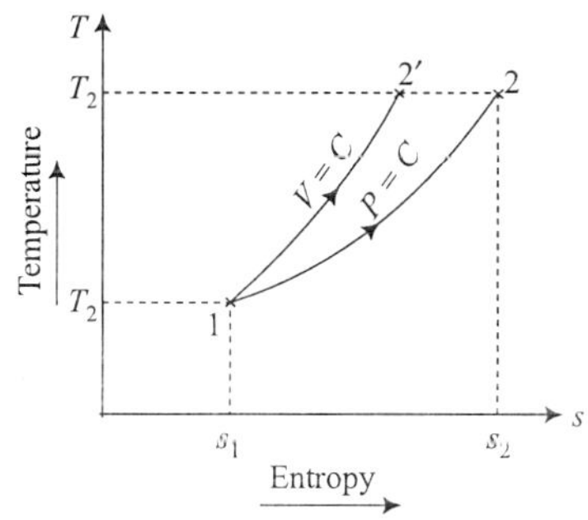

Fig. 4.4 Constant pressure process

$$dQ = mCpdT$$

$$\therefore \quad \frac{dQ}{T} = mCp\frac{dT}{T}$$

$$dS = mCp\left(\frac{dT}{T}\right)$$

$$\int_{S_1}^{S_2} dS = mCp \int_{T_1}^{T_2} \frac{dT}{T}$$

$$S_2 - S_1 = mCp \ln\frac{T_2}{T_1}$$

For $p = C$,
$$\frac{T_2}{T_1} = \frac{V_2}{V_1}$$

$$\therefore \quad S_2 - S_1 = mCp \ln\left(\frac{V_2}{V_1}\right)$$

This equation is valid for both reversible and irreversible processes.

Again,
$$dS = mCp\frac{dT}{T}$$

$$\therefore \qquad \frac{dT}{dS} = \frac{T}{Cp} = \text{slope of process line 1–2}$$

For constant volume process,

$$\frac{dT}{dS} = \frac{T}{Cv}$$

$$\because \qquad Cv < Cp$$

$$\therefore \qquad \frac{1}{Cv} > \frac{1}{Cp}$$

$$\therefore \qquad \frac{T}{Cv} > \frac{T}{Cp}$$

The slope of constant volume process is higher than that for constant pressure process.

3. *Isothermal process*

An isothermal process 1–2 is shown on *T-s* diagram. For isothermal process,

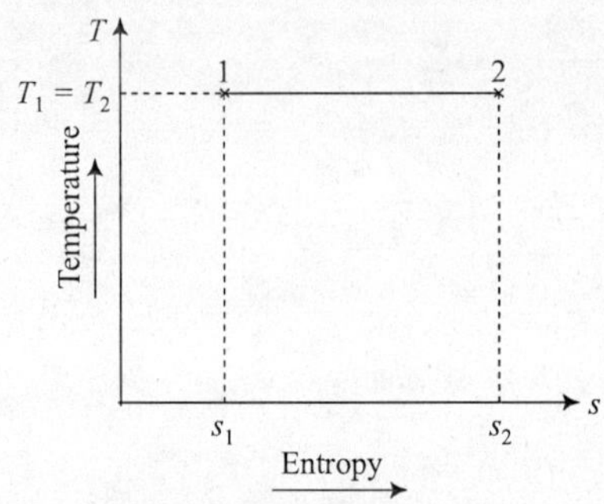

Fig. 4.5 Isothermal process

$$Q_{1-2} = W_{1-2} = mRT \ln\left(\frac{V_2}{V_1}\right)$$

$$\therefore \qquad S_2 - S_1 = \frac{mRT}{T} \ln\left(\frac{V_2}{V_1}\right)$$

$$= mR \ln\left(\frac{V_2}{V_1}\right)$$

$$= m(Cp - Cv) \ln\left(\frac{V_2}{V_1}\right)$$

For $T = C$, $$\frac{V_2}{V_1} = \frac{p_1}{p_2}$$

$\therefore$ $$S_2 - S_1 = mR \ln\left(\frac{p_1}{p_2}\right)$$

or $$S_2 - S_1 = m(Cp - Cv) \ln\left(\frac{p_1}{p_2}\right)$$

The equation is valid for reversible and irreversible processes.

4. *Adiabatic process*

A reversible adiabatic process 1–2 is shown on *T*-*s* diagram whereas irreversible adiabatic process 1–2 is shown a dotted line.

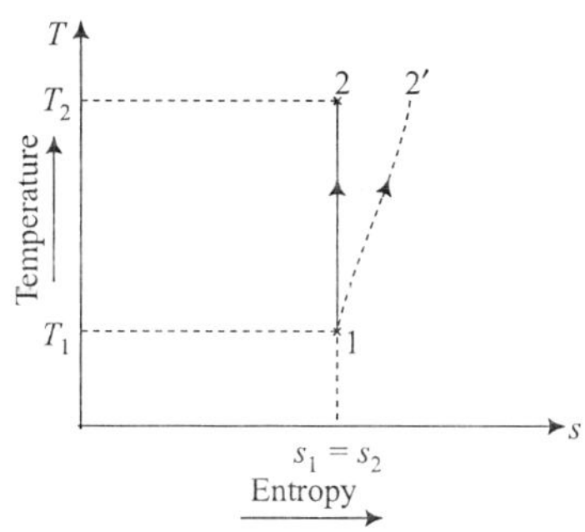

Fig. 4.6 Isentropic process

For adiabatic process,

$$dQ = 0$$

$\therefore$ $$dS = 0 \text{ for reversible process.}$$

For irreversible adiabatic process, there is increase in entropy due to internal heating.

5. *Polytropic process*

A polytropic process 1–2 is shown on *T*-*s* diagram. For a polytropic process,

$$dQ = \frac{\gamma - n}{\gamma - 1} dW$$

Now, $$dW = pdV$$

$\therefore$ $$dQ = \frac{\gamma - n}{\gamma - 1} pdV$$

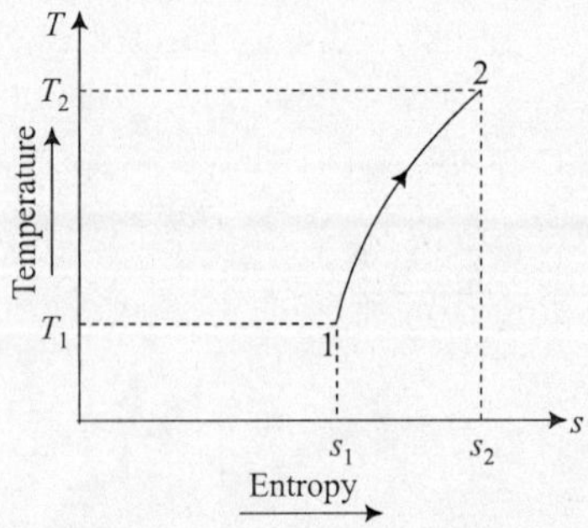

Fig. 4.7 Polytropic process

$$\frac{dQ}{T} = \frac{\gamma - n}{\gamma - 1}\frac{pdV}{T}$$

$$\therefore \qquad dS = \frac{\gamma - n}{\gamma - 1} mR \frac{dV}{V}$$

$$\therefore \qquad pdV = mRT$$

$$\therefore \qquad \frac{p}{T} = \frac{mR}{V}$$

Integrating,

$$S_2 - S_1 = \frac{\gamma - n}{\gamma - 1} mR \ln\left(\frac{V_2}{V_1}\right)$$

6. *Carnot cycle*

Carnot cycle is represented on *T-s* diagram.

Process 1–2: Isothermal expansion
Process 2–3: Isentropic expansion
Process 3–4: Isothermal compression
Process 4–1: Isentropic compression

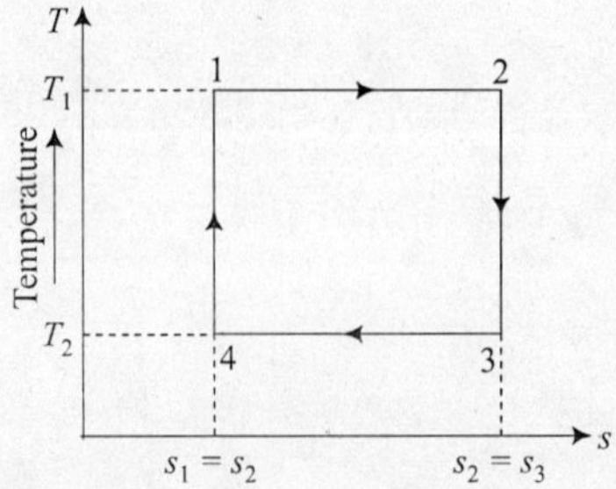

Fig. 4.8 Carnot cycle

Heat added, $Q_1 = T_1(s_2 - s_1)$

Heat rejected, $Q_2 = T_2(s_4 - s_3)$

$$= -T_2(s_3 - s_4)$$

$$s_2 - s_1 = s_3 - s_4$$

Work done during cycle,

$$W = Q_1 - Q_2 = (T_1 - T_2)\,(s_2 - s_1)$$

Thermal efficiency,

$$\eta = \frac{W}{Q_1} = \frac{(T_1 - T_2)(s_2 - s_1)}{T_1(s_2 - s_1)} = \frac{T_1 - T_2}{T_1}$$

4.4 ENTROPY CHANGE FOR IDEAL GASES

Changes of entropy of a certain quantity, m of a perfect gas during heating by any thermodynamic process from state 1 ($p_1V_2T_1$) to state 2 ($p_2V_2T_2$) can be found out as follows:

1. *In terms of volume and temperature*

General energy equation of 1st law of thermodynamics:

$$dQ = dU + dW = m\,Cv\,dT + pdV$$

$$\therefore \qquad \frac{dQ}{T} = m\,Cv\,\frac{dT}{T} + \frac{pdV}{T}$$

From equation of state, $pV = mRT$

$$\frac{p}{T} = \frac{mR}{V}$$

$$\therefore \qquad \frac{dQ}{T} = dS = m\,Cv\frac{dT}{T} + mR\frac{dV}{V}$$

$$\int_{s_1}^{s_2} dS = m\,Cv\int_{T_1}^{T_2} \frac{dT}{T} + mR\int_{V_1}^{V_2}\frac{dV}{V}$$

$$\therefore \qquad S_2 - S_1 = mCv\ln\left(\frac{T_2}{T_1}\right) + mR\ln\left(\frac{V_2}{V_1}\right)$$

$$= m\left[Cv\ln\frac{T_2}{T_1} + (Cp - Cv)\ln\left(\frac{V_2}{V_1}\right)\right]$$

2. *In terms of pressure and temperature*

From equation of state,

$$\frac{p_1 V_1}{T_1} = \frac{p_2 V_2}{T_2}$$

$$\therefore \quad \frac{V_2}{V_1} = \frac{p_1 T_2}{p_2 T_1}$$

$$S_2 - S_1 = mCv \ln\left(\frac{T_2}{T_1}\right) + mR\left(\frac{p_1}{p_2} \times \frac{T_2}{T_1}\right)$$

$$= mCv \ln\left(\frac{T_2}{T_1}\right) + mR \ln\left(\frac{p_1}{p_2}\right) + mR \ln\left(\frac{T_2}{T_1}\right)$$

$$= m(Cv + R) \ln\left(\frac{T_2}{T_1}\right) + mR \ln\left(\frac{p_1}{p_2}\right)$$

Now, $R = Cp - Cv$

$$Cv + R = Cp$$

$$\therefore \quad S_2 - S_1 = mCp \ln\left(\frac{T_2}{T_1}\right) + m(Cp - Cv) \ln\left(\frac{p_1}{p_2}\right)$$

$$= m\left[Cp \ln\left(\frac{T_2}{T_1}\right) + (Cp - Cv) \ln\left(\frac{p_1}{p_2}\right)\right]$$

3. *In terms of pressure and volume*

$$\frac{p_1 V_1}{T_1} = \frac{p_2 V_2}{T_2}$$

$$\therefore \quad \frac{T_2}{T_1} = \frac{p_2}{p_1} \times \frac{V_2}{V_1}$$

$$S_2 - S_1 = mCv \ln\left(\frac{T_2}{T_1}\right) + mR \ln\left(\frac{V_2}{V_1}\right)$$

$$= mCv \ln\left(\frac{p_2}{p_1} \times \frac{V_2}{V_1}\right) + m(Cp - Cv) \ln\left(\frac{V_2}{V_1}\right)$$

$$= mCv \ln\left(\frac{p_2}{p_1}\right) + mCv \ln\left(\frac{V_2}{V_1}\right) + m(Cp - Cv) \ln\left(\frac{V_2}{V_1}\right)$$

$$= mCv \ln\left(\frac{p_2}{p_1}\right) + mCv \ln\left(\frac{V_2}{V_1}\right) + mCp \ln\left(\frac{V_2}{V_1}\right) - mCv \ln\left(\frac{V_2}{V_1}\right)$$

$$S_2 - S_1 = m\left[Cv \ln\frac{p_2}{p_1} + Cp \ln\frac{V_2}{V_1}\right]$$

4.5 ENTROPY CHANGE DURING STEAM GENERATION

(i) 1 kg of water at temperature T_1 is heated to boiling temperature T_2.

Heat supplied,

$$Q_{1-2} = h_f - h_1 = C_w(T_2 - T_1)$$

where, h_f = enthalpy of saturated water,

C_w – specific heat of water

$= 4.186$ kJ/kg-K

Change of entropy,

$$S_2 - S_1 = \int_{T_1}^{T_2} \frac{dQ}{T} = \int_{T_1}^{T_2} \frac{C_w\, dT}{T} = C_w \ln\left(\frac{T_2}{T_1}\right)$$

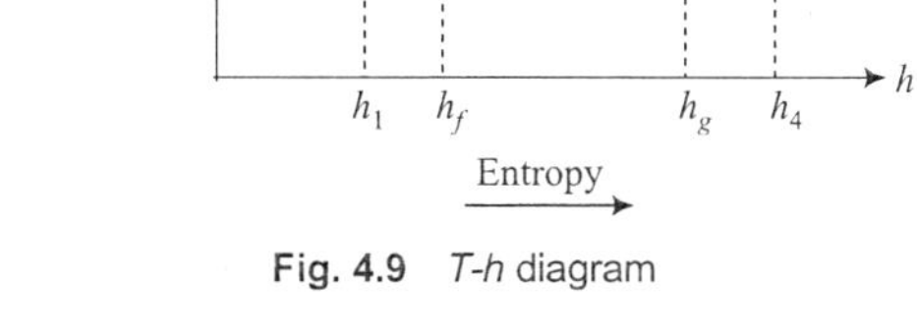

Fig. 4.9 *T-h* diagram

(ii) During process 2–3 there is evaporation of water. The temperature is constant at saturation temperature T_s.

$$S_3 - S_2 = \int_2^3 \frac{dQ}{T} = \frac{Q_{2-3}}{T_S} = \frac{h_{fg}}{T_S}$$

(iii) During process 3–4, the steam is superheated. The heat added

$$dQ = mC_S\, dT$$

where, C_S = specific heat of steam

$$= 2.1 \text{ kJ/kg-K}$$

The entropy change,

$$S_4 - S_3 = \int_{T_S}^{T_4} \frac{dQ}{T} = m \int_{T_S}^{T_4} \frac{C_S\, dT}{T} = mC_S \ln\left(\frac{T_4}{T_S}\right)$$

4.6 ENTROPY CHANGE OF THERMAL RESERVOIRS

A thermal reservoir (source or sink) has infinite capacity of heat energy and its temperature remains constant during heat exchange.

$$dS = -\frac{dQ}{T}$$

$$S_2 - S_1 = \int -\frac{dQ}{T} = -\frac{Q_{1-2}}{T}$$

4.7 ENTROPY CHANGE OF THE UNIVERSE

The entropy change of the universe is the algebraic sum of the change of entropy of the system and change of entropy of the surrounding.

$$(\Delta S)_{\text{universe}} = (\Delta S)_{\text{system}} - (\Delta S)_{\text{surrounding}}$$

For a reversible process,

$$(\Delta S)_{\text{system}} = (\Delta S)_{\text{surrounding}}$$

But in practice, all processes are irreversible. The change of entropy of the system undergoing irreversible process will always be greater than change of entropy of the surrounding where the process is reversible.

$$(\Delta S)_{\text{universe}} \geq 0$$

The entropy of the universe goes on increasing. This is called *principle of increase of entropy*.

4.8 AVAILABLE AND UNAVAILABLE HEAT ENERGY

The heat supplied Q_1 to a heat engine can be divided into two parts, the work done W and heat rejected Q_2. The maximum work done by an engine is called available heat energy (AHE) and the minimum heat rejected is called unavailable heat energy (UHE).

$$dQ = \text{AHE} + \text{UHE}$$

$$= \text{Work done} + \text{Heat rejected}$$

$$\eta = \frac{dW}{dQ} = \left(1 - \frac{T_2}{T_1}\right)$$

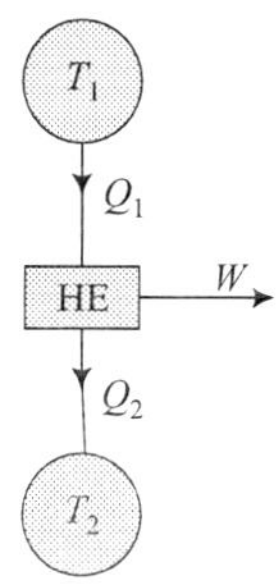

Fig. 4.10 Available energy

$$\therefore \quad dW_{max} = dQ\left(1 - \frac{T_2}{T_1}\right)$$

But $T_2 = T_\infty$ = Temperature of surrounding for minimum heat rejection

$$\therefore \quad dW_{max} = dQ\left(1 - \frac{T_\infty}{T_1}\right) = dQ - dQ \times \frac{T_\infty}{T_1}$$

$\therefore$ Unavailable heat energy = Heat rejected

$$= dQ\frac{T_\infty}{T_1} = T_\infty \frac{dQ}{T_1} = T_\infty dS$$

Change in entropy is a measure of unavailable heat energy or irreversibility of a process.

4.9 IMPORTANT MATHEMATICAL RELATIONS OF ENTROPY

1. $dS = \frac{dQ}{T}$
2. $dQ = TdS$
3. $TdS = dU + pdV$
4. Unavailable heat energy, UHE = $T_\infty dS$.
   where, T_∞ = Ambient temperature
5. For reversible process, $dS = \frac{dQ}{T}$
6. For irreversible process, $dS > \frac{dQ}{T}$
7. For universe (isolated system), $ds \geq 0$

8. For reversible cyclic process, $ds = 0$
9. For irreversible cyclic process, $ds > 0$
10. For a constant pressure process,

$$S_2 - S_1 = mCp \ln \frac{V_2}{V_1} = mCp \ln \frac{T_2}{T_1}$$

4.10 CARNOT CYCLE

Sadi Carnot proposed the concept of Carnot engine in 1824 which works on reversible cycle called Carnot cycle.

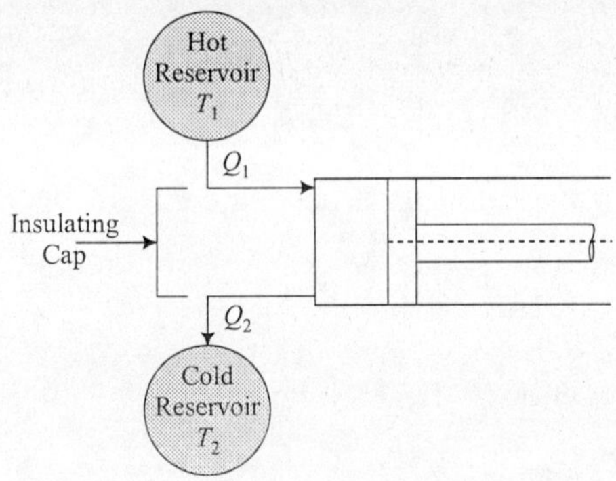

Fig. 4.11 Carnot engine

The *Carnot engine* has air enclosed in a cylinder. The cylinder receives heat Q_1 from hot reservoir and rejects heat Q_2 to cold reservoir. The cylinder head can be covered with insulating cap at will. The following assumptions are made:

1. The working substance is air.
2. Piston-cylinder arrangement is weightless and frictionless.
3. The heat transfer takes place with the help of heat reservoirs.
4. The walls of cylinder and piston are insulated.

4.10.1 Carnot Cycle Processes

Carnot cycle consists of two isothermals and two reversible adiabatic or isentropic processes. The *p-v* diagram and *T-s* diagram of the cycle are shown in Fig. 4.23.

1. *Process 1–2: Isothermal expansion*

The cylinder is brought in contact with hot reservoir. The air expands at constant temperature ($T_1 = T_2$) from volume V_1 to V_2. The heat supplied to the air is utilized in doing external work.

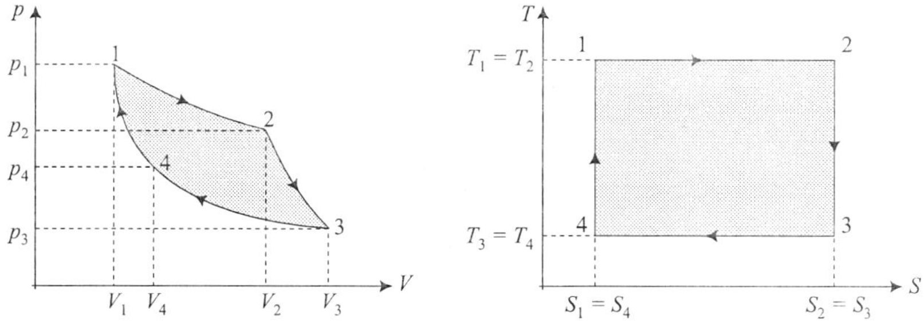

Fig. 4.12 Carnot cycle

Applying first law of thermodynamics,

$$Q_{1-2} = (U_2 - U_1) + W_{1-2}$$

$\therefore \quad T_1 = T_2$

$\therefore \quad U_2 - U_1 = 0$

$$Q_{1-2} = W_{1-2} = p_1 V_1 \ln\left(\frac{V_2}{V_1}\right)$$

$\because \quad p_1 V_1 = mRT_1$

$\therefore \quad Q_{1-2} = mRT_1 \ln\left(\frac{V_2}{V_1}\right)$

Let $\quad \dfrac{V_2}{V_1} = r =$ expansion ratio

$\therefore \quad Q_{1-2} = mRT_1 \ln r$

Also, $\quad Q_{1-2} = T_1(S_2 - S_1)$

2. *Process 2–3: Reversible Adiabatic Expansion or Isentropic Expansion*

The cylinder is disconnected from hot reservoir and insulation cap is put on. The air expands from V_2 to V_3 and T_3.

$$Q_{2-3} = 0$$

Applying first law of thermodynamics,

$$Q_{2-3} = (U_3 - U_2) + W_{2-3} = 0$$

$\therefore \quad (U_3 - U_2) = \dfrac{p_2 V_2 - p_3 V_3}{\gamma - 1}$

$$pV = mRT$$

$$\therefore \qquad (U_3 - U_2) = \frac{mRT_2 - mRT_3}{\gamma - 1}$$

$$= \frac{mR(T_1 - T_3)}{\gamma - 1} \qquad [\because T_1 = T_2]$$

Also, $\qquad S_2 = S_3.$

3. *Process 3–4: Isothermal Expansion*

Insulating cap is removed and the cylinder head is brought in contact with cold reservoir. The air is compressed at constant temperature ($T_3 = T_4$) from volume V_3 to V_4.

Applying 1st law of thermodynamics,

$$Q_{3-4} = (U_4 - U_3) + W_{3-4}$$

$$\therefore \qquad T_3 = T_4$$

$$\therefore \qquad U_4 - U_3 = 0$$

$$\therefore \qquad Q_{3-4} = W_{3-4} = p_3 V_3 \ln \frac{V_3}{V_4}$$

$$= mRT_3 \ln \frac{V_3}{V_4} \qquad [\because pV = mRT]$$

$$= mRT_3 \ln r$$

where, $r = \text{compression ratio} = \dfrac{V_3}{V_4} = \dfrac{V_2}{V_1}$

Also, $\qquad Q_{3-4} = T_3(S_3 - S_4).$

4. *Process 4–1: Reversible Adiabatic or Isentropic Compression*

The cylinder head is disconnected from cold reservoir and covered with insulating cap. The air is allowed to be compressed isentropically from V_4 to V_1.

$$Q_{4-1} = 0$$

Applying 1st law of thermodynamics,

$$Q_{4-1} = (U_4 - U_1) + W_{4-1}$$

$$\therefore \qquad (U_4 - U_1) = \frac{p_1 V_1 - p_2 V_2}{\gamma - 1}$$

$$= \frac{mR(T_1 - T_2)}{\gamma - 1}$$

But, $\qquad T_3 = T_4$

$$\therefore \qquad U_4 - U_1 = \frac{mR(T_1 - T_3)}{\gamma - 1}$$

$$\therefore \qquad U_3 - U_2 = U_4 - U_1$$

Also, $S_4 = S_1$.

4.10.2 Efficiency of Carnot Cycle

Work done,

$$W = \text{Heat supplied} - \text{Heat rejected}$$

$$= mRT_1 \ln r - mRT_3 \ln r$$

$$= mR \ln r(T_1 - T_3)$$

Efficiency,

$$\eta = \frac{\text{Work done}}{\text{Heat supplied}} = \frac{mR\ln r(T_1 - T_3)}{mR\ln rT_1} = \frac{T_1 - T_3}{T_1}$$

From *T-s* diagram,

$$W = T_1(S_2 - S_1) - T_3(S_3 - S_4)$$

$$\eta = \frac{T_1(S_2 - S_1) - T_3(S_3 - S_4)}{T_1(S_2 - S_1)}$$

But, $S_3 = S_2$ and $S_1 = S_4$

$$\therefore \qquad \eta = \frac{T_1 - T_3}{T_1}$$

For process 2–3, $\frac{T_2}{T_3} = \left(\frac{V_3}{V_2}\right)^{\gamma - 1}$

But, $T_2 = T_1$

$$\therefore \qquad \frac{T_1}{T_3} = \left(\frac{V_3}{V_2}\right)^{\gamma - 1}$$

For process 4–1, $\frac{T_1}{T_4} = \left(\frac{V_4}{V_1}\right)^{\gamma - 1}$

But, $T_4 = T_3$

$$\therefore \qquad \frac{T_1}{T_3} = \left(\frac{V_4}{V_1}\right)^{\gamma-1}$$

$$\therefore \qquad \frac{V_3}{V_2} = \frac{V_4}{V_1}$$

$$\text{or} \qquad \frac{V_2}{V_1} = \frac{V_3}{V_4} = r$$

$$\therefore \qquad \frac{T_1}{T_3} = (r)^{\gamma-1}$$

$$\therefore \qquad \eta = 1 - \frac{T_3}{T_1} = 1 - \left(\frac{1}{r}\right)^{\gamma-1}$$

$$= 1 - \frac{1}{r^{\gamma-1}}$$

4.10.3 Observations

1. Carnot efficiency, $\eta = 1 - \dfrac{T_2}{T_1}$.

   Thermal efficiency, $\eta = \dfrac{\text{Work done}}{\text{Heat supplied}} = \dfrac{W}{Q_1} = \dfrac{Q_1 - Q_2}{Q_1} = 1 - \dfrac{Q_2}{Q_1}$.

   $$\therefore \qquad \frac{Q_2}{Q_1} = \frac{T_2}{T_1}$$

   $$\text{or} \qquad \frac{Q_1}{T_1} = \frac{Q_2}{T_2} = \text{constant.}$$

   The heat transfer from a heat reservoir is proportional to its temperature. This is called *Carnot principle*.
2. Carnot efficiency is independent of working fluid.
3. It is a function of absolute temperature of heat reservoirs.
4. Carnot efficiency increases with the decrease of sink temperature T_2.

   If $T_2 = 0$, η is 100%. It violates Kelvin-Planck statement as no cold reservoir is required.
5. The sink temperature T_2 cannot be lower than atmospheric temperature or temperature of sea or river or lake which can act as heat sink.

6. Carnot efficiency increases with the increase of source temperature, T_1. This requires high temperature resisting materials which are very costly and not easily available. Therefore, T_1 is restricted by metallurgical considerations and is called *metallurgical limitation*.
7. Carnot efficiency depends upon temperature difference $(T_1 - T_2)$. Higher the temperature difference between source and sink, higher the Carnot efficiency.
8. For same degree increase of source temperature or decrease in sink temperature, Carnot efficiency is more sensitive to change in sink temperature.

4.10.4 Impracticability of Carnot Engine (Carnot Cycle)

Although Carnot cycle is the most efficient, it is not practical and Carnot engine is only a hypothetical device. The main reasons for impracticability of Carnot cycle are listed bolow:

1. All the processes are taken as reversible in a Carnot cycle. Practically processes are not reversible due to the presence of internal fluid friction of working substance and mechanical friction between piston and cylinder.
2. There is also irreversibility due to heat absorption and rejection taking place with finite temperature difference between reservoir and working substance.
3. The cycle demands that the engine speed should be very slow to ensure isothermal process and very fast for adiabatic process. Such a large variation of piston speed is not practical.
4. The slope difference of isothermal and adiabatic processes is very small on *p-v* diagram. The area of *p-v* diagram and, hence, work done is very small per cycle. Therefore, piston-cylinder arrangement has to be very big and bulky to get sufficient work output as the engine has to work with large range of pressures and volumes.

Though, it is not practicable to build an engine which can work on Carnot cycle, however, it serves as a standard for comparison and improvement of practical cycles. The Carnot cycle has maximum efficiency for given temperatures of source and sink.

Example 4.1 5 kg of ice at –10°C is kept in atmosphere which is at 30°C. Calculate the change in entropy of universe when it melts and comes into thermal equilibrium with atmosphere. Take latent heat of fusion as 335 kJ/kg and specific heat of ice is half that of water.

Solution

$$m = 5 \text{ kg}$$

$$T_1 = -10°\text{C} + 273 = 263 \text{ K}$$

$$T_S = 0°C + 273 = 273 \text{ K}$$

$$T_2 = 30°C + 273 = 303 \text{ K}$$

$$h_{\text{fusion}} = 335 \text{ kJ/kg}$$

$$C_i = \frac{C_w}{2} = \frac{4.186}{2} = 2.093 \text{ kJ/kg-K}$$

The heat transfer takes place in two stages:

1. Sensible heating of ice.
2. Latent heat of fusion.

Total heat transfer, $Q_{1-2} = mC_i(T_S - T_1) + mh_{\text{fusion}}$

$$= 5 \times 2.093(273 - 263) + 5 \times 335$$

$$= 104.65 + 1675$$

$$= 1779.65 \text{ kJ}$$

Change of entropy of the surrounding at constant temperature 303 K,

$$(\Delta S)_{\text{surrounding}} = -\frac{Q_{1-2}}{303} = \frac{-1779.65}{303} = -5.8734 \text{ kJ}$$

Change of entropy of the system,

$$= mC_i \ln\left(\frac{T_S}{T_1}\right) + \frac{mh_{\text{fusion}}}{T_S}$$

$$= 5 \times 2.093 \ln\left(\frac{273}{263}\right) + \frac{5 \times 335}{273}$$

$$= 0.39 + 6.1355$$

$$= 6.5255 \text{ kJ}$$

$$(\Delta S)_{\text{universe}} = (\Delta S)_{\text{surrounding}} + (\Delta S)_{\text{system}}$$

$$= -5.8734 + 6.5255 = \mathbf{0.652 \text{ kJ}} \quad \textbf{Ans.}$$

Example 4.2 0.25 kg/s of water is heated from 30°C to 60°C by hot gases that enter at 180° and leave at 80°C. Calculate the mass flow rate of gases when its Cp = 1.08 kJ/kg-K. Find the entropy change of water and of hot gases. Take the specific heat of water as 4.186 kJ/kg-K. Also find the increase of unavailable energy if the ambient temperature is 27°C.

Solution

Data on water,

$$\dot{m}_w = 0.25 \text{ kg/s}$$

$$T_{w1} = 30°C + 273 = 303 \text{ K}$$
$$T_{w2} = 60°C + 273 = 333 \text{ K}$$
$$C_w = 4.186 \text{ kJ/kg-K}$$

Data on gas

$$T_{g1} = 180°C + 273 = 453 \text{ K}$$
$$T_{g2} = 80°C + 273 = 353 \text{ K}$$
$$C_g = 1.08 \text{ kJ/kg-K}$$
$$T_\infty = 27°C + 273 = 300 \text{ K}$$

1. *Mass flow rate of gases*

   Heat gained by water = Heat lost by gases

$$\dot{m}_w C_w (T_{w2} - T_{w1}) = \dot{m}_g C_g (T_{g1} - T_{g2})$$
$$0.25 \times 4.186(333 - 303) = \dot{m}_g \times 1.08\,(453 - 353)$$
$$\dot{m}_g = \frac{0.25 \times 4.186(333 - 303)}{1.08(453 - 353)} = 0.29 \text{ kg/s}$$

2. *Increase of entropy of water,*

   Water is heated at constant pressure

$$S_2 - S_1 = m_w C_w \ln \frac{T_{w2}}{T_{w1}} = 0.25 \times 4.186 \ln\left(\frac{333}{303}\right)$$
$$= 0.0988 \text{ kJ}$$

3. *Decrease in entropy of gas,*

$$S_2 - S_1 = m_g C_g \ln \frac{T_{g2}}{T_{g1}}$$
$$= 0.29 \times 1.08 \ln\left(\frac{353}{453}\right)$$
$$= -\,0.078 \text{ kJ}$$

4. *Loss of available energy of gas,*

   UHE $= T_\infty\, dS = 300 \times 0.078 =$ **23.436 kJ** **Ans.**

4.11 AVAILABILITY

Available energy of a system in a given state is the maximum useful work that can be obtained by reducing the system from its initial state to the dead state, i.e., a state in equilibrium with the earth and the atmosphere.

Availability is the maximum work that can be done by a system in any given state.

4.11.1 Availability of a Work Reservoir

The work reservoir has infinite capacity and never reaches the dead state.

The availability of work reservoir is the maximum useful work obtained from a given quantity of energy withdrawn from the reservoir.

Availability,

$$A = \Delta W$$

where, ΔW = energy withdrawn from the reservoir.

4.11.2 Availability of a Heat Reservoir

Heat ΔQ is withdrawn from a heat reservoir at constant temperature T. The work of the cycle will be maximum when,

(i) the cycle is reversible

(ii) the heat rejection occurs at atmospheric pressure, T_∞

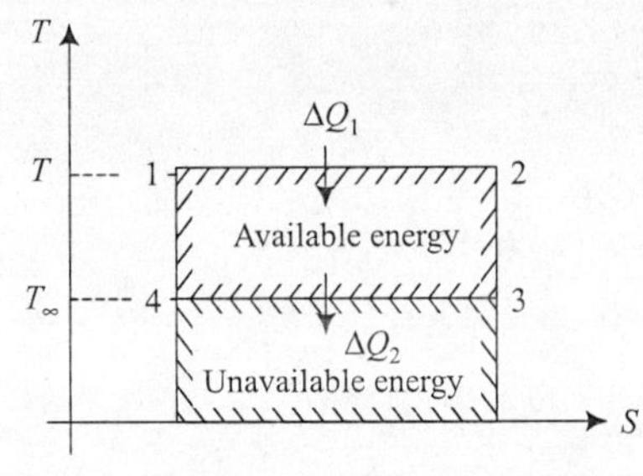

Fig. 4.13 Availability of a heat reservoir

$$(\Delta W)_{max} = \Delta Q_1 - \Delta Q_2$$

$$= \int_1^2 TdS - \int_3^4 T_\infty dS$$

$$= T(S_2 - S_1) - T_\infty(S_3 - S_4)$$

$$= \Delta Q - T_\infty \Delta S$$

$$= \Delta Q\left(\frac{T - T_\infty}{T}\right)$$

$$= \Delta Q\left(1 - \frac{T_\infty}{T}\right)$$

Therefore, availability of heat ΔQ,

$$A = \Delta Q - T_\infty \Delta S$$

$$= \Delta Q\left(1 - \frac{T_\infty}{T}\right)$$

4.11.3 Availability of a Process

For process 1-2, heat ΔQ is withdrawn from a source of finite size.

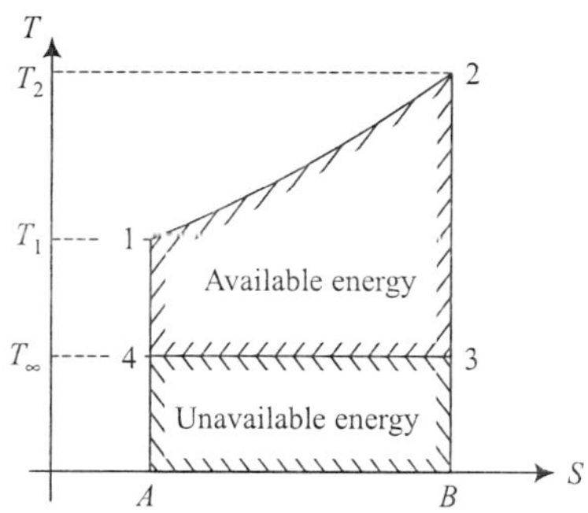

Fig. 4.14 Availability of a process

The work done during the cycle,

$$(\Delta W)_{\text{cycle}} = [\text{Heat supplied}] - [\text{Heat rejected}]$$

$$= \Delta Q - T_\infty (S_B - S_A)$$

$$= \Delta Q - T_\infty \Delta S$$

Availability, $\quad A = \Delta Q - T_\infty \Delta S$

Unavailability energy, $= T_\infty \Delta S$

where, $\quad \Delta S = S_2 - S_1$

The values of $S_2 - S_1$ will depend upon the type of process as given below.

Process	*Change of entropy* $(S_2 - S_1)$[kJ/kg-K]
1. Isochoric ($v = c$)	$C_V \ln \frac{T_2}{T_1}$
2. Isobaric ($p = c$)	$C_p \ln \frac{T_2}{T_1}$
3. Isothermal ($t = c$)	$R \ln \frac{v_2}{v_1}$

Contd.

Process	$(S_2 - S_1)$
4. Isentropic ($s = c$) or reversible adiabatic ($pv^{\gamma} = c$)	0,
5. Polytropic ($pv^n = c$)	$C_V\left(\frac{n-\gamma}{n-1}\right)\ln\frac{T_2}{T_1}$

4.11.4 Availability of a Closed System

A fluid expands reversibly from initial conditions of p_1, T_1 to the final atmospheric condition p_{∞}, T_{∞}. The system is represented by a cylinder-piston arrangement which supplies heat ΔQ to a reversible heat engine (RHE). The heat engine can be represented by an infinite number of reversible heat engines arranged in parallel and each operating on a Carnot cycle. Each engine receives heat at constant temperature and rejects heat at T_{∞}. The scheme of engine and T-s diagram are shown in Fig. 4.15.

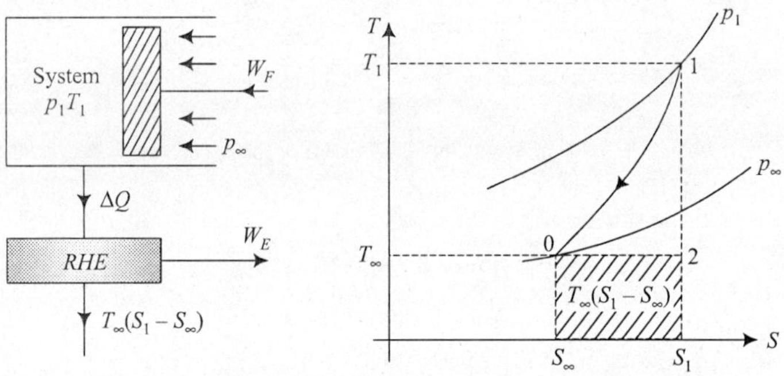

Fig. 4.15 Availability of a closed system

The work done by the engine,

$$W_E = \text{Heat supplied} - \text{Heat rejected}$$
$$= \Delta Q - T_{\infty}(S_1 - S_{\infty})$$

The fluid in the cylinder is undergoing a process 1 to 0 and rejects heat ΔQ.

$$\therefore \quad -\Delta Q = (U_{\infty} - U_1) + W_F$$

$\therefore$ Work done by fluid,

$$W_F = (U_1 - U_{\infty}) - \Delta Q$$

$$\therefore \quad W_F + W_E = [(U_1 - U_{\infty}) - \Delta Q] + [\Delta Q - T_{\infty}(S_1 - S_{\infty})]$$
$$= (U_1 - U_{\infty}) - T_{\infty}(S_1 - S_{\infty})$$

Work done on the atmosphere,

$$W_{\infty} = p_{\infty}(U_{\infty} - U_1)$$

Maximum work available,

$$W_{max} = (U_1 - u_\infty) - T_\infty(S_1 - S_\infty) - p_\infty(V_\infty - V_1)$$
$$= (U_1 + p_\infty V_1 - T_\infty S_1) - (U_\infty + p_\infty V_\infty - T_\infty S_\infty)$$
$$= A_1 - A_\infty$$

The availability per unit mass,

$$a = (u + p_\infty v - T_\infty s)$$

a is called non-flow availability function.

4.11.5 Availability of a Steady Flow Process

A steady flow process is shown in Fig. 4.16 in which fluid is flowing with a velocity V_1 from a reservoir at p_1 and T_1.

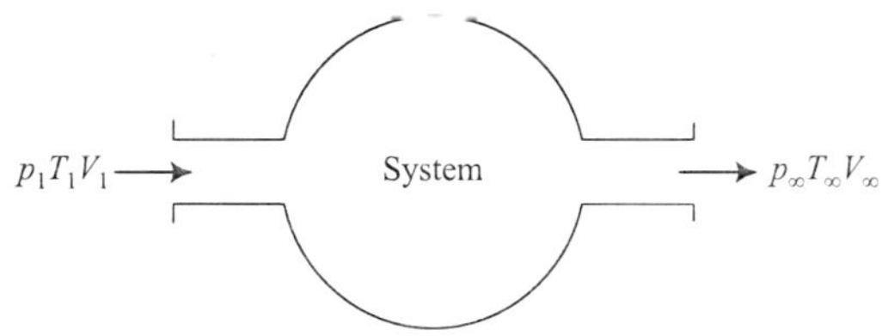

Fig. 4.16 Availability of a steady flow process

Maximum work done per unit mass,

$$w_{max} = \left(h_1 + \frac{V_1^2}{2} + gZ_1\right) - [h_\infty - T_\infty(s_1 - s_\infty)]$$

where at earth's surface, $V_\infty = 0$ and $Z_\infty = 0$.

The heat supplied from the system to an engine will be rejecting heat $T_\infty(s_1 - s_\infty)$.

If $ke = 0$ and $pe = 0$

$$w_{max} = (h_1 - h_\infty) - T_\infty(s_1 - s_\infty)$$
$$= (h_1 - T_\infty s_1) - (h_\infty - T_\infty s_\infty)$$
$$= b - b_\infty$$

where b is called steady flow availability function per unit mass. This is also called Keenan function and is a composite property of system and its environment.

The Keenan function per unit mass,

$$b = (h - T_\infty s)$$

4.11.6 Helmholtz and Gibb's Function

The work done per unit mass in a non-flow reversible system.

$$
\begin{aligned}
w &= q - (U_\infty - U_1) \\
&= Tds - (U_\infty - U_1) \\
&= T(s_\infty - s_1) - (U_\infty - U_1) \\
&= (U_1 - Ts_1) - (U_\infty - Ts_\infty) \\
&= f_1 - f_\infty
\end{aligned}
$$

f is called Helmholtz function

$$f = (u - Ts)$$

For total flow,

$$F = (U - TS)$$

F gives the maximum possible work output for a given heat input ΔQ from a heat reservoir at constant temperature.

The work done against atmosphere,

$$w_\infty = p_\infty (u_\infty - u_1)$$

Therefore, maximum work done,

$$
\begin{aligned}
w_{max} &= (w - w_\infty) \\
&= (u_1 - Ts_1) - (u_\infty - Ts_\infty) - p_\infty(v_\infty - v_1) \\
&= (u_1 + p_\infty v_1 - Ts_1) - (u_\infty + p_\infty u_\infty - Ts_\infty) \\
&= (h_1 - Ts_1) - (h_\infty - Ts_\infty) \\
&= g_1 - g_\infty
\end{aligned}
$$

g is called free energy function or Gibb's function per unit mass.

$$g = (h - Ts)$$

Gibb's function for total flow,

$$G = (H - TS)$$

The maximum possible available work when system changes from state 1 to 2, in a closed system:

$$
\begin{aligned}
W_{max} &= (G_1 - G_\infty) - (G_2 - G_\infty) \\
&= (G_1 - G_2)
\end{aligned}
$$

The maximum work available during steady flow process

$$W_{max} = (G_1 - G_2) + (KE_1 - KE_2) + (PE_1 - PE_2)$$

The Gibb's function, $g = (h - Ts)$, is a property of the system.

The availability function, $a = (u + p_\infty v - T_\infty s)$, is a composite property of the system and its surrounding

Now,

$$a = (u + p_\infty v - T_\infty s)$$

$$b = (u + pv - T_\infty s)$$

$$g = (u + pv - Ts)$$

When the system proceeds from state 1 to dead state (zero state),

$$a = b = g$$

4.12 IRREVERSIBILITY AND EFFECTIVENESS

4.12.1 Irreversibility

The difference between the maximum useful work and actual work of a system is called irreversibility of the process.

The irreversibility,

$$I = W_{max} - W_{actual}$$

The irreversibility is also called dissipation or degradation of energy or work.

A. Non-flow process

When the system exchanges heat with the environment only, the irreversibility per unit mass,

$$i = [(u_1 - u_2) - T_\infty(s_1 - s_2)] - [(u_1 - u_2) + \Delta Q]$$

$$= T_\infty (s_2 - s_1) - \Delta Q$$

$$= T_\infty(\Delta s)_{system} + T_\infty(\Delta s)_{surrounding}$$

$$= T_\infty[(\Delta s)_{system} + (\Delta s)_{surrounding}]$$

Irreversibility for a reversible process is zero and is always positive for irreversible process.

$$i \geq 0.$$

B. Steady-flow process

The irreversibility,

$$I = W_{max} - W_{actual}$$

For unit mass, irreversibility,

$$i = w_{max} - w_{actual}$$

$$= \left[\left(b_1 + \frac{V_1^2}{2} + gZ_1\right) - \left(b_2 + \frac{V_2^2}{2} + gZ_2\right)\right]$$

$$- \left[\left(h_1 + \frac{V_1^2}{2} + gZ_1\right) - \left(h_2 + \frac{V_2^2}{2} + gZ_2\right)\right] + \Delta q$$

$$= T_\infty(s_2 - s_1) - \Delta Q$$

$$= T_\infty[(\Delta s)_{system} + (\Delta s)_{surrounding}]$$

Irreversibility is also called increase in unavailable energy or energy.

4.12.2 Effectiveness

Effectiveness of a system is the ratio of actual useful work and maximum useful work.

Effectiveness,

$$\varepsilon = \frac{W_{\text{useful}}}{W_{\text{max useful}}}$$

Effectiveness can also be expressed in terms of increase of availability of surrounding.

$$\varepsilon = \frac{\text{Increase of availability of surrounding}}{\text{Loss of availability of surrounding}}$$

The effectiveness of an actual process is always less than unity.

The effectiveness of a process is the measure of the extent to which advantage has been taken of an opportunity to obtain useful work.

For compression or heating process,

$$\varepsilon = \frac{\text{Increase of availability of the system}}{\text{Loss of availability of the surrounding}}$$

4.13 PROPERTY RELATIONSHIPS

4.13.1 Mathematical Conditions for Exact Differential

Thermodynamic properties are exact differentials. There are certain mathematical conditions for exact differentials.

Theorem 1 If variables x, y and z are correlated, then z can be expressed as a function of x and y.

$$z = f(x, y)$$

$$\therefore \qquad dz = \left(\frac{\partial z}{\partial x}\right)_y dx + \left(\frac{\partial z}{\partial y}\right)_x dy$$

Let $$\left(\frac{\partial z}{\partial x}\right)_y = M \text{ and } \left(\frac{\partial z}{\partial y}\right)_x = N$$

$$\therefore \qquad dz = Mdx + Ndy$$

where z, M and N are functions of x and y.

Differentiating M partially with respect to y and N with respect to x,

$$\left(\frac{\partial M}{\partial y}\right)_x = \frac{\partial^2 z}{\partial x\,\partial y}$$

$$\left(\frac{\partial N}{\partial x}\right)_y = \frac{\partial^2 z}{\partial y . \partial x}$$

$$\therefore \qquad \left(\frac{\partial M}{\partial y}\right)_x = \left(\frac{\partial N}{\partial x}\right)_y$$

This is the *condition of perfect (or exact) differential.*

Theorem 2 If a quantity f is a function of x, y and z and variables x, y and z are correlated, then f is a function of any two of x, y, and z. If any one of x, y, or z is a function of f and any one of x, y, or z.

If $\qquad x = x(f, y)$

$$dx = \left(\frac{\partial x}{\partial f}\right)_y df + \left(\frac{\partial x}{\partial y}\right)_f dy \qquad \ldots(1)$$

Similarly, if $\qquad y = y(f, z)$

$$dy = \left(\frac{\partial y}{\partial f}\right)_z df + \left(\frac{\partial y}{\partial z}\right)_f dz \qquad \ldots(2)$$

Substituting the value of dy (equation 2) into equation (1)

$$dx = \left(\frac{\partial x}{\partial f}\right)_y df + \left(\frac{\partial x}{\partial y}\right)_f \left[\left(\frac{\partial y}{\partial f}\right)_z df + \left(\frac{\partial y}{\partial z}\right)_f dz\right]$$

$$= \left[\left(\frac{\partial x}{\partial f}\right)_y + \left(\frac{\partial x}{\partial y}\right)_f \left(\frac{\partial y}{\partial f}\right)_z\right] df + \left(\frac{\partial x}{\partial y}\right)_f \left(\frac{\partial y}{\partial z}\right)_f dz$$

Again, $\qquad dx = \left(\frac{\partial x}{\partial f}\right)_z df + \left(\frac{\partial x}{\partial z}\right)_f dz$

$$\therefore \qquad \left(\frac{\partial x}{\partial z}\right)_f = \left(\frac{\partial x}{\partial y}\right)_f \left(\frac{\partial y}{\partial z}\right)_f$$

$$\therefore \qquad \left(\frac{\partial x}{\partial y}\right)_f \left(\frac{\partial y}{\partial z}\right)_f \left(\frac{\partial z}{\partial x}\right)_f = 1$$

This is also the condition of exact differential.

Theorem 3 Any one variable among x, y and z may be considered a function of the other two.

If $$x = x(y, z)$$

$$\therefore \quad dx = \left(\frac{\partial x}{\partial y}\right)_z dy + \left(\frac{\partial x}{\partial z}\right)_y dz$$

Similarly $$dz = \left(\frac{\partial z}{\partial x}\right)_y dx + \left(\frac{\partial z}{\partial y}\right)_z dy$$

Substituting the value of dz,

$$dx = \left(\frac{\partial x}{\partial y}\right)_z dy + \left(\frac{\partial x}{\partial z}\right)_y \left[\left(\frac{\partial z}{\partial x}\right)_y dx + \left(\frac{\partial z}{\partial y}\right)_z dy\right]$$

$$= \left[\left(\frac{\partial x}{\partial y}\right)_z + \left(\frac{\partial x}{\partial z}\right)_y \left(\frac{\partial z}{\partial y}\right)_x\right] dy + \left(\frac{\partial x}{\partial z}\right)_y \cdot \left(\frac{\partial z}{\partial x}\right)_y dx$$

$$= \left[\left(\frac{\partial x}{\partial y}\right)_z + \left(\frac{\partial x}{\partial z}\right)_y \cdot \left(\frac{\partial z}{\partial y}\right)_x\right] dz + dx$$

$$\therefore \quad \left(\frac{\partial x}{\partial y}\right)_z + \left(\frac{\partial x}{\partial z}\right)_y \cdot \left(\frac{\partial z}{\partial y}\right)_x = 0$$

or $$\left(\frac{\partial x}{\partial y}\right)_z \left(\frac{\partial z}{\partial x}\right)_y \cdot \left(\frac{\partial y}{\partial z}\right)_x = -1$$

This relation can be applied to thermodynamic properties p, V and T

$$\therefore \quad \left(\frac{\partial p}{\partial V}\right)_T \left(\frac{\partial V}{\partial T}\right)_p \left(\frac{\partial T}{\partial p}\right) = -1$$

4.13.2 Maxwell's Relation

Any pure substance existing in a single phase has only two independent variables. Any one property may be expressed as function of any other two properties. The eight thermodynamic properties are:

p = Pressure

V = Volume

T = Temperature

S = Entropy

U = Internal energy
H = Enthalpy
F = Helmholtz function
G = Gibb's function

Now U, H, F and G are exact differentials.

The following mathematical conditions hold good for exact differential

$$dz = Mdx + Ndy$$

$$\therefore \qquad \left(\frac{\partial M}{\partial y}\right)_z = \left(\frac{\partial N}{\partial x}\right)_y$$

If a pure substance undergoes a slow reversible process, then

$$dU = TdS - pdV \qquad \ldots(1)$$

$$dH = dU + pdV + Vdp \qquad \ldots(2)$$

$$dF = dU - TdS - SdT = -pdV - SdT \qquad \ldots(3)$$

$$dG = dH - TdS - SdT = Vdp - SdT \qquad \ldots(4)$$

Applying the condition of exact differentials to the four equations given above, we get

$$\left(\frac{\partial T}{\partial V}\right)_S = -\left(\frac{\partial P}{\partial S}\right)_V \qquad \ldots(5)$$

$$\left(\frac{\partial T}{\partial p}\right)_S = \left(\frac{\partial V}{\partial S}\right)_p \qquad \ldots(6)$$

$$\left(\frac{\partial p}{\partial T}\right)_V = \left(\frac{\partial S}{\partial V}\right)_T \qquad \ldots(7)$$

$$\left(\frac{\partial V}{\partial T}\right)_p = -\left(\frac{\partial S}{\partial p}\right)_T \qquad \ldots(8)$$

Equations (5), (6), (7), and (8) are known as *Maxwell's equations.*

4.13.3 *TdS* Equations

If entropy S is taken as function of T and V, then

$$dS = \left(\frac{\partial S}{\partial T}\right)_V dT + \left(\frac{\partial S}{\partial V}\right)_T dV$$

$$\therefore \qquad TdS = T\left(\frac{\partial S}{\partial T}\right)_V dT + T\left(\frac{\partial S}{\partial V}\right)_T dV$$

Now, $$T\left(\frac{\partial S}{\partial T}\right)_V = C_V \text{ (Heat capacity at constant volume)}$$

$$\left(\frac{\partial S}{\partial V}\right)_T = \left(\frac{\partial P}{\partial T}\right)_V \quad \text{(Maxwell's equation)}$$

∴ $$TdS = C_V dT + T\left(\frac{\partial P}{\partial T}\right)_V dV$$

This is called *first TdS equation.*

If $$S = S(T, p)$$

$$dS = \left(\frac{\partial S}{\partial T}\right)_p dT + \left(\frac{\partial S}{\partial p}\right)_T dp$$

∴ $$TdS = T\left(\frac{\partial S}{\partial T}\right)_p dT + T\left(\frac{\partial S}{\partial p}\right)_T dp$$

Now, $$T\left(\frac{\partial S}{\partial T}\right)_p = C_p$$

and $$\left(\frac{\partial S}{\partial p}\right)_T = -\left(\frac{\partial V}{\partial p}\right)_p$$

∴ $$TdS = C_p dT - T\left(\frac{\partial V}{\partial T}\right)_p dp$$

This is called *second TdS equation.*

4.13.4 Isothermal Compressibility

The *TdS* equations are

$$TdS = C_V dT + T\left(\frac{\partial p}{\partial T}\right)_V dV$$

and $$TdS = C_p dT - T\left(\frac{\partial V}{\partial T}\right)_p dp$$

Equating the two equations,

$$C_p dT - T\left(\frac{\partial V}{\partial T}\right)_p dp = C_V dT + T\left(\frac{\partial p}{\partial T}\right)_V dV$$

$$(C_p - C_V)dT = T\left(\frac{\partial p}{\partial T}\right)VdV + T\left(\frac{\partial V}{\partial T}\right)_p dp$$

$$dT = \frac{T\left(\frac{\partial P}{\partial T}\right)V}{(C_p - C_V)}dV + \frac{T\left(\frac{\partial V}{\partial T}\right)_p}{(C_p - C_V)}dp$$

Also, $$dT = \left(\frac{\partial T}{\partial V}\right)_p dV + \left(\frac{\partial T}{\partial p}\right)_V dp$$

$\therefore$ $$\frac{T\left(\frac{\partial p}{\partial T}\right)V}{C_p - C_V} = \left(\frac{\partial T}{\partial V}\right)_p$$

and $$\frac{T\left(\frac{\partial V}{\partial T}\right)_p}{C_p - C_V} = \left(\frac{\partial T}{\partial p}\right)_V$$

$\therefore$ $$C_p - C_V = T\left(\frac{\partial p}{\partial T}\right)_V \cdot \left(\frac{\partial V}{\partial T}\right)_p$$

But, $$\left(\frac{\partial p}{\partial T}\right)_V \left(\frac{\partial T}{\partial V}\right)_p \cdot \left(\frac{\partial V}{\partial p}\right)_T = -1$$

$\therefore$ $$C_p - C_V = -T\left(\frac{\partial V}{\partial T}\right)_p^2 \left(\frac{\partial P}{\partial V}\right)_T \qquad \ldots(1)$$

Following conclusions can be made from the above equation.

1. $\left(\frac{\partial V}{\partial T}\right)_p^2$ is always +ve, $\left(\frac{\partial P}{\partial V}\right)_T$ is –ve, therefore, $(C_p - C_V)$ is always positive. C_p is always greater than C_V.
2. $C_p \rightarrow C_V$ as $T \rightarrow$ OK
   At $T = OK$, $C_p = C_V$
3. $\left(\frac{\partial V}{\partial T}\right)_p = 0$ for water at 4°C and $C_p = C_V$.
4. $C_p - C_V = R$

Let us define *volume expansivity* as

$$\beta = \frac{1}{V}\left(\frac{\partial V}{\partial T}\right)_p$$

and *isothermal compressibility as*

$$k_T = -\frac{1}{V}\left(\frac{\partial V}{\partial T}\right)_T$$

From equation (1)

$$C_p - C_V = \frac{TV\left[\frac{1}{V}\left(\frac{\partial V}{\partial T}\right)_p\right]^2}{-\frac{1}{V}\left(\frac{\partial V}{\partial P}\right)_T}$$

$\therefore$ $$C_p - C_V = \frac{TV\beta^2}{k_T}$$

4.13.5 Adiabatic Compressibility

The TdS equation for S = constant are

$$C_p(dT)_S = T\left(\frac{\partial V}{\partial T}\right)_p (dp)_S$$

and $$C_V(dT)_S = -T\left(\frac{\partial p}{\partial T}\right)_V (dV)_S$$

$\therefore$ $$\frac{C_p}{C_V} = -\left(\frac{\partial V}{\partial T}\right)_p \cdot \left(\frac{\partial T}{\partial p}\right)_V \left(\frac{\partial p}{\partial V}\right)_S$$

$$= \frac{\left(\frac{\partial p}{\partial V}\right)_S}{\left(\frac{\partial p}{\partial V}\right)_T} = \gamma$$

Now, $$\gamma > l$$

$\therefore$ $$\left(\frac{\partial p}{\partial V}\right)_S = \left(\frac{\partial p}{\partial V}\right)_T$$

The slope of an isentrope is greater than that of isotherm on $p - v$ diagram as shown.

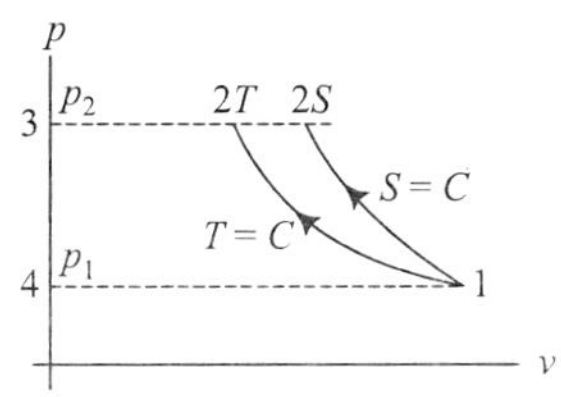

Fig. 4.17 Compression work

For isentropic or reversible adiabatic process, work done is

$$w_s = (h_{2s} - h_1)$$

$$= \int_1^{2s} v dp = \text{Area } 1 - 2s - 3 - 4 - 1$$

For reversible isothermal compression process, work done is

$$w_T = (h_{2T} - h_1)$$

$$= \int_1^{2T} v dp = \text{Area } 1 - 2T - 3 - 4 - 5$$

$$\therefore \qquad w_T < w_s.$$

For n between 1 and γ, work done will be in between w_T and w_s.

Let us define *adiabatic compressibility* as

$$k_s = -\frac{1}{V}\left(\frac{\partial V}{\partial p}\right)_s$$

$$\therefore \qquad \frac{C_p}{C_V} = \gamma = \frac{-\frac{1}{V}\left(\frac{\partial V}{\partial p}\right)_T}{-\frac{1}{V}\left(\frac{\partial V}{\partial p}\right)_S}$$

$$\therefore \qquad \gamma = \frac{k_T}{k_s}.$$

Example 4.3 5 kg of air at 550 K and 4 bar is enclosed in a closed vessel.

(i) Determine the availability of the system if the surrounding pressure and temperature are 1 bar and 290 K.

(ii) If the air is cooled at constant pressure to the atmospheric temperature, determine the availability and effectiveness.

Solution

$$m = 5 \text{ kg}$$

$$T_1 = 550 \text{ K}$$

$$p_1 = 4 \text{ bar} = 4 \times 10^5 \text{ N/m}^2$$

$$T_\infty = 290 \text{ K}$$

$$p_\infty = 1 \text{ bar} = 1 \times 10^5 \text{ N/m}^2.$$

(i) Availability of the system,

$$A = m[(u_1 - u_\infty) - T_\infty(s_1 - s_\infty)]$$

$$= m[C_V(T_1 - T_\infty) - T_\infty(\Delta s)]$$

$$\Delta s = C_p \ln \frac{T_1}{T_\infty} - R \ln \frac{p_1}{p_\infty}$$

$$= 1.005 \ln \frac{550}{290} - 0.287 \ln \frac{4}{1}$$

$$= 0.246 \text{ kJ/kgK}$$

$$\therefore \quad A = 5[0.718(550 - 290) - 290 \times 0.246]$$

$$= \mathbf{576.7 \text{ kJ Ans.}}$$

(ii) Heat transfer during cooling upto T_∞

$$\Delta Q = mC_p (T_1 - T_\infty)$$

$$= 5 \times 1.005 (550 - 290)$$

$$= 1306.5 \text{ kJ}$$

$$\Delta S = mC_p \ln \frac{T_1}{T_\infty}$$

$$= 5 \times 1.005 \ln \frac{550}{290}$$

$$= 3.216 \text{ kJ/kgK}$$

$$\text{Available energy} = \Delta Q - T_\infty \Delta s$$

$$= 1306.5 - 290 \times 3.216$$

$$= 373.86 \text{ kJ}$$

$$\text{Effectiveness,} \quad \varepsilon = \frac{\text{Available energy}}{\text{Availability of the system}}$$

$$= \frac{373.86}{526.7} = 0.648 \text{ or } \mathbf{64.8\%} \quad \textbf{Ans.}$$

Example 4.4 Determine the reversible work and irreversibility associated with a process in which a metallic block of 2 kg mass is cooled from 300°C to 150°C by keeping it in the open at 20°C. Specific heat of metal may be taken as 0.45 kJ/kgK.

Solution

$$m = 2 \text{ kg}$$

$$T_1 = 300°\text{C} + 273 = 573 \text{ K}$$

$$T_2 = 150°\text{C} + 273 = 423 \text{ K}$$

$$T_\infty = 20°\text{C} + 273 = 293 \text{ K}$$

$$C_p = 0.45 \text{ kJ/kgK}$$

$$(\Delta W)_{\text{rev.}} = \Delta Q - T_\infty \Delta s$$

$$= mC_p(T_2 - T_1) - T_\infty\, mC_p \ln \frac{T_2}{T_1}$$

$$= 2 \times 0.45\,(573 - 423) - 293 \times 2 \times 0.45 \ln \frac{423}{573}$$

$$= \mathbf{143\ kJ} \quad \textbf{Ans.}$$

Irreversibility,

$$I - T_\infty\,[(\Delta s)_{\text{system}} + (\Delta s)_{\text{surr.}}]$$

$$= 293\left[mC_p \ln \frac{T_2}{T_1} + \frac{mC_p(T_2 - T_1)}{T_\infty}\right]$$

$$= 293\left[2 \times 0.45 \ln \frac{423}{577} + \frac{2 \times 0.45(573 \times 423)}{293}\right]$$

$$= 293\,[-0.273 + 0.46]$$

$$= \mathbf{55\ kJ} \quad \textbf{Ans.}$$

QUESTION BANK 4

1. What is entropy? Show that entropy is a point function.
2. Derive the formula used for change of entropy of an ideal gas when its state changes from 1 to 2.
3. Explain available and unavailable energy.
4. Are the following statements true or false?
   (i) The process that violates the second law also violates the first law of thermodynamics.
   (ii) The change in entropy of a closed system must be greater than or equal to zero.

   (iii) The entropy of an isolated system always increases.
   (iv) Entropy increases in the process of mixing hot water with cold water.

5. Define the term, availability, explaining the concept of available and unavailable energy. Also define the term irreversibility and show how it is applied for a closed system.
6. What is the importance of Maxwell's relations in thermodynamics? Define the terms, Helmholtz and Gibb's function and with their help derive the Maxwell relation for a simple compressible substance.
7. Explain the following terms:
   (i) Keenan function
   (ii) Isothermal compressibility
   (iii) Adiabatic compressibility
   (iv) Effectiveness
   (v) Tds equations

PART II

PROPERTIES OF GASES AND VAPOURS

CHAPTER 5

Gases and Gas Mixtures

5.1 INTRODUCTION

There is a relationship between the pressure, temperature and specific volume for all substances. For solids and liquids, changes in pressure and temperature have negligible effect on the specific volume. However, pressure as well as temperature changes produce appreciable changes in the specific volume of a gas. The products pv of all real gases at a given temperature approach the same value as the pressure approaches zero. In other words, at the state of zero pressure (or vanishing density), all real gases behave in a similar manner which may be called ideal behaviour and the state of zero pressure as an ideal state.

$$pv = (RT)_{p \to 0}$$

or

$$p = \left(\frac{RT}{v}\right)_{p \to 0}$$

An *ideal gas* is a hypothetical gas that will behave in an ideal manner at all pressures, and obey the following laws:

1. The equation of state

$$pv = RT$$

2. The internal energy of an ideal gas is a function of temperature only

$$u = u(T)$$

5.2 LAWS OF IDEAL GASES

The ideal gases obey the following laws:

1. *Boyle's Law*

The volume of a given gas is inversely proportional to its pressure at constant temperature.

$$v \propto \left(\frac{1}{p}\right)_{T=\text{const}}$$

or $$p_1 v_1 = p_2 v_2 = pv \text{ at } T = \text{constant}$$

2. *Charle's Law* or *Gay-Lussac's Law*

(i) The volume of a given mass of gas varies directly with the absolute temperature at constant pressure.

$$v \propto (T)_{p = \text{constant}}$$

or $$\frac{v_2}{v_1} = \frac{T_2}{T_1} \text{ at } p = \text{constant}$$

(ii) The pressure of a given mass of gas varies directly with absolute temperature at constant volume.

$$p \propto (T)_{v = \text{constant}}$$

or $$\frac{p_2}{p_1} = \frac{T_2}{T_1} \text{ at } v = \text{constant}$$

3. *Avogadro's Hypothesis*

"Equal volumes of all gases under identical conditions of temperature and pressure, contain same number of molecules."

When $$p = 760 \text{ mm Hg } (1.013 \times 10^5 \text{ N/m}^2)$$

$$T = 273.15 \text{ K}$$

$$\bar{v} = 22.4 \text{ m}^3/\text{kmol}$$

$\therefore$ $$\bar{R} = \frac{1.013 \times 10^5 \times 22.4}{273.15} = 8.3143 \text{ kJ/kmol K}$$

$\bar{R}$ is called universal gas constant or molar constant and its value is same for all gases.

4. *Equation of state*

Combining the above laws, the ideal gas equation may be obtained as:

$$pv = RT$$

$$pV = mRT$$

$$p\bar{v} = \bar{R}T$$

5. *Thermodynamic properties*

The thermodynamic properties of an ideal gas are:

Internal energy, $u = u(T)$

Enthalpy, $h = h(T)$

Specific heat at constant volume,

$$Cv = \frac{du}{dT}$$

Specific heat at constant pressure,

$$C_p = \frac{dh}{dT}$$

$$C_p - C_V = R$$

$$\frac{C_p}{C_V} = \gamma.$$

6. *Thermodynamic processes*

Various thermodynamic processes for ideal gas are discussed in Chapter 1

5.3 DEVIATION OF REAL GASES FROM IDEAL GAS

The ideal gas behaviour is not sufficiently accurate for real gases at very high pressures and low temperatures. The real gases show marked deviation from the ideal gas laws.

The equation of state, $pv = RT$, holds good for a gas under the following assumed conditions.

1. Any finite volume of a gas consists of a very large number of molecules.
2. The molecules are separated by distances large compared to molecular dimensions and are in a state of continuous motion.
3. Molecules do not exert any force on one another except when they collide.
4. Collisions of molecules with one another and with the walls of the container are perfectly elastic.

The real gases deviate from the behaviour of an ideal gas when the above conditions are not valid. The main reasons for deviation are:

1. *Intermolecular Forces:* The molecules of real gases have finite volume and there are interacting forces among them.
2. *Electrical Forces:* Non-permanent electrical forces such as electric moments arise during collision or near collision positions.
3. *Hydrogen Bonding:* Forces due to hydrogen bonding usually occur.
4. *Quantum Effects:* Molecules of some gases such as H_2, He, Ne have quantum effects at low temperature.

5.4 EQUATIONS OF STATE FOR REAL GASES

At high pressures, the intermolecular forces of attraction and repulsion increase, and also the volume of the molecules becomes appreciable compared to the total gas volume. The real gases deviate considerably from the ideal gas equation. Different scientists have introduced correction factors to take care of reasons of deviation.

5.4.1 Van der Waals Equation of State

Van der Waals used laws of mechanics and introduced two corrections in the equation of ideal gas. The van der Waals equation of state is:

$$\left(p+\frac{a}{v^2}\right)(v-b)=RT$$

where, $\frac{a}{v^2}$ = force of cohesion to account for the existence of mutual attraction between the molecules.

b = co-volume to account for the volumes of the molecules.

Van der Waals equation holds good for all gases at high pressure but not in all ranges of pressures and temperatures.

5.4.2 Compressibility Factor

Important equations of state apart from van der Waals equation are:

(i) Berthelot: $p=\frac{RT}{v-b}-\frac{a}{Tv^2}$

(ii) Dieterici: $p=\frac{RT}{v-b}e^{\frac{-a}{RTv}}$

(iii) Redlick-Kwong: $p=\frac{RT}{v-b}-\frac{a}{T^{1/2}v(v+b)}$

The constants a and b for van der Waals and other equations are evaluated from the critical data. The equation of state can be written as:

$$pv = ZRT$$

where, Z = compressibility factor

For a given gas, Z is a function of p and T which can be found out from a graph as shown in Fig. 5.1. For any value of p and T, the value of v can be calculated.

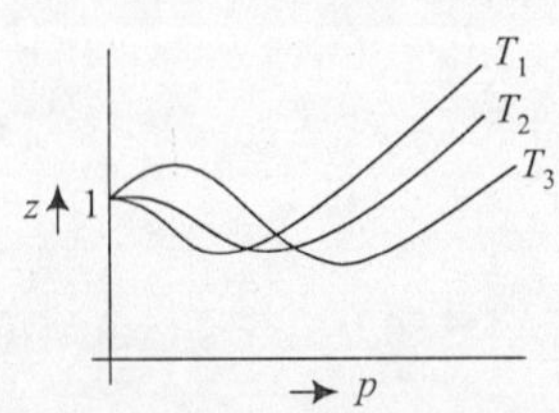

Fig. 5.1 Compressibility chart

There is a compressibility factor chart for every gas.

5.4.3 Generalized Compressibility Chart

The reduced properties are the ratios of existing property to the critical property.

$$p_r = \frac{p}{p_c};\ T_r = \frac{T}{T_c};\ v_r = \frac{v}{v_c}$$

where subscript, r = reduced property

c = property at critical state

It has been found from experimental data that the reduced volume of different gases is approximately same at the same reduced pressure and reduced temperature.

$$v_r = f(p_r, T_r)$$

Now,
$$v_r = \frac{v}{v_c} = \frac{ZRTp_c}{Z_cRT_cp} = \frac{Z}{Z_c}\cdot\frac{T_r}{p_r}$$

$$Z_c = \frac{p_cv_c}{RT_c}$$

where, Z_c = critical compressibility factor

= constant

$$\therefore \quad Z = f(p_r, T_r)$$

The generalized compressibility chart can be plotted between Z and p_r for different values of T_r.

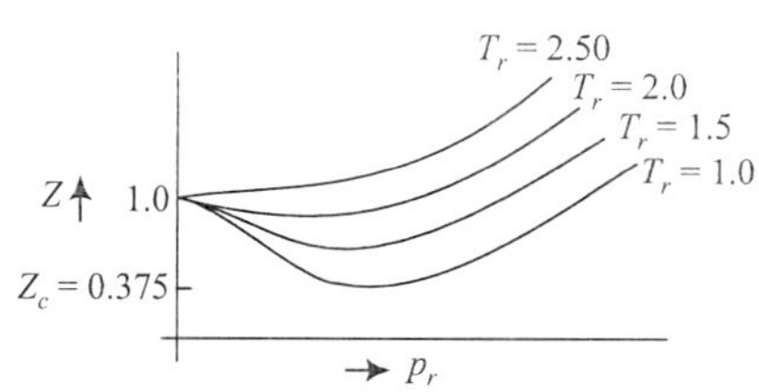

Fig. 5.2 Generalized compressibility chart

5.4.4 Corresponding States

Law of corresponding states is the relation among the reduced properties; p_r, T_r and v_r.

The van der Waals equation in reduced parameters becomes:

$$p_r v_r = \frac{8 T_r v_r}{3 v_r - 1} - \frac{3}{v_r}$$

where, $$Z_C = 0.375$$

$$R = \frac{8}{3} \frac{p_c v_c}{T_c}$$

5.5 NON-REACTIVE MIXTURES OF GASES AND VAPOURS

Non-reactive gas mixture is a mixture of two or more pure substances (gases and vapours) where there is no chemical reaction between the components. The components are considered inert gases or vapours.

There are two types of mixtures of pure substances:

1. *Mixture of ideal gases*

Example 1 Products of combustion of I.C. engines and boilers, i.e., CO_2, SO_2, NO_x, etc.

Example 2 Air as mixture of nitrogen, oxygen, etc.

2. *Mixture of ideal gases and vapours.*

Example Atmospheric air as a mixture of dry air and water vapour.

3. *Total volume of ideal gas mixture*

$$V = V_a + V_b + V_c$$

Volume fraction (Z) is the ratio of the volume occupied by a constituent gas (if theoretically assumed to be separated at the pressure and temperature of the mixture) to the total volume of ideal gas mixture.

$$Z_a = \frac{V_a}{V}$$

$$Z_b = \frac{V_b}{V}$$

$$Z_c = \frac{V_c}{V}$$

$$\therefore \quad Z_a + Z_b + Z_c = \frac{V_a}{V} + \frac{V_b}{V} + \frac{V_c}{V} = 1$$

4. *Total pressure of ideal gas mixture*

The total pressure of mixture,

$$p = p_a + p_b + p_c$$

The pressure of the mixture of the ideal gases is equal to the sum of the partial pressures of individual constituent gases.

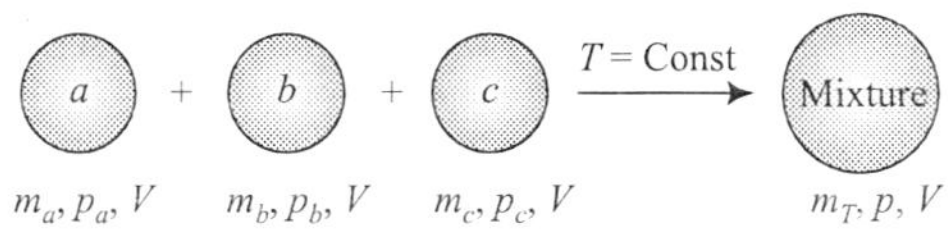

Fig. 5.3 Pressure of gas mixture

p_a, p_b, and p_c are the partial pressures of gas a, b and c respectively.

The partial pressure of any component in a mixture of gases is the pressure of the constituent if it alone occupied the entire volume of the mixture at the same temperature.

5. *Internal Energy of the Ideal Gas Mixture*

It is the sum of internal energies of the individual constituents at the pressure and temperature of the mixture.

$$U = U_a + U_b + U_c.$$

5.5.1 Ideal Gas Mixtures

A homogeneous mixture of two or more ideal gases is called an ideal gas mixture.

A mixture of three ideal gases, a, b, and c at a pressure p and temperature t occupies a volume V.

1. Total mass of ideal gas mixture

$$m_T = m_a + m_b + m_c$$

Mass fraction (y) is the ratio of the mass of a constituent to that of the total mass of the mixture.

$$y_a = \frac{m_a}{m_T}$$

$$y_b = \frac{m_b}{m_T}$$

$$y_c = \frac{m_c}{m_T}$$

$$\therefore \quad y_a + y_b + y_c = \frac{m_a}{m_T} + \frac{m_b}{m_T} + \frac{m_c}{m_T}$$

$$= \frac{m_a + m_b + m_c}{m_T} = 1$$

2. Total number of moles of the gas mixture

Mole is the quantity of a substance equal in mass to the molecular mass of the substance.

Examples 1 kg mol (kmol) of $H_2 = 2$ kg

1 kgmol (kmol) of $N_2 = 28$ kg

1 kgmol (kmol) of $CO_2 = 44$ kg

The total number of moles of gas mixture,

$$N = n_a + n_b + n_c$$

Mole fraction (x) is the ratio of number of moles of a constituent gas to the total number of moles of the ideal gas mixture.

$$\therefore \quad x_a = \frac{n_a}{N}$$

$$x_b = \frac{n_b}{N}$$

$$x_c = \frac{n_c}{N}$$

$$\therefore \quad x_a + x_b + x_c = \frac{n_a}{N} + \frac{n_b}{N} + \frac{n_c}{N} = \frac{n_a + n_b + n_c}{N} = 1.$$

3. Entropy of the ideal gas mixture

The entropy of the mixture of ideal gases is the sum of entropies of the constituent gases as they exist in the mixture.

$$S = S_a + S_b + S_c$$

4. Dalton's law of partial pressures states that in a mixture of ideal gases, any individual gas constituent behaves as *vacuum* to other gas constituents of the mixture

$$p = p_a + p_b + p_c + \dots$$

The final pressure is the sum of partial pressures of individual constituents. The mixture of gases is assumed to behave as an ideal gas.

5. Gibb's-Dalton Law also includes the laws of total internal energy and total entropy of the mixture as sum of these properties of individual gas constituents.

$$U = U_a + U_b + U_c$$

and

$$S = S_a + S_b + S_c$$

5.5.2 Important Relationships of Ideal Gas Mixtures

1. *Amagats Leduc Law*

The *volume* of a mixture of ideal gases is equal to the sum of *partial volumes* or *separated volumes* of the constituent gases when determined at the pressure and temperature of the mixture.

$$V_{\text{mixture}} = V_a + V_b + V_c$$

2. *Partial Pressure Ratio*

The partial pressure ratios of constituent gases in a mixture of ideal gases are equal to mole fraction of corresponding constituent gases.

$$\frac{p_a}{p} = \frac{n_a}{N}; \frac{p_b}{p} = \frac{n_b}{N}; \frac{p_c}{p} = \frac{n_c}{N}.$$

Similarly, volume fraction of each constituent gas of the mixture is equal to mole fraction of corresponding constituent gases.

$$\frac{V_a}{V} = \frac{n_a}{N}; \frac{V_b}{V} = \frac{n_b}{N}; \frac{V_c}{V} = \frac{n_c}{N}$$

Therefore, for each constituent gas in a mixture of ideal gases, the mole fraction, volume fraction and partial pressure ratios are all equal.

3. *Gas Constant of the Mixture*

The equations of state for the mixture and constituent gases are:

$$pV = m_T R_{\text{mix}} T$$

$$p_a V = m_a R_a T$$

$$p_b V = m_b R_b T$$

$$p_c V = m_c R_c T$$

The law of partial pressures:

$$p = p_a + p_b + p_c$$

$$\therefore \quad \frac{m_T (R_{\text{mix}}) T}{V} = \frac{m_a R_a T}{V} + \frac{m_b R_b T}{V} + \frac{m_c R_c T}{V}$$

or

$$R_{\text{mix}} = \frac{m_a}{m_T} Ra + \frac{m_b}{m_T} Rb + \frac{m_c}{m_T} Rc$$

$$= y_a R_a + y_b R_b + y_c R_c$$

where y is mass fraction.

4. *Molecular Mass of the Mixture*

The universal gas constant or molar constant ($\overline{R}$) of a gas is the product of the characteristic gas constant (R) and the molecular mass (M) of the gas.

$$\overline{R} = MR$$

$$\frac{\overline{R}}{M_{\text{mix}}} = y_a \frac{\overline{R}}{M_a} + y_b \frac{\overline{R}}{M_b} + y_c \frac{\overline{R}}{M_c}$$

$$\therefore \quad M_{\text{mix}} = \frac{1}{(y_a / M_a) + (y_b / M_b) + (y_c / M_c)}$$

5. *Specific Heats of Gas Mixture*

The internal energy of gas mixture,

$$U = U_a + U_b + U_c$$

$$\therefore \quad m_T u = m_a\, u_a + m_b\, u_b + m_c\, u_c$$

$$m_T (C_V)_{\text{mix}}\, dT = m_a (C_V)_a\, dT + m_b (C_V)_b\, dT + m_c (C_V)_c\, dT$$

$$\therefore \quad (C_V)_{\text{mix}} = \frac{m_a}{m_T}(C_V)_a + \frac{m_b}{m_T}(C_V)_b + \frac{m_c}{m_T}(C_V)_c$$

$$(C_V)_{\text{mix}} = y_a (C_V)_a + y_b (C_V)_b + y_c (C_V)_c$$

The molar specific heat,

$$(\overline{C}_V)_{\text{mix}} = x_a (\overline{C}_V)_a + x_b (\overline{C}_V)_b + x_c (\overline{C}_V)_c$$

The specific heat at constant pressure,

$$(C_p)_{\text{mix}} = y_a (C_p)_a + y_b (C_p)_b + y_c (C_p)_c$$

The molar specific heat,

$$(\overline{C}_p)_{\text{mix}} = x_a (\overline{C}_p)_a + x_b (\overline{C}_p)_b + x_c (\overline{C}_p)_c$$

6. *Entropy Increase of gas Mixture*

If n_a moles of gas a at pressure p and n_b moles of gas b at pressure p mix adiabatically and isothermally, the entropy change of gas a,

$$\Delta S_a = (S_2 - S_1)_a = -n_a\, \overline{R}\, \ln \left(\frac{p_a}{p} \right)$$

and entropy change of gas b,

$$\Delta S_b = (S_2 - S_1)_b = -n_b\, \overline{R}\, \ln \left(\frac{p_b}{p} \right)$$

$$\therefore \quad \Delta S_a + \Delta S_b = -\overline{R} \left[n_a \ln \frac{p_a}{p} + n_b \ln \frac{p_b}{p} \right]$$

$$= -\, \overline{R}\, [n_a \ln x_a + n_b \ln x_b]$$

Observations

1. The increase of entropy of a gas mixture depends only on the number of moles of constituent gases.
2. The increase of entropy does not depend upon the constituent gases.
3. The entropy increases only when different gases mix.

5.5.3 Mixture of Gas and Vapour

If one of the constituent of the mixture is a vapour which may have a very low partial pressure, the vapour can be treated as an ideal gas. The general rules and properties as applicable to mixture of ideal gases are also applicable to such gas and vapour mixture. Atmospheric air (humid) is a mixture of dry air and water vapour at very low partial pressure. The science which deals with processes of air-water mixture is called *psychrometrics*. It will be dealt with separately.

There are significant difference in the behaviour of gas-vapour mixtures as compared to ideal gas mixtures.

The composition of gas-vapour mixtures may change on heating and cooling. The vapour constituent of the mixture may condense or solidify at low temperatures resulting in the change of composition.

5.5.4 Properties of Atmospheric Air

Dry air is a mechanical mixture of many gases mainly consist of 21% O_2 and 79% N_2 by volume and 23% O_2 and 77% N_2 by mass. However, dry air does not exist and atmospheric air is a mixture of dry air and water vapour.

1. The pressure of atmospheric air can be found out by Dalton's law of partial pressures.

$$p = p_a + p_w$$

where, p = atmospheric pressure (atm)

p_a = partial pressure of dry air (atm)

p_w = partial pressure of water vapour (atm)

2. Mole fraction of dry air,

$$x_a = \frac{p_a}{p} = p_a \qquad (\because \quad p = 1 \text{ atm})$$

Mole fraction of water vapour

$$x_w = \frac{p_w}{p} = p_w$$

Since p_w is very small, the saturation temperature of water-vapour at p_w is less than atmospheric temperature, t_{atm} and water vapour exists in the superheated state. The air is said to be unsaturated.

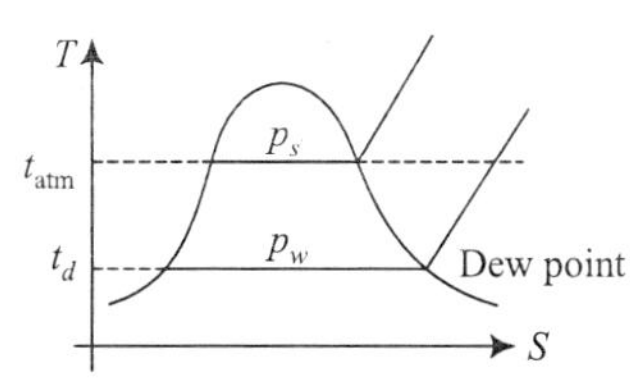

Fig. 5.4 State of water vapour in mixture

The properties of atmospheric air can be described by the following terms:

1. Relative Humidity (ϕ) is the ratio of partial pressure of water vapour in a mixture to the saturation pressure of pure water at the temperature of the mixture.

   Relative humidity

$$\phi = \frac{p_w}{p_s}$$

where, p_w = partial pressure of water vapour in the mixture.

p_s = saturation pressure at the mixture temperature

$$\phi = \frac{p_w}{p_s} = \frac{m_w}{m_s} = \frac{n_w}{n_s} = \frac{x_w}{x_s}$$

2. Specific humidity or humidity ratio (W) is the mass of water vapour per unit mass of dry air in a mixture of air and water vapour.

$$W = \frac{m}{G}$$

where, m = mass of water vapour

G = mass of dry air

Specific humidity is maximum when air is saturated at atmospheric temperature.

$$W_{max} = \frac{m_s}{G}$$

Equation of state for dry air,

$$p_a V = G R_a T$$

Equation of state for water vapour (assumed ideal gas)

$$p_w V = m R_w T$$

$$\therefore \quad W = \frac{m}{G} = \frac{p_w}{p_a} \cdot \frac{R_a}{R_w} = \frac{p_w}{p - p_w} \frac{8.3143/28.96}{8.3143/18}$$

$$= 0.622 \frac{p_w}{p - p_w}$$

$$W_s = 0.622 \frac{p_s}{p - p_s}$$

3. The degree of saturation (μ) is the ratio of actual specific humidity and the saturated specific humidity both at the same temperature.

$$\mu = \frac{W}{W_s} = \frac{0.622\dfrac{p_w}{p-p_w}}{0.622\dfrac{p_s}{p-p_s}}$$

$$= \frac{p_w}{p_s} \cdot \frac{p-p_s}{p-p_w}$$

$\mu = 0$ for dry air

$\mu = 1$ for saturated air

Example 5.1 Calculate the pressure exerted by 1 kg of CO_2 at 100°C if the specific volume is 1 m^3/kg.

Compare the result if pressure is calculated by the use of van der Waals equation with $a = 362.85 \times 10^3$ Nm^4/kg mol^2 and $b = 0.0423$ m^3/kg mol.

Solution (i) $pv = RT$

$$R = \frac{\bar{R}}{M} = \frac{8.3143}{44} \text{ kJ/kgK}$$

$$pV = mRT$$

$$p = \frac{RT}{V} = \frac{8.3143}{44} \times \frac{(100+273)}{1} = 6845 \text{ kN/m}^2$$

(ii) Van der Waals equation

$$p = \frac{RT}{v-b} - \frac{a}{v^2}$$

$$p = \frac{8314.3 \times 373}{44 - 0.0423} - \frac{362.85 \times 10^3}{44^3}$$

$$= 68260.2 \text{ N/m}^2 = 68.26 \text{ kN/m}^2.$$

Example 5.2 A mixture of carbon monoxide and oxygen is to be prepared in the proportion of 7 kg to 4 kg in a vessel of 0.3 m^3 capacity. If the temperature of the mixture is 15°C, determine the pressure to which the vessel is subjected. If the temperature is raised to 40°C, what will then be the pressure in the vessel?

Solution (i) The partial pressures of constituent gases can be calculated from equation of state.

$$(p)_{CO} = \frac{m\bar{R}T}{VM_{CO}} = \frac{7 \times 8314.3 \times (273 + 15)}{0.3 \times 28}$$

$$= 199.54 \times 10^4 \text{ N/m}^2$$

$$(p)_{O_2} = \frac{m\bar{R}T}{VM_{O_2}} = \frac{4 \times 8314.3 \times 288}{0.3 \times 32}$$

$$= 99.77 \times 10^4 \text{ N/m}^2$$

The pressure of mixture in the vessel,

$$p = (p)_{CO} + (p)_{O_2} = 199.54 \times 10^4 + 99.77 \times 10^4$$

$$= 299.31 \times 10^4 \text{ N/m}^2$$

$$= 29.93 \text{ bar}$$

(ii) For $T = 273 + 40 = 313$ K

$$(p)_{CO} = \frac{199.54 \times 10^4 \times 313}{288} = 216.88 \times 10^4 \text{ N/m}^2$$

$$(p)_{O_2} = \frac{99.77 \times 10^4 \times 313}{288} = 108.43 \times 10^4 \text{ N/m}^2$$

The pressure of the mixture,

$$p = (p)_{CO} + (p)_{O_2} = 216.88 \times 10^4 + 108.43 \times 10^4$$

$$= 325.3 \times 10^4 \text{ N/m}^2$$

$$= 32.53 \text{ bar}$$

QUESTION BANK NO. 5

1. What is an ideal gas? What are the main reasons of deviation of real gas from ideal gas?
2. What is compressibility factor? What is the difference between compressibility chart and generalized compressibility chart?
3. Explain Dalton's law of partial pressures. How is it different from Gibb's-Dalton law?
4. Define the following terms related to atmospheric air.
   (a) Relative humidity
   (b) Specific humidity
   (c) Degree of saturation

5. Explain the following terms for ideal gas mixture
   (a) Mass fraction
   (b) Mole fraction
   (c) Volume fraction
   (d) Partial pressure ratio
6. Derive the formula for the following properties for ideal gas mixture
   (a) Gas constant and molecular mass
   (b) Specific heats
   (c) Entropy increase

CHAPTER 6

Properties of Steam

6.1 INTRODUCTION

When water is heated in a boiler or steam generator at constant pressure, its heat content called enthalpy increases and its physical state changes. With the addition of heat, the temperature of water rises and vaporization takes place. The vapour so formed is called steam. Steam does not obey laws of perfect gases. However, in a highly superheated state, its behaviour approaches that of a perfect gas.

The different stages of steam generation are:

1. **Compressed and subcooled water:** Water (feedwater) is pumped to boiler pressure in the state of below saturation temperature. Feedwater is called compressed and subcooled water.
2. **Saturated water:** The subcooled water is heated upto saturation temperature in the economizer, a part of steam generator. Saturation temperature is the temperature at which it starts to evaporate. It is a function of water pressure.
3. **Saturated steam:** The saturated water is evaporated into saturated steam in the boiler part of steam generator. Saturated steam is the steam in a state where it starts condensing on cooling.
4. **Wet steam:** It is a mixture of dry steam and water particles. Steam passes through a state of wet steam between saturated water and saturated steam.
5. **Superheated steam:** The dry and saturated steam can be heated to higher temperatures in the superheater. Steam above saturation temperature is called superheated steam.

6.2 APPLICATIONS OF STEAM

Steam is perhaps the most important working fluid used in the industry. The various applications of steam are:

1. Process heating
2. Space heating

3. Thermo-compression and vacuum generation
4. Distillation and stripping
5. Cleaning and fire-fighting
6. Displacement of liquids and gases
7. Refrigeration
8. Steam jet liquid pumping
9. Evaporation
10. Steam reforming
11. Steam locomotion and marine propulsion
12. Power generation in thermal, nuclear and geothermal power plants.

Steam finds applications in many types of industries. Some of these industries are:

1. Central power plants (thermal and nuclear)
2. Industrial power plants (combined-cycle and co-generation plants)
3. Chemical and petrochemical industries
4. Oil industries (refineries)
5. Sugar industries
6. Dairy plants
7. Distilleries
8. Fertilizer plants
9. Textile industries
10. Pharmaceutical plants
11. Food processing industries

6.3 STEAM GENERATION

Steam is generated by heating water at constant pressure in a boiler. Water passes through different states and same is shown in various property diagrams.

Process *A* – *B*: Feed water at pressure p_A temperature t_A is heated. The temperature will go on increasing till boiling temperature t_B is reached. This process is called sensible heating of water. The boiling point is called saturation temperature, t_s. The saturation temperature varies with pressure of water. The specific heat of water is 4.186 kJ/kg-K. At $p_s = 1.01325$ bar, $t_s = 100°\text{C}$.

Enthalpy change,

$$(dh)_{AB} = C\,dt = 4.186\,(t_B - t_A)$$

Process *B* – *C*: Water is evaporated and a two-phase mixture of liquid water and vapour is formed. There is no increase in temperature and heat is utilized to break up molecular bonds of water.

$$(dh)_{BC} = \text{latent heat of vaporization}$$

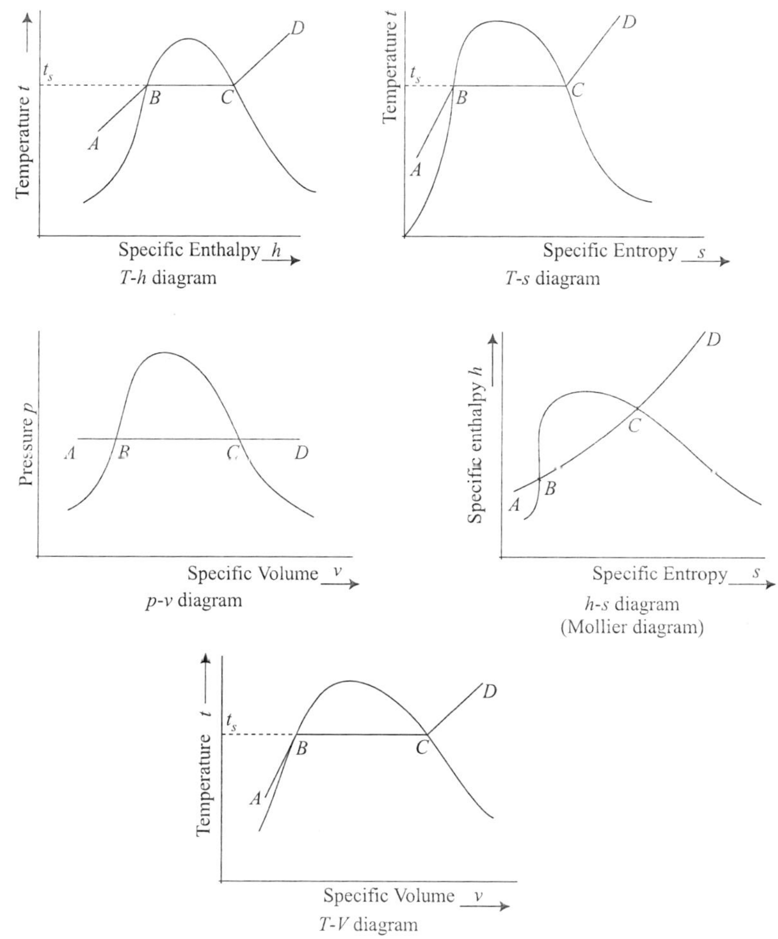

Fig. 6.1 Steam generation on different property diagrams

$$= h_C - h_B = h_g - h_f = h_{fg}$$

where, h_f = specific enthalpy of saturated water (kJ/kg)

h_g = specific enthalpy of saturated steam (kJ/kg)

h_{fg} = latent heat or specific enthalpy of vaporization (kJ/kg)

The mixture of water and steam is called wet steam.

Process *C* – *D*: The steam first becomes completely dry and at point *C*, it is called dry and saturated steam. The temperature of steam increases on further heating and the process is called superheating.

$$(dh)_{CD} = C_s\, dt = 2.1\,(t_D - t_C)$$

$$C_s = 2.1 \text{ kJ/kg-K} = \text{specific heat of steam.}$$

6.4 IMPORTANT CONSTANTS

1. Triple point: Ice, water and water vapour coexist at the triple point of water. It is the point of intersection of sublimation line, melting line and vaporization line on p-t diagram. On $p-v$ diagram, it is represented by triple phase line as shown in Fig. 6.3.

The triple point for water:

Pressure, $p = 0.611$ kPa

Temperature, $T = 273.16$ K or 0.01°C

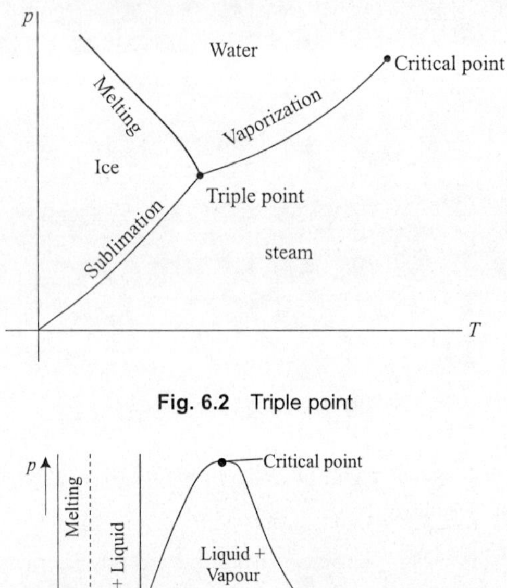

Fig. 6.2 Triple point

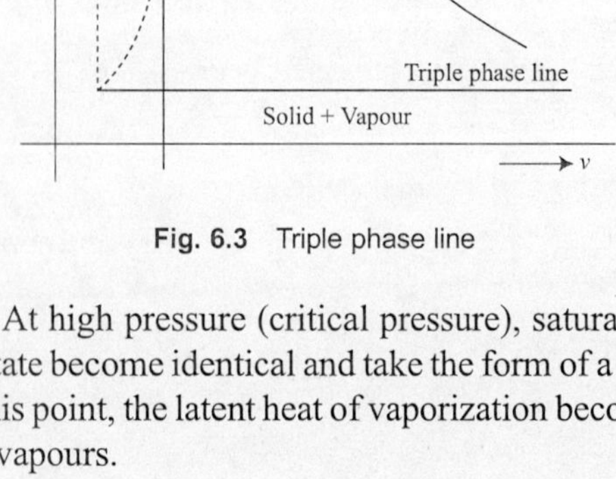

Fig. 6.3 Triple phase line

2. Critical point: At high pressure (critical pressure), saturated liquid state and saturated vapour state become identical and take the form of a point. This is called critical point. At this point, the latent heat of vaporization becomes zero and water directly flashes to vapours.

The critical point for water:

Pressure, $p_{cr} = 22.09$ MPa

Temperature, $t_{cr} = 274.14°C$

Specific volume, $v_{cr} = 0.003155\ m^3/kg$

The steam at pressure and temperature above critical point is called supercritical steam.

3. Dryness fraction: The wet steam is a mixture of dry steam and saturated water. The vapour fraction in the wet steam is called dryness fraction (X).

$$X = \frac{\text{Mass of actual dry steam in the mixture}}{\text{Total mass of wet steam}}$$

$$= \frac{m_g}{m_g + m_f}$$

where, m_g = mass of dry and saturated steam

m_f = mass of saturated water

$m_f + m_g$ = total mass of wet steam

The dryness fraction plays an important role in the determination of thermodynamic properties of wet steam. Experimentally, it can be determined by using a throttling calorimeter or separating calorimeter.

4. Specific heat: It is the quantity of heat required to raise the temperature of one kilogram of a substance through one degree

Specific heat of ice = 2.09 kJ/kg-K.

Specific heat of water = 4.1868 kJ/kg-K.

Specific heat of superheated steam = 2.1 kJ/kg-K

5. Latent heat: It is the heat required to completely change the phase of a substance.

Latent heat of vaporization of water = 2256.9 kJ/kg at 1 bar.

Heat of fusion of ice = 335.7 kJ/kg at 1 bar.

6. Miscellaneous: Mass density of water, $\rho = 1000\ kg/m^3$ at 1 bar.

Polytropic index of superheated steam, $n = 1.3$

Gas constant of superheated steam,

$$R = \frac{\bar{R}}{M} = \frac{8.3143}{18}$$

$$= 0.462 \text{ kJ/kg-K.}$$

6.5 THERMODYNAMIC PROPERTIES OF STEAM

Important thermodynamic properties of steam along with their symbols and units are given in Table 6.1.

Table 6.1 Thermodynamic properties of steam

S. No.	*Property*	*Symbol*	*Unit*
1.	Absolute pressure	p	bar
2.	Temperature	t	°C
3.	Saturation temperature	t_s	°C
4.	Specific volume of saturated water	v_f	m^3/kg
5.	Specific volume of saturated steam	v_g	m^3/kg
6.	Specific enthalpy of saturated water	h_f	kJ/kg
7.	Specific enthalpy of saturated steam	h_g	kJ/kg
8.	Specific enthalpy of vaporization ($h_g - h_f$)	h_{fg}	kJ/kg
9.	Specific entropy of saturated water	s_f	kJ/kg K
10.	Specific entropy of saturated steam	s_g	kJ/kg K
11.	Specific entropy of vaporization ($s_g - s_f$)	s_{fg}	kJ/kg K
12.	Specific internal energy	u	kJ/kg

The properties of steam are required for the design of plant and equipment and study and analysis of various processes based on steam. These properties based on extensive study by different scientists are available in steam tables and Mollier diagram. Steam tables have been prepared for different states of water and steam.

1. Compressed and subcooled water: The properties of water at high pressure with temperature below saturation point are available in separate tables. The values v, u, h, s are recorded as a function of pressure and temperature.

2. Saturated water and saturated steam: The properties of saturated water v_f, h_f, and s_f are available in tables with varying temperatures and varying pressures. Similarly, in the same tables properties of saturated steam v_g, h_g, h_{fg}, s_g and s_{fg} are available.

The values of internal energy can be calculated as follows:

Internal energy of saturated water,

$$u_f = h_f - pv_f \times 10^2 \;\; \text{[kJ/kg]}$$

where, p is in bar.

Internal energy of saturated steam,

$$u_g = h_g - pv_g \times 10^2 \text{ (kJ/kg)}$$

3. Properties of wet steam: The properties of wet steam with dryness fraction, X can be calculated by using steam tables for saturated water and saturated steam.

(i) Specific volume

$$v = x\,v_g + (1 - x)\,v_f.$$
$$= v_f + x\,(v_g - v_f)$$

The values of v_g and v_f are read from steam tables.

(ii) Specific enthalpy,

$$h = x\,h_g + (1 - x)\,h_f$$
$$= h_f + x\,(h_g - h_f)$$
$$= h_f + x\,h_{fg}.$$

The values of h_g and h_{fg} are read from steam tables.

(iii) Specific entropy,

$$s = xs_g + (1 - x)s_f$$
$$= s_f + x\,s_{fg}$$

The values of s_g and s_{fg} are read from steam tables.

(iv) Specific internal energy,

$$v = h - pv \times 10^2$$

where, h = specific enthalpy of wet steam

$= h_f + xh_{fg}$

v = specific volume of wet steam

$= v_f + x\,(v_g - v_f)$

p = pressure of steam in bar.

4. Superheated steam: All power plants and industries use superheated steam for the following advantages:

i. More work is obtained per kg of steam without increasing the pressure.
ii. The plant or process efficiency increases due to higher temperature of steam.
iii. There is less corrosion and erosion of equipment due to absence of moisture in the steam.
iv. Superheating is mostly done from waste heat of boiler without additional cost of fuel.

The properties of steam can be found out:

(1) From tables for superheated steam
(2) Mollier diagram
(3) From tables for saturated steam.

(i) Specific volume of superheated steam at pressure p and temperature T

$$v = v_g \times \frac{T}{T_s}\ (\text{m}^3/\text{kg})$$

$T_s = t_s + 273$ = saturated temperature of steam (K)

v_g can be read from steam tables.

(ii) Specific enthalpy of superheated steam at pressure p and temperature t.

$$h = h_g + C_s (T - T_s)$$

where, C_s = specific heat of superheated steam

= 2.1 kJ/kg-K or as given

h_g can be read from steam tables for saturated steam.

(iii) Entropy of superheated steam at pressure p and temperature t,

$$s = s_g + C_s \ln\left(\frac{T}{T_s}\right)$$

(iv) Internal energy of superheated steam at pressure p and temperature t.

$$v = h - pv \times 10^2 \text{ (kJ/kg)}$$

where, h = specific enthalpy of superheated steam

p = pressure in bar

v = specific volume of superheated steam

5. Supercritical steam: The thermodynamic properties of supercritical steam at parameters above critical point (221.2 bar and 374.15°C) are read from steam tables for supercritical steam for varying pressure and temperature.

6. Linear interpolation of properties: The steam tables are sufficiently elaborate, the values of properties are given for each degree of change in temperature and 1 bar increase in pressure in saturated steam tables. But in case of superheated steam tables, values for a given pressure and temperature may not be given. The principle of linear interpolation can be used.

Example 6.1 Determine the enthalpy, volume, internal energy and entropy of superheated steam at 15 bar pressure and a temperature of 220°C. The volume of water may be neglected. Take specific heat of superheated steam equal to 2.2 kJ/kg-K.

Solution p =15 bar

t = 220°C.

From steam tables for saturated steam, read the following against a pressure of 15 bar,

h_f = 844.7 kJ/kg

h_{fg} = 1945.3 kJ/kg

t_s = 198.3°C

v_g = 0.13167 m^3/kg

$$s_f = 2.314 \text{ kJ/kg-K}$$
$$s_{fg} = 4.127 \text{ kJ/kg-K}$$

(i) specific enthalpy of superheated steam,

$$h = h_f + h_{fg} + C_s\,(t - t_s)$$
$$= 844.7 + 1945.3 + 2.2\,(220 - 198.3)$$
$$= 2837.54 \text{ kJ/kg.}$$

(ii) specific volume of superheated steam,

$$v = v_g \frac{T}{T_s} = 0.13167 \frac{(220 \times 273)}{(198.3 + 273)}$$
$$= 0.1377 \text{ m}^3\text{/kg.}$$

(iii) internal energy,

$$u = h - pv \times 10^2$$
$$= 2837.54 - 15 \times 10^2 \times 0.1377$$
$$= 2630.94 \text{ kJ/kg}$$

(iv) specific entropy of superheated steam,

$$s = s_f + s_{fg} + C_s \text{ lsn } \frac{T}{T_s}$$
$$= 2.314 + 4.127 + 2.2 \text{ lsn } \frac{(220 \times 273)}{(273 + 198.3)}$$
$$= 6.84 \text{ kJ/kg-K.}$$

Example 6.2 5 kg of steam is generated at a pressure of 10 kg bar from feedwater at a temperature of 25°C. Starting from the basic principles and taking the help of steam tables only, calculate the enthalpy and entropy of steam, if

(i) steam is dry and saturated

(ii) steam is superheated upto a temperature of 300°C. Take C_p for steam as 2.1 kJ/kg-K and C_p for water as 4.187 kJ/kg-K.

Solution

(i) Mass of steam $m = 5$ kg

$p = 10$ bar

Final condition = steam dry and saturated.

Steam tables are prepared with datum temperature at 0°C.

Enthalpy of water at 25°C $= m\, Cp_w\, (t - 0)$

$= 5 \times 4.187 \times 25$

$= 523.375$ kJ

Entropy of water at $25°C = m\, Cp_w \ln \dfrac{T_2}{T_1}$

$$= 5 \times 4.187 \ln \frac{273 + 25}{273}$$

$= 1.83436$ kJ/kg-K.

The effect of pressure is neglected.

Enthalpy of saturated water $h_f = 762.8$ kJ/kg at 10 bar from steam tables.

Entropy of saturated water $s_f = 2.139$ kJ/kg-K.

For dry and saturated steam at 10 bar, $h_g = 2778.1$ kJ/kg

∴ Enthalpy of saturated steam $= m\, h_g$

$= 5 \times 2778.1$

$= 1389.5$ kJ

Entropy of dry saturated steam at 10 bar $= \text{m}\, s_g = 5 \times 6.587$

$= 32.935$ kJ/K.

(ii) Specific enthalpy of superheated steam,

$$h = h_g + C_s\,(t - t_s)$$

From steam tables at 10 bar, $h_g = 2778.1$ kJ/kg

$t_s = 179.7°C$

∴ $h = 2778.1 + 2.1(300 - 179.7)$

$= 3030.73$ kJ/kg

Total enthalpy $= mh = 5 \times 3030.73 = 15153.65$ kJ.

Entropy of superheated steam

$$s = s_g + C_s \ln \frac{T}{T_s} = 6.587 + 2.1 \ln \frac{300 + 273}{179.7 + 273}$$

$= 7.0818777$ kJ/kg-K

Total entropy $= ms = 35.41$ kJ/K

6.6 MOLLIER DIAGRAM

The work done during a process can be found out by the area under the curve on a *p-v* diagram. Similarly, the heat supplied during a process can be found out by the area under the process curve on a *T-s* diagram. In 1904, Dr. Mollier developed Mollier diagram (*h-s* diagram) where the work done and heat supplied can be measured directly as a length of process line. It is also widely used to obtain steam properties for various engineering devices and processes. The engineering devices can be nozzles, turbines, steam compressors, condensers, boilers, throttling calorimeter, etc.

The important processes involved in steam engineering are:

1. Heating and cooling at constant volume in vessels.
2. Heating at constant pressure in economizers, boilers and superheaters.
3. Cooling at constant pressure in condensers.
4. Isentropic (or irreversib2le) expansion of steam in nozzles, turbines, etc.
5. Throttling of steam in calorimeters, valves, pipes, etc.

A skeleton diagram of Mollier chart is shown in Fig. 6.4.

1. The constant enthalpy lines are horizontal and constant entropy lines are vertical.
2. The constant temperature lines are straight in wet steam region and horizontal in superheated steam region.
3. The constant pressure lines do not change shape in wet steam region and superheated steam region.
4. The constant volume lines are shown as broken lines in wet steam region and superheated steam region. The slope of constant specific volume lines is more than that of pressure lines.
5. Constant dryness fraction lines or quality lines (x) are also shown in wet steam region.

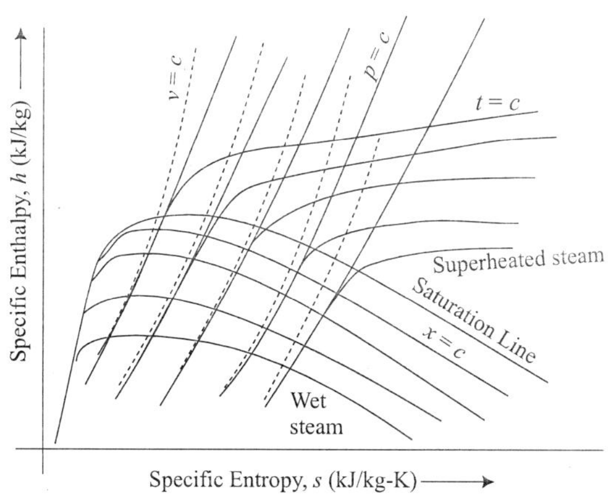

Fig. 6.4 Mollier diagram

Example 6.3 Show the following process on a Skeleton Mollier diagram.

(i) Constant pressure cooling
(ii) Adiabatic compression with friction
(iii) Isothermal process
(iv) Constant volume heating process.

Solution The process lines are shown on Skeleton Mollier diagram,

(i) Line 1–2 is constant pressure cooling process

$$p_1 = p_2$$
$$h_2 < h_1$$

(ii) Line 3–4 is isentropic compression process.

$$s_3 - s_4$$

Dotted line 3–4′ is adiabatic compression with friction

$$s'_4 > s_4$$

(iii) Line 5–6 is isothermal process

$$t_5 = t_6$$

(iv) Line 7–8 is constant volume heating process.

$$v_7 = v_8$$
$$h_8 > h_7$$

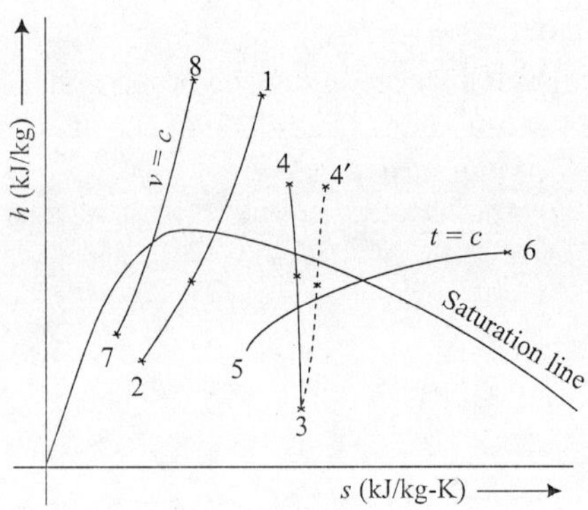

Example 6.4 For estimating the quality of steam at 3 MPa in a boiler, its sample is throttled in a throttling calorimeter where its pressure and temperature were found to be 1 bar and 140°C. What is the quality of steam?

Solution

(a) Using Mollier Diagram

1. Locate point 2 on the intersection of 1 bar and 140°C.
2. Draw line 2 – 1 horizontally so that point 1 lies on 30 bar line.
3. Read dryness fraction X_1,

$$X_1 = 0.97.$$

(b) Using Steam Tables

Steam at point 2 is superheated.

From steam table for superheated steam.

For pressure 1 bar and temperature.

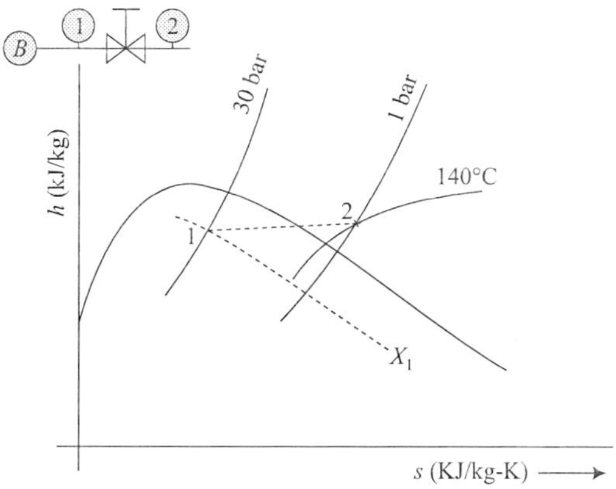

t	100°C	150°C	50°C	140°C
h	2676.2	2776.4	100.2	$2676.2 + \dfrac{100.2 \times 40}{50} = 2756.36$

For throttling process, $h_2 = h_1$

$$\therefore \quad h_1 = h_{f1} + x_1 \, h_{fg1}$$

From saturated steam tables, for a pressure of 30 bar

$$h_{f1} = 1008.5 \text{ kJ/kg}$$

$$h_{fg1} = 1795.7 \text{ kJ/kg}$$

$$\therefore \quad X_1 = \frac{h_1 - h_{g1}}{h_{fg1}} = \frac{2756.36 - 1008.5}{1795.7} = 0.973.$$

6.7 VAPOUR PROCESSES

Water vapour or steam can be subjected to heating, cooling, expansion or compression but the results obtained are different from these processes for gases. The quality of steam continuously changes.

Thermodynamic laws are applicable to steam but gas laws and gas equations do not hold good for steam.

For closed system or non-flow process:

$$dq = du + pdv.$$

For open system and steady flow process.

$$dq = dh + v\,dp$$

Also,

$$dq = Tds$$

6.7.1 Constant Pressure Process

A constant pressure heating process from state 1, wet steam (p_1, x_1) to superheated steam state 2 $(p_2 = p_1, t_2)$ is shown in Fig. 6.5. The initial and final values of specific enthalpy, specific volume and specific entropy can be found out from steam tables or Mollier diagram.

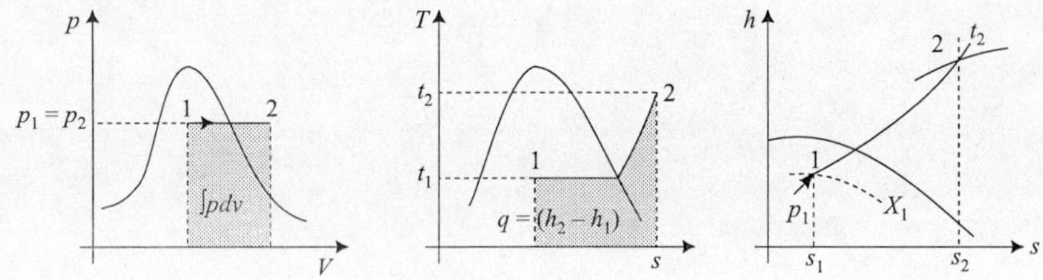

Fig. 6.5 Constant pressure process

(a) Closed system

(i) Work done,

$$w_{1-2} = \int_1^2 pdv = p(v_2 - v_1)$$

If v is in m^3/kg and p in bar,

$$w_{1-2} = p \times 10^2 (v_2 - v) \quad \text{[kJ]}$$

$$= (p_2 v_2 - p_1 v_1) \times 10^2 \qquad [\because \quad p_1 = p_2]$$

(ii) Change in internal energy,

$$(u_2 - u_1) = (h_2 - p_2 v_2) - (h_1 - p_1 v_1)$$

$$= (h_2 - h_1) - (p_2 v_2 - p_1 v_1)$$

$$v_1 = x_1 v_{g1}$$

$$v_2 = v_{g2} \frac{T_2}{T_{s_2}}$$

(iii) Heat transfer,

Using 1st law of thermodynamics,

$$q_{1-2} = w_{1-2} + (u_2 - u_1)$$

$$= (p_2 v_2 - p_1 v_1) + [(h_2 - h_1) - (p_2 v_2 - p_1 v_1)]$$

$$= (h_2 - h_1)$$

(b) Open system (Steady flow process)

(i) Work done,

$$w_{1-2} = \int_1^2 -vdp$$

But, $dp = 0$

$\therefore$ $w_{1-2} = 0$

(ii) Heat transferred,

$$q_{1-2} = w_{1-2} + (h_2 - h_1) = (h_2 - h_1)$$

6.7.2 Constant Volume Process

Heating or cooling of steam in a closed vessel is the example of constant volume process. 1 kg of wet steam (p_1, X_1) at state 1 is heated at constant volume to superheated state 2 (p_2, t_2). The process is shown in Fig. 6.6.

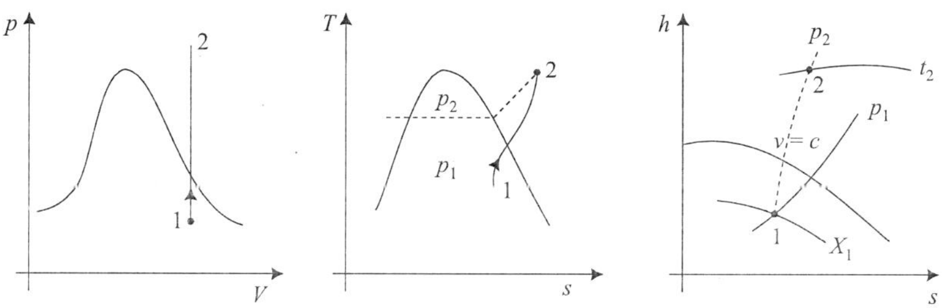

Fig. 6.6 Constant volume process

The initial volume,

$$v_1 = x_1 v_{g1}$$

The final volume,

$$v_2 = v_1$$

(i) If $v_2 < v_{g2}$: steam is wet

$\therefore$ $v_2 = x_2 v_{g2}$.

(ii) If $v_2 = v_{g2}$: steam is dry and saturated

(iii) If $v_2 > v_{g2}$: steam is superheated

$$v_2 = v_{g2} \frac{T_2}{T_{s_2}}$$

where, T_{s2} is saturation temperature at p_2.

(a) Closed system

(i) Work done,

$$w_{1-2} = \int_1^2 pdv = 0 \quad [\because dv = 0]$$

(ii) Change in internal energy

$$u_2 - u_1 = (h_2 - p_2 v_2) - (h_1 - p_1 v_1)$$
$$= (h_2 - h_1) - 10^2 (p_2 v_2 - p_1 v_1)$$

where, p is in bar.

(iii) Heat transfer,

$$q_{1-2} = w_{1-2} + (u_2 - u_1)$$
$$= 0 + [(h_2 - h_1) - 10^2 (p_2 v_2 - p_1 v_1)]$$

(b) Open system (Steady flow process)

(i) Work done, $w_{1-2} = \int_1^2 -vdp = -v \int_1^2 = -v(p_2 - p_1) = (p_1 v_1 - p_1 v_2)$

$$[\because \quad v_1 = v_2 = v]$$

(ii) Change in internal energy,

$$u_2 - u_1 = (h_2 - h_1) - (p_2 v_2 - p_1 v_1)$$
$$= (h_2 - h_1) - 10^2 (p_2 v_2 - p_1 v_1) \qquad \text{[If } p \text{ is in bar]}$$

(iii) Heat transfer,

From 1st law of thermodynamics,

$$q_{1-2} = w_{1-2} + (h_2 - h_1)$$
$$= (p_1 v_1 - p_2 v_2) + [(u_2 + p_2 v_2) - (u_1 + p_1 v_1)]$$
$$= u_2 - u_1.$$

6.7.3 Constant Temperature Process

1 kg of wet steam at state 1 (p_1, X_1) is heated at constant temperature to a state 2 (p_2). The heating of steam in wet region takes place simultaneously at constant pressure and temperature as clear from curve 1 – *a* in $T - s$ diagram. Heating of superheated steam at constant pressure is accompanied by decrease in pressure. It is not practical to heat steam at constant temperature with simultaneous expansion.

(a) Heat transfer,

$$q_{1-2} = T_1(s_2 - s_1)$$

(b) Change in internal energy

$$u_2 - u_1 = (h_2 - p_2 v_2) - (h_1 - p_1 v_1)$$

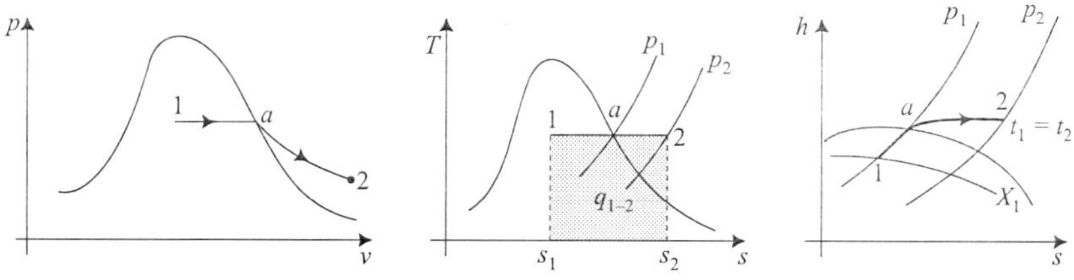

Fig. 6.7 Constant temperature process

(c) Work done,

From 1st law of thermodynamics,

$$q_{1-2} = w_{1-2} + (u_2 - u_1)$$

$$\therefore \quad w_{1-2} = q_{1-2} - (u_2 - u_1)$$

A constant temperature process for steam is same as constant pressure process in wet region and is similar to hyperbolic process in superheated region since steam behaves like a gas in this region.

6.7.4 Hyperbolic Process ($pv = c$)

1 kg of wet steam at state 1 (p_1, X_1) is hyperbolically expanded to state 2 (P_2, t_2) as shown in Fig. 6.8.

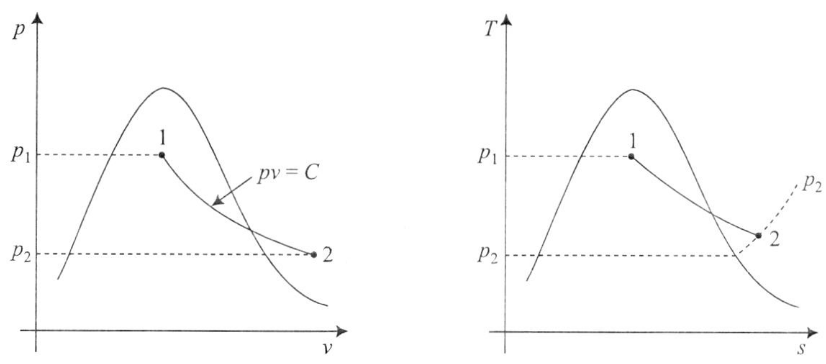

Fig. 6.8 Hyperbolic process

$$v_1 = x_1 v_{g1}$$

$$v_2 = v_{g2} \cdot \frac{T_2}{T_{s_2}} \text{ (if final state of steam is superheated)}$$

$$v_2 = x_2 v_{g2} \text{ (if final state of steam is wet)}$$

For hyperbolic process,

$$p_1v_1 = p_2v_2$$

$$\therefore \quad p_1v_1 = p_2(X_2v_{g2}) \text{ for wet steam}$$

$$X_2 = \frac{p_1\, x_1\, v_{g1}}{p_2v_{g2}}$$

X_2 can be determined.

For superheated steam at final state,

$$p_1X_1v_{g1} = p_2.v_{g2}\,\frac{T_2}{T_{s_2}}$$

T_2 can be determined.

(a) Closed system

(i) Work done,

$$w_{1-2} = p_1v_1 \ln\frac{v_2}{v_1}$$

(ii) Change in internal energy,

$$u_2 - u_1 = (h_2 - h_1) - (p_2v_2 - p_1v_1)$$

$$p_2v_2 = p_1v_1$$

$$\therefore \quad v_2 - v_1 = h_2 - h_1$$

(iii) Heat removed,

From 1st law of thermodynamics,

$$q_{1-2} = w_{1-2} + u_2 - u_1$$

$$= p_1v_1 \ln\frac{v_2}{v_1} + (h_2 - h_1)$$

(b) Open system

(i) Work done,

$$w_{1-2} = \int -vdp = p_1v_1 \ln\frac{v_2}{v_1}$$

(ii) Heat removed,

$$q_{1-2} = w_{1-2} + (h_2 - h_1)$$

$$= p_1v_1 \ln\frac{v_2}{v_1} + (h_2 - h_1)$$

6.7.5 Polytropic Process ($pv^n = c$)

1 kg of superheated steam at state 1 (p_1, t_1) is expanded polytropically to a pressure of p_2. The process is shown in Fig. 6.9.

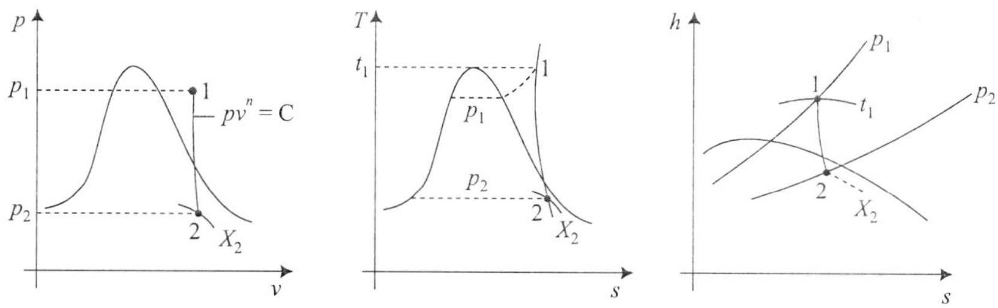

Fig. 6.9 Polytropic expansion process

v_1 can be determined from superheated steam tables against pressure p_1 and temperature t_1,

$$p_1 v_1^n = p_2 v_2^n$$

$$\therefore \qquad v_2 = \left(\frac{p_1}{p_2}\right)^{\frac{1}{n}} . v_1$$

Find v_{g2} against pressure p_2 from superheated steam tables

If $v_2 < v_{g2}$: steam is wet.

$$v_2 = x_2 \, v_{g2}$$

If $v_2 = v_{g2}$: steam is dry and saturated

If $v_2 > v_{g2}$: steam is superheated

$$v_2 = v_{g2} . \frac{T_2}{T_{s_2}} .$$

(a) Closed system

(i) Work done,

$$w_{1-2} = \frac{p_1 v_1 - p_2 v_2}{(n-1)}$$

(ii) Change in internal energy,

$$u_2 - u_1 = (h_2 - h_1) - (p_2 v_2 - p_1 v_1)$$

(iii) Heat transfer,

$$q_{1-2} = w_{1-2} + (u_2 - u_1)$$

$$= \frac{p_1 v_1 - p_2 v_2}{(n-1)} + [(h_2 - h_1) - (p_2 v_2 - p_1 v_1)]$$

(b) Open system

(a) Work done,

$$W_{1-2} = \left(\frac{n}{n-1}\right)(p_1v_1 - p_2v_2)$$

(b) Heat transfer,

$$q_{1-2} = w_{1-2} + (h_2 - h_1)$$

$$= \frac{n}{n-1}(p_1v_1 - p_2v_2) + (h_2 - h_2).$$

6.7.6 Isentropic Process (Reversible Adiabatic)

Conditions of isentropic process:

1. $q_{1-2} = 0$: No heat transfer.
2. $s_2 = s_1$: No change of entropy
3. $pv^n = C$: $n = 1.135$ for wet or saturated steam initially
 : $n = 1.3$ for superheated steam initially.

One kg of steam from state 1 (p_1, t_1) expands isentropically to final pressure p_2. The process is shown in Fig. 6.10.

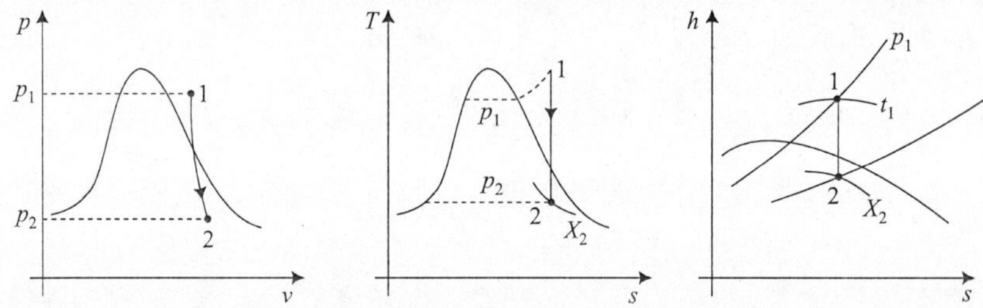

Fig. 6.10 Isentropic expansion process

Superheated steam tables can be used to find h_1 and s_1.

Now, $$s_1 = s_2 = s_{f2} + x_2 s_{fg2}$$

$\therefore$ $$x_2 = \frac{s_1 - s_{f2}}{s_{fg2}}$$

s_{f2} and s_{fg2} can be found from saturated steam tables against a pressure p_2.

$\therefore$ $$h_2 = h_{f2} + x_2 h_{fg2}$$

Again h_{f_2} and h_{fg_2} are available from saturated steam tables against pressure p_2.

(a) Closed system

(i) Heat transfer, $q_{1-2} = 0$

(ii) Change in internal energy,

$$u_2 - u_1 = (h_2 - p_2v_2) - (h_1 - p_1v_1)$$

(iii) Work done,

$$q_{1-2} = w_{1-2} + (u_2 - u_1) = 0$$

$\therefore$ $$w_{1-2} = (u_1 - u_2) = (h_2 - p_2v_2) - (h_1 - p_1v_1)$$

(b) Open system (Steady flow process)

(i) Heat transfer,

$$q_{1-2} = 0$$

(ii) Work done,

$$q_{1-2} = w_{1-2} + (h_2 - h_1) = 0$$

$\therefore$ $$w_{1-2} = (h_1 - h_2)$$

6.7.7 Throttling Process

When steam is passed through a partly opened valve or a restriction, it is throttled to a lower pressure without any work being done. Steam from state 1 (p_1, x_1) is throttled to state 2 (p_2, t_2). Applying steady flow energy equation:

$$q_{1-2} - w_{1-2} = (h_2 - h_1) + \frac{V_2^2 - V_1^2}{2} + g(Z_2 - Z_1)$$

Here, $V_1 = V_2,\quad Z_1 = Z_2,\quad q_{1-2} = 0,\quad w_{1-2} = 0$

$\therefore$ $(h_2 - h_1) = 0$

or $h_1 = h_2 = \text{constant.}$

In a throttling process, enthalpy of steam remains constant. The process is shown in Fig. 6.11.

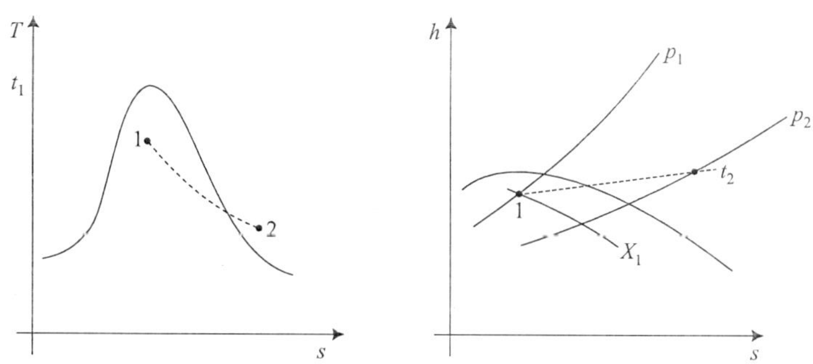

Fig. 6.11 Throttling process

The problem can be solved with the help of Mollier diagram or steam tables. Find h_2 from steam tables (usually superheated steam tables) against a pressure p_2 and temperature t_2.

$$h_2 = h_1 = h_{f1} + x_1 h_{fg1}$$

Find h_{f1} and h_{fg_1} from saturated steam tables against pressure p_1.

∴ Quality of steam at inlet of throttling valve,

$$x_1 = \frac{h_2 - h_{f1}}{h_{fg1}}$$

Example 6.5 2 kg of steam occupying 0.3 m^3 at 15 bar is expanded according to law $pv^{1.3}$ = constant to a pressure of 1.5 bar.

Calculate: (a) work done, and

(b) condition of steam after expansion.

Solution

$$p_1 = 15 \text{ bar}$$

$$V_1 = 0.3 \text{ m}^3$$

$$m = 2 \text{ kg}$$

$$n = 1.3$$

$$p_2 = 1.5 \text{ bar}$$

$$v_1 = \frac{V_1}{m} = \frac{0.3}{2} = 0.15 \text{ m}^3/\text{kg}$$

Now, $$p_1 v_1^n = p_2 v_2^n$$

∴ $$v_2 = \left(\frac{p_1}{p_2}\right)^{\frac{1}{n}} v_1$$

$$= \left(\frac{15}{1.5}\right)^{\frac{1}{1.3}} \times 0.15 = 0.8817 \text{ m}^3/\text{kg}$$

(b) From saturated steam tables for p_2 = 1.5 bar

$$v_{g2} = 1.1635 \text{ m}^3/\text{kg}.$$

Since, $v_2 < v_{g2}$: steam is wet.

Now, $$v_2 = x_2\, v_{g2}$$

∴ $$x_2 = \frac{v_2}{v_{g2}} = \frac{0.8817}{1.1635} = 0.8595$$

(a) Work done,

$$W_{1-2} = m\left[\frac{p_1 v_1 - p_2 v_2}{n-1}\right] = 2\left[\frac{15\times 10^5 \times 0.15 - 1.5\times 10^5 \times 0.8817}{(1.3-1)}\right]$$

$$= 6.183 \times 10^5 \text{ J}$$

$$= \mathbf{618.3 \text{ kJ}} \quad \textbf{Ans.}$$

Example 6.6 A turbine in a steam power plant operating under steady state conditions, receives superheated steam at 3 MPa and 350°C at the rate of one kg/s and with a velocity of 50 m/s at an elevation of 2 m above the ground level. The steam leaves the turbine at 10 kPa with a quality of 0.95 at an elevation of 5 m above the ground level. The exit velocity of steam is 120 m/s. The energy losses on heat from the turbine are estimated as 5 kJ/s. Estimate the power output of the turbine. How much error will be introduced if the kinetic energy and potential energy terms are ignored?

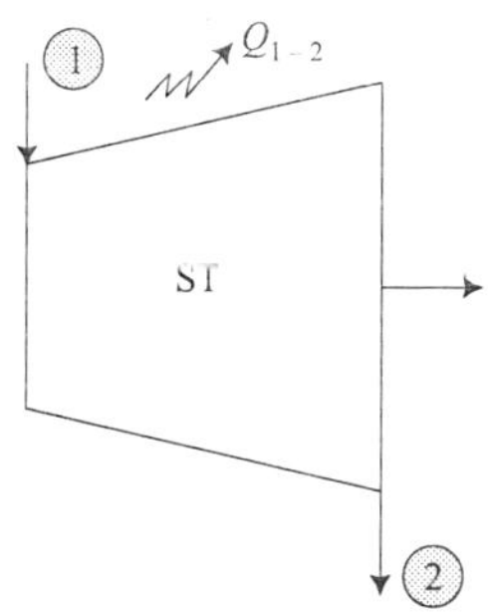

Solution At turbine inlet,

$$p_1 = 3 \text{ MPa} = 3 \times 10^3 \text{ kN/m}^2$$
$$t_1 = 350°\text{C}$$
$$V_1 = 50 \text{ m/s}$$
$$z_1 = 2 \text{ m}$$
$$m = 1 \text{ kg/s}$$

At turbine output,

$$p_2 = 10 \text{ kN/m}^2$$
$$X_2 = 0.95$$
$$V_2 = 120 \text{ m/s}$$
$$z_2 = 5 \text{ m}$$
$$Q_{1-2} = -5 \text{ kJ/s}$$

From superheated steam tables at 30 bar (3×10^3 kN/m^2) and 350°C

$$h_1 = 3115.25 \text{ kJ/kg}$$

From saturated steam tables at 0.1 bar (10 kN/m^2)

$$h_{f2} = 191.81 \text{ kJ/kg}$$

$$h_{fg2} = 2392.83 \text{ kJ/kg}$$

$$\therefore \quad h_2 = h_{f2} + x_2 h_{fg2} = 191.81 + 0.95(2392.82)$$

$$= 2465 \text{ kJ/kg}$$

Applying steady flow energy equation,

$$q_{1-2} - w_{1-2} = (h_2 - h_1) + \frac{V_2^2 - V_1^2}{2\times 10^3} + \frac{g}{10^3}(z_2 - z_1)$$

$$-5 - w_{1-2} = (2465 - 3115.25) + \frac{120^2 - 50^3}{2\times 10^3} + \frac{9.81}{10^3}(5-2)$$

$$w_{1-2} = 639 \text{ kW}$$

Neglecting the effect of *ke* and *pe*.

$$-5 - w_{1-2} = (2465 - 3115.25)$$

$$w_{1-2} = 645.25 \text{ kW}$$

$$\text{Absolute error} = 639 - 645.25 = -6.25 \text{ kW}$$

$$\%\text{ age error} = \frac{6.25}{639} \times 100 = \mathbf{0.978\%\ Ans.}$$

Example 6.7 Steam at 90 kPa and 150°C enters a subsonic diffuser with a velocity of 150 m/s and leaves the diffuser at 190 kPa and a velocity of 55 m/s. During the process, 1.5 kJ/kg of heat is lost to the surrounding. Determine:

(i) The final temperature.

(ii) The mass flow rate

(iii) The exit diameter assuming the inlet diameter as 10 cm and steady flow.

Solution

Inlet conditions:

$$p_1 = 90 \text{ kPa} = 90 \text{ kN/m}^2$$

$$T_1 = 150 + 273 = 423 \text{ K}$$

$$V_1 = 150 \text{ m/s}$$

Outlet conditions:

$$p_2 = 190 \text{ kN/m}^2$$

$$V_2 = 55 \text{ m/s}$$

Applying steady flow energy equation for unit mass:

$$q_{1-2} - w_{1-2} = (h_2 - h_1) + \frac{V_2^2 - V_1^2}{2} + g(Z_2 - Z_1)$$

For given diffuser,

$$Z_2 = Z_1 \text{ and } w_{1-2} = 0$$

$$\therefore \quad q_{1-2} = (h_2 - h_1) + \frac{V_2^2 - V_1^2}{2}$$

$$\therefore \quad (h_2 - h_1) = -q_{1-2} - \frac{V_2^2 - V_1^2}{2}$$

$$\therefore \quad h_1 - h_2 = q_{1\ 2} + \frac{V_2^2 - V_1^2}{2}$$

$$= 1.5 + \frac{55^2 - 150^2}{2 \times 10^3} = 11.2375 \text{ kJ/kg}$$

From superheated steam tables against $p_1 = 90$ kPa (90×10^{-2} bar) and 150°C,

$$h_1 = 2777.1 \text{ kJ/kg}$$

$$\therefore \quad h_2 = 2777.1 - 11.2375 = 2765.86 \text{ kJ/kg.}$$

(i) Again from superheated steam tables, against $p_2 = 190$ kN/m^2 (1.9 bar) and

$$h_2 = 2765.86 \text{ kJ/kg}$$

$$t_2 = 154°\text{C}.$$ **Ans.**

(ii)

$$v_1 = \frac{RT_1}{p_1}$$

$$R = \frac{\bar{R}}{M} = \frac{8.314}{18} = 0.4619 \text{ kJ/kg-K}$$

$$v_1 = \frac{0.4619 \times 423}{90} = 2.17 \text{ m}^3\text{/kg}$$

$$v_2 = \frac{RT_2}{p_2} = \frac{0.4619 \times (154 + 273)}{190} = 1.038 \text{ m}^3\text{/kg}$$

$$\therefore \quad \dot{m} = \frac{A_1 V_1}{v_1} = \frac{\pi/4 (0.10)^2 \times 150}{2.17} = \mathbf{0.543 \text{ kg/s}}$$

(iii) Exit area,

$$A_2 = \frac{\dot{m}\, v_2}{V_2} = \frac{0.543 \times 1.038}{55} = 0.010248 \text{ m}^2$$

$$\therefore \quad d_2 = \sqrt{\frac{A_2 \times 4}{\pi}} = \sqrt{\frac{0.010248 \times 4}{\pi}} = 0.1142 \text{ m}$$

$$= \mathbf{11.42\ cm} \quad \textbf{Ans.}$$

Example 6.8 For estimating the quality of steam at 3 MPa in a boiler, its sample is throttled in a throttling calorimeter where its pressure and temperature were found to be 1 bar and 140°C. What is the quality of steam?

Solution

$$p_1 = 3 \text{ MPa} = 30 \text{ bar}$$

$$p_2 = 1 \text{ bar}$$

$$t_2 = 140°\text{C}$$

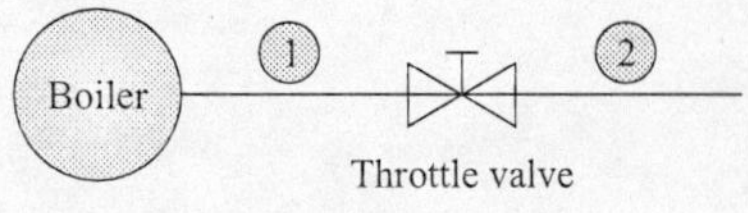

From steam tables for superheated steam against a pressure of 1 bar and temperature of 140°C,

$$h_2 = 2756.3 \text{ kJ/kg}$$

For throttling process,

$$h_2 = h_1 = h_{g1} + x_1 h_{fg1}$$

$$\therefore \quad X_1 = \frac{h_2 - h_{f1}}{h_{fg1}}$$

From steam tables against a pressure of 30 bar,

$$h_{f1} = 1008.5 \text{ kJ/kg}$$

$$h_{fg1} = 1795.7 \text{ kJ/kg}$$

$$\therefore \quad X_1 = \frac{2756.5 - 1008.5}{1795.7} = \mathbf{0.97} \quad \textbf{Ans.}$$

QUESTION BANK NO. 6

1. With the help of $(T - v)$ and $(T - s)$ diagram, explain the difference between 'wet', 'dry saturated' and 'superheated steam'.
2. How can properties of steam be calculated with the help of Steam Tables and Mollier Diagram?
3. Discuss the generation of steam at constant pressure. Show various processes on $(T - v)$, $(p - v)$ and $(T - s)$ diagram.
4. For superheated steam, the isobaric and isothermal lines diverge on 'Mollier chart". Explain the statement.
5. Explain the following:
   (a) Triple point of water
   (b) Critical point
   (c) Dryness fraction

6. What are the main uses of Mollier diagram? Explain with suitable examples.
7. Show the following processes on Mollier diagram:
   (i) Isentropic expansion
   (ii) Adiabatic compression with fraction
   (iii) Throttling process
   (iv) Constant pressure heat addition

CHAPTER 7

Refrigerants

7.1 INTRODUCTION

Refrigeration is used to remove heat from a system or space and maintain it at a low temperature lesser than the surrounding. Refrigeration and air-conditioning industry is the second biggest industry in the world after automobile industry. There are many refrigeration systems.

1. Ice refrigeration system
2. Air refrigeration system
3. Mechanical vapour compression refrigeration system
4. Vapour absorption refrigeration system
5. Steam jet refrigeration system
6. Thermoelectric refrigeration system
7. Other special refrigeration systems

Mechanical vapour compression refrigeration is most widely used. Cooling effect is produced by vaporization of a liquid at appropriate pressure. The liquid is called refrigerant. Different refrigerants are used to maintain various systems at different temperature to as low a temperature as –250°C.

Refrigerant alternates between liquid phase and gas phase.

1. Refrigerant in the form of low pressure gas is received from the system and compressed to an appropriate pressure by a compressor.
2. The high pressure gas is condensed into liquid by removing its latent heat in an air cooled or water cooled heat exchanger. The liquid is stored in a receiver.
3. The high pressure liquid is expanded in a valve or capillary tube by throttling process. A very low temperature liquid is received.
4. The low temperature liquid is used to cool and maintain the space at low temperature or cool air for air-conditioning. The liquid is evaporated into low pressure gas to complete the cycle in a closed loop.

7.2 APPLICATIONS OF REFRIGERATION

The main applications of refrigeration are listed below:

1. Ice making.
2. Cold storages for storage of food products, vegetables and fruits.
3. Chilling (freezing) storage and transportation of milk and dairy products, foodstuffs, fruits, vegetables, fruit juices, fish, meat, poultry products.
4. Processing of food products, farm products, beverages.
5. Chemical and pharmaceutical industries for processing and storage of drugs, tobacco, petroleum and other chemical products.
6. Manufacturing and treatment of metals.
7. Oil refining and synthetic rubber manufacturing.
8. Building construction, plumbing, cooling of concrete for dams.
9. Processing of textiles, printing work, photographic materials.
10. Preservation of archaeological documents.
11. Medical and surgical aids.
12. Cryogenics, production of dry ice.
13. Industrial air-conditioning.
14. Comfort air-conditioning of residences, hospitals, malls, commercial offices, hotels, restaurants, theatres, etc.

7.3 SELECTION OF REFRIGERANTS

The selection of refrigerant depends upon the application, temperature range, capital and running cost of the system, thermodynamic and thermophysical properties of the refrigerant. No refrigerant may be ideal for all jobs but must possess as many qualities as possible from the following list:

1. Low boiling point
2. High critical temperature
3. High latent heat of vaporization
4. Low specific heat of liquid
5. Low specific heat of vapour
6. Non-corrosive to metal
7. Non-inflammable and non-explosive
8. Non-toxic and harmless to human beings
9. Easy to liquefy at moderate pressure and temperature
10. Ease to leak detection by odour or suitable indicators.
11. Low cost and easy availability
12. Inert to lubricants and oils
13. Low specific volume of vapour

14. Satisfactory heat transfer and viscosity coefficients
15. No pollution effect on environment

7.4 PROPERTIES OF REFRIGERANTS

Experimental thermodynamic and thermophysical properties of different refrigerants are available in the form of property tables and p-H charts.

7.4.1 Property Tables

Following property tables are available for every refrigerant.

1. *Thermodynamic properties* of saturated liquid and vapor. The tables include the following properties for different temperatures.

S.No.	Property	Symbol	Units
1.	Temperature	t	°C
2.	Saturation pressure	p	bar
3.	Specific volume of liquid	v_f	m^3/kg
4.	Specific volume of vapour	v_g	m^3/kg
5.	Specific enthalpy of liquid	h_f	kJ/kg
6.	Specific enthalpy of vapour	h_g	kJ/kg
7.	Specific enthalpy of vaporization	h_{fg}	kJ/kg
8.	Specific entropy of liquid	s_f	kJ/kgK
9.	Specific entropy of vapour	s_g	kJ/kgK

2. *Thermodynamic properties* of superheated vapour

The following properties of superheated vapour are given against various pressures and degrees of superheate.

S.No.	Property	Symbol	Units
1.	Pressure	p	bar
2.	Saturation temp.	t_s	°C
3.	Degree of superheat	T_s	°C
4.	Specific volume	v	m^3/kg
5.	Specific enthalpy	h	kJ/kg
6.	Specific entropy	s	kJ/kg-K

3. *Thermophysical properties*

The following thermophysical properties for saturated liquid, saturated vapour and gas at 0 atm pressure are given for different temperatures.

S.No.	Property	Symbol	Units
1.	Temperature	T	K
2.	Dynamic viscosity	μ	Pa-s
3.	Kinematic viscosity	ν	m^2/s
4.	Thermal conductivity	K	W/mK
5.	Specific heat at constant pressure	C_p	kJ/kgK
6.	Specific heat at constant volume	C_v	kJ/kgK
7.	Density	ρ	kg/m^3
8.	Prandtl Number	P_r	–

7.4.2 *p-h* Chart

Pressure-enthalpy charts have been prepared for refrigerants which give their various thermodynamic properties. Enthalpy scale is most convenient to use for refrigeration problems, pressure on vertical ordinates and enthalpy on horizontal ordinates have been selected for displaying the various thermodynamic properties of refrigerants. The following values are selected as datum in these charts.

Temperature, $t = -40°C$

Specific enthalpy, $h = 0$ kJ/kg

Specific entropy, $s = 0$ kJ/kg

A typical *p-h* chart is shown in Fig. 7.1. The saturated liquid line and the saturated vapour line merge into one another at the critical point. The space to the left of the saturated liquid line is the sub-cooled liquid region. The space between the saturated liquid and vapour line is called wet vapour region and to the right of the saturated vapour line is a superheated vapour region. A few important lines of the complete chart are shown.

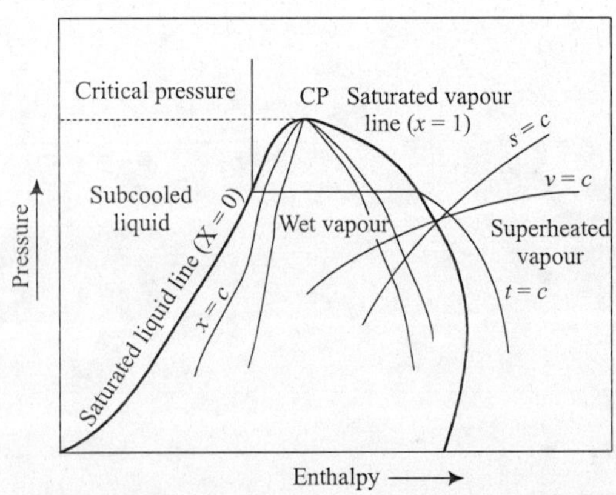

Fig. 7.1 Phase diagram with constant property lies on *p-h* diagram

7.5 COMMON REFRIGERANTS

Common refrigerants are a group of halogenated hydrocarbons under various proprietary names such as Freon, genetron, areton, isotron, figen, mafron, etc. These are either methane-based or ethane-based compounds where hydrogen atoms are replaced by chlorine or fluorine atoms.

Examples Refrigerant – 12 (*R* – 12) is CCl_2F_2 and is the compound dichloro-difluoro methane used for the design of domestic refrigerator, window air-conditioners.

Refrigerant – 22 (*R* – 22) is $CHClF_2$ and is the compound monochloro-difluoro methane used for the design of comfort air-conditioning plants.

The use of these refrigerants is now discouraged. These are largely insoluble in (rain) water, move up, react with ozone in the ozone layer (which protects the earth from pernicious ultraviolet rays) and deplete it.

The important refrigerants used are discussed below:

1. Ammonia (NH_3)

It has a boiling point of –333°C and liquid specific gravity of 0.684. It is widely used in large industrial and commercial mechanical refrigeration systems. It is also widely used as a refrigerant in vapour absorption system.

It is highly toxic and inflammable and therefore, not used in domestic refrigeration and air-conditioning plants.

2. Carbon dioxide (CO_2)

It is colourless and odourless gas heavier than atmospheric air. It has a boiling point of –77.6°C and a liquid specific gravity of 1.56 at atmospheric pressure. It is non-toxic and non-flammable. It has low critical temperature but high critical pressure. It is used only in low temperature refrigeration system.

It has high operating pressure (70 bar) and therefore, high power requirements per tonne of refrigeration.

3. Sulphur dioxide (SO_2)

It is colourless as liquid and gas and non-inflammable and non-explosive. It has a boiling point of –10.5°C and liquid specific gravity of 1.36. It is extremely toxic and has pungent irritating smell. Therefore, it is used in small commercial machines with hermetically sealed units.

4. Freon –12

The entire Freon group is odourless and white in colour. Freon–12 has a boiling point of –30°C and corresponding pressure of 8 bar. It is non-inflammable and non-toxic. It is widely used for domestic refrigerators.

QUESTION BANK NO. 7

1. What is a refrigerant? What are the applications of refrigeration?
2. State the properties of a good refrigerant.
3. What are the normal refrigerants used?
4. Write notes on:
   (a) Refrigerant Property Tables
   (b) Pressure-enthalpy Chart

CHAPTER 8

Products of Combustion

8.1 INTRODUCTION

Solid, liquid and gaseous fuels are burnt in the furnaces of boilers, combustion chambers of gas turbines or cylinders of I.C. engine to generate heat by combustion. Combustion is the chemical combination of hydrocarbons of fuel with oxygen of atmospheric air. A large amount of heat is produced which is converted into mechanical work by the heat engines.

The products of combustion of a fuel form a mixture of gases and vapours. The composition of products depends upon the composition of fuel, degree of completion of reaction and amount of excess (or deficient) air used. The main gases are CO_2, CO, SO_2, N_2, and O_2. The water vapour (H_2O) is also part of products of combustion. The composition of the products can be expressed volumetrically or gravimetrically (mass basis) and are called volumetric analysis and gravimetric analysis respectively. These analysis are mutually convertible using the laws of mixtures of gases and vapours. The analysis of products of combustion is used for finding out the following:

1. Composition of fuel (if unknown)
2. Air-fuel ratio and excess air used.
3. Amount of gases by volume and weight
4. Enthalpy of formation or heat of reaction
5. Heat of combustion and heating value of fuel
6. Combustion temperatures and dew point temperature

The above information is required to control the combustion process, design and performance analysis of combustion equipment, fuel supply system and air supply systems, design of emission system and control of environmental pollution.

8.2 COMBUSTION REACTIONS

There are two methods to determine the analysis of products of combustion of fuels.

1. Experimental method by *Orsat* apparatus: It gives the volumetric analysis of the dry products of combustion. The products are cooled to room temperature which is below the dew point temperature of the gases. All the water vapour is condensed. The dry sample is passed through the apparatus and volume of each constituent of exhaust gases is measured.
2. Analytical method by use of combustion equation. Combustion reactions are expressed as chemical equations in terms of conservation of atoms. The total mass of reactants must be equal to total mass of products.

   For ideal gases, the mole may also be regarded as a unit of volume. It is not necessary that the number of moles (or volumes) of reactants be equal to the number of moles (or volumes) of the products.

8.2.1 Combustion Equations of Atoms

Following equations are important for the combustion of fuels.

1. When carbon atoms of fuel burn in sufficient quantity of oxygen of atmospheric air, carbon dioxide is produced along with a release of large amount of heat of reaction called heat of formation. This process is represented by the following chemical equation.

$$\begin{array}{lccccccl} & C & + & O_2 & = & CO_2 & & \\ & 1\text{ mol} & + & 1\text{ mol} & = & 1\text{ mol} & & \text{(by volume)} \\ & 12\text{ kg} & + & 32\text{ kg} & = & 44\text{ kg} & & \text{(by mass)} \\ \text{or} & 1\text{ kg} & + & \frac{32}{12}\text{ kg} & = & \frac{44}{12} & & \text{kg} \end{array}$$

1 kg of carbon requires $\frac{8}{3}$ kg of oxygen for its complete combustion and produces 11/3 kg of carbon dioxide gas.

2. When sufficient quantity of oxygen is not available, the combustion of carbon is incomplete producing carbon monoxide.

$$\begin{array}{cccccl} 2C & + & O_2 & = & 2CO & \\ 2\text{ mol} & + & 1\text{ mol} & = & 2\text{ mol} & \text{(by volume)} \\ 2\times 12\text{ kg} & + & 2\times 16\text{ kg} & = & 2\times 28\text{ kg} & \text{(by mass)} \\ 1\text{ kg} & + & \frac{32}{24}\text{ kg} & = & \frac{56}{24}\text{ kg} & \\ 1\text{ kg} & + & \frac{4}{3}\text{ kg} & = & \frac{7}{3}\text{ kg} & \end{array}$$

1 kg of carbon requires $\frac{4}{3}$ kg of oxygen to produce $\frac{7}{3}$ kg of carbon monoxide.

3. When CO is burnt into CO_2.

$$\begin{array}{ccccccc} 2CO & + & O_2 & = & 2CO_2 & \\ \left\{\begin{array}{l} 2\text{ mol} \\ 2m^3 \end{array}\right. & \begin{array}{c} + \\ + \end{array} & \begin{array}{c} 1\text{ mol} \\ 1m^3 \end{array} & \begin{array}{c} = \\ = \end{array} & \left.\begin{array}{r} 2\text{ mol} \\ 2m^3 \end{array}\right\} & \text{(by volume)} \\ 2 \times 28\text{ kg} & + & 2 \times 16\text{ kg} & = & 2 \times 44\text{ kg} & \text{(by mass)} \\ 1\text{ kg} & + & \frac{4}{7}\text{ kg} & = & \frac{11}{7}\text{ kg} & \end{array}$$

1 kg of CO requires $\frac{4}{7}$ kg of O_2 to produce $\frac{11}{7}$ kg of CO_2.

4. When hydrogen atoms of fuel burn to form H_2O.

$$\begin{array}{ccccccc} 2H_2 & + & O_2 & = & 2H_2O & \\ \left\{\begin{array}{l} 2\text{ mol} \\ 2m^3 \end{array}\right. & \begin{array}{c} + \\ + \end{array} & \begin{array}{c} 1\text{ mol} \\ 1m^3 \end{array} & \begin{array}{c} = \\ = \end{array} & \left.\begin{array}{r} 2\text{ mol} \\ 2m^3 \end{array}\right\} & \text{(by volume)} \\ 2 \times 2\text{ kg} & + & 2 \times 16\text{ kg} & = & 2 \times 18\text{ kg} & \text{(by mass)} \\ 1\text{ kg} & + & 8\text{ kg} & = & 9\text{ kg} & \end{array}$$

1 kg of hydrogen requires 8 kg of oxygen to produce 9 kg of H_2O.

5. When sulphur atoms of fuel burn to form sulphur dioxide.

$$\begin{array}{ccccccc} S & + & O_2 & = & SO_2 & \\ 1\text{ mol} & + & 1\text{ mol} & = & 1\text{ mol} & \text{(by volume)} \\ 32\text{ kg} & + & 2 \times 16\text{ kg} & = & 64\text{ kg} & \text{(by mass)} \\ 1\text{ kg} & + & 1\text{ kg} & = & 2\text{ kg} & \end{array}$$

1 kg of sulphur requires 1 kg of oxygen to produce 2 kg of sulphur dioxide.

8.2.2 Complete Combustion Equations

The complete combustion equations of fuels can be written by balancing carbon atoms, hydrogen atoms and oxygen atoms of reactants and products.

Example Combustion of heptane, C_7H_{16}.

$$C_7H_{16} + xO_2 = aCO_2 + bH_2O$$

1. Balancing carbon atoms of reactants and products

$$7C = aC$$

$$\therefore \quad a = 7$$

2. Balancing hydrogen atoms of reactants and products

$$16H = 2bH$$

$$\therefore \quad b = 8$$

$$\therefore \quad C_7H_{16} + xO_2 = 7CO_2 + 8H_2O$$

3. Balancing oxygen atoms of reactants and products

$$2x = (2 \times 7) + (8 \times 1) = 22$$

$$\therefore \quad x = 11$$

Final combustion equation is

$$C_7H_{16} + 11\ O_2 = 7\ CO_2 + 8\ H_2O$$

4. For most combustion processes, oxygen is supplied as part of atmospheric air.

   Composition of air:

   (a) On mass basis:

   $$O_2 = 23.15\%$$
   $$N_2 = 76.85\%$$

   (b) On volume basis:

   $$O_2 = 21\%$$
   $$N_2 = 79\%$$

   $$\text{Amount of } N_2 \text{ per kg of } O_2 \text{ supplied} = \frac{76.85}{23.15} = 3.32 \text{ kg}$$

   $$\text{Amount of air per kg of } O_2 \text{ supplied} = \frac{100}{23.15} = 4.32 \text{ kg}$$

   $$\text{or, Moles of } N_2 \text{ per mole of } O_2 \text{ supplied} = \frac{79}{21} = 3.76 \text{ mole}$$

   $$\text{Moles of air per mole of } O_2 \text{ supplied} = \frac{100}{21} = 4.76 \text{ mole}$$

5. Complete combustion equation of heptane.

$$C_7H_{16} + 11\ O_2 + (11 \times 3.76)N_2 = 7\ CO_2 + 8\ H_2O + (11 \times 3.76)N_2$$

or $$C_7H_{16} + 11\ O_2 + 41.63\ N_2 = 7\ CO_2 + 8\ H_2O + 41.36\ N_2$$

8.2.3 Air Fuel Ratio

Air fuel ratio is the ratio of mass of air to the mass of fuel in the combustion reaction.

Air fuel ratio,

$$AF = \frac{\text{mass of air}}{\text{mass of fuel}} = \frac{m_a}{m_f}$$

Fuel air ratio,

$$\text{FA} = \frac{\text{mass of fuel}}{\text{mass of air}} = \frac{m_f}{m_a}$$

For the above example,

$$\text{AF ratio} = \frac{(\text{mass of 11 mole } O_2) + (\text{mass of 41.36 mole } N_2)}{(\text{mass of 1 mole } C_7H_{16})}$$

$$= \frac{(11 \times 32) + (41.36 \times 28)}{(1 \times 100)} = 15.10$$

$$\text{FA ratio} = \frac{\text{mass of } (11 + 41.36) \text{ mole air}}{\text{mass of 1 mole } C_7H_{16}}$$

$$= \frac{52.36 \times 29}{1 \times 100} = 15.18$$

1. *Stoichiometric* or theoretical quantity of air required for complete combustion of 1 kg of fuel. There is no oxygen in the products of combustion.

   Excess air = (Actual amount of air supplied)
   – (Theoretical amount of air required for complete combustion)
2. Excess air is shown in the products of combustion as nitrogen and oxygen.
3. Insufficient air = (Theoretical amount of air required for complete combustion) – (Actual amount of air supplied)
4. Chemically correct air fuel ratio shows deficiency or excess amount of air supplied.

Example 8.1 The Orsat apparatus readings of products of combustion are:

$CO_2 = 8.0\%$, $CO = 0.9\%$, $O_2 = 8.8\%$, $N_2 = 82.3\%$

Determine: (a) composition of fuel

(b) air fuel ratio

(c) per cent excess air supplied

Solution The chemical reaction can be written as:

$$C_x H_y + a\, O_2 + 3.76a\, N_2 = 8\, CO_2 + 0.9\, CO + 8.8\, O_2 + 82.3\, N_2 + \frac{y}{2}\, H_2O$$

Using oxygen balance:

$$2a = 8 \times 2 + 0.9 + 8.8 \times 2 + \frac{y}{2} \times 1 \qquad \ldots(1)$$

Using nitrogen balance:

$$3.76 \times 2a = 82.3 \times 2 \qquad \ldots(2)$$

$$\therefore \qquad a = 21.89$$

From equation (1), $\quad y = 18.5$

Using carbon balance:

$$x = 8 + 0.9 = 8.9$$

(a) Chemical formula of fuel is $C_{8.9}H_{18.5}$.

The fuel composition is:

$$\%\text{ age carbon} = \frac{8.9 \times 12}{8.9 \times 12 + 18.5 \times 1} \times 100 = 85.23\%$$

$$\%\text{ age hydrogen} = 100 - 85.23 = 14.77\%$$

(b) Air fuel ratio, $AF = \dfrac{32a + 3.76a \times 28}{12x + 1y}$

$$= \frac{32 \times 21.89 + 3.76 \times 21.89 \times 28}{12 \times 8.9 + 18.5 \times 1}$$

$$= 24$$

(c) Excess air supplied $= \dfrac{8.8 \times 32}{21.89 \times 32 - 8.8 \times 32} \times 100$

$$= 67.22\%$$

8.3 MUTUAL CONVERSION OF VOLUMETRIC AND GRAVIMETRIC ANALYSIS

1. The volumetric composition of product can be converted into gravimetric composition as follows:
   (i) Proportional mass of the constituent gas can be found out by product of volume and own molecular mass.
   (ii) Percentage analysis by mass is found out by dividing the proportional mass by total mass multiplied by 100.
2. The gravimetric (mass) analysis of product gas mixture can be converted into volumetric analysis as follows:
   (i) Proportional volume of constituent gas can be found out by dividing the percentage mass by its own molecular mass.
   (ii) Percentage analysis by volume is found out by dividing the proportional value of constituent gases by total volume multiplied by 100.

Example 8.2 The volumetric analysis of product of composition is: $CO_2 = 14\%$, $CO = 1\%$, $O_2 = 5\%$, $N_2 = 80\%$. Calculate the gas composition by mass.

Solution The solution is given in Table 8.1.

Table 8.1 Gravimetric analysis

Constituent gas	*Volume in $1m^3$ of gas mixture*	*Molecular mass*	*Proportional mass*	*Mass per kg of gas mixture*	*Mass per cent*
0	*(1)*	*(2)*	*(3) = (1 × 2)*	$(4) = \frac{(3)}{\Sigma(3)}$	*(5) = (4) × 100*
CO_2	0.14	44	6.16	$\frac{6.16}{30.44} = 0.202$	20.2%
CO	0.01	28	0.28	$\frac{0.28}{30.44} = 0.009$	0.9%
O_2	0.05	32	1.60	$\frac{1.60}{30.44} = 0.053$	5.3%
N_2	0.80	28	22.40	$\frac{22.40}{30.44} = 0.736$	73.6%
Σ	1.00		30.44	1.000	100.0

Result: $CO_2 = 20.2\%$; $CO = 0.9\%$; $O_2 = 5.3\%$; $N_2 = 73.6\%$

Example 8.3 Convert the following gravimetric analysis into volumetric anaylsis:

$$CO_2 = 13.3\%, CO = 0.95\%, O_2 = 8.35\%, N_2 = 77.4\%$$

Solution The solution is given in Table 8.2.

Table 8.2 Volumetric analysis

Constituent gas	*Mass per cent*	*Molecular mass*	*Proportional volume*	*Volume per m^3 of gas mixture*	*Volume percent*
0	*(1)*	*(2)*	$(3) = \frac{(1)}{(2)}$	$(4) = \frac{(3)}{\Sigma(3)}$	*(5) = (4) × 100*
CO_2	13.3	44	$\frac{13.3}{44} = 0.302$	$\frac{0.302}{3.357} = 0.090$	9.00
CO	0.95	28	$\frac{0.95}{44} = 0.034$	$\frac{0.034}{3.357} = 0.010$	1.00
O_2	8.35	32	$\frac{8.35}{32} = 0.261$	$\frac{0.261}{3.357} = 0.078$	7.80
N_2	77.4	28	$\frac{774}{28} = 2.76$	$\frac{2.76}{3.357} = 0.822$	82.20
Σ	100.00		3.357	1.000	100.00

Volumetric composition of gas mixture: $CO_2 = 9\%$; $CO = 1\%$; $O_2 = 7.8\%$; $N_2 = 82.2\%$

8.4 HEAT OF COMBUSTION

A chemical reaction can be exothermic when heat is evolved or endothermic when heat is absorbed. Combustion is an exothermic reaction because energy of products of reaction is more than the energy of reactants when evaluated even at the same temperature.

8.4.1 Enthalpy of Reaction in an Open System

The first law equation for steady-state steady-flow system (SSSF) may be written for a chemically reactive system where combustion process is taking place. Ignoring the changes in kinetic and potential energies:

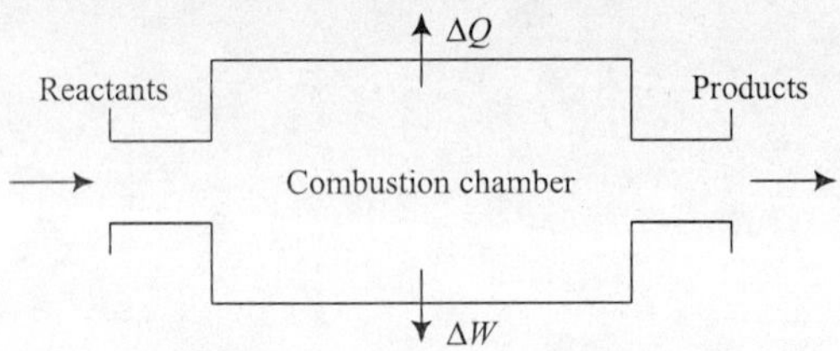

Fig. 8.1 Combustion process

$$H_R - \Delta Q = H_p + \Delta W$$

where,
H_R = Enthalpy of reactants (kJ)
H_p = Enthalpy of products (kJ)
ΔQ = Heat transferred to the surrounding (kJ)
ΔW = Work done by the system
= 0 for combustion process.

The enthalpy of reactants and products are evaluated at ambient condition T_∞ and p_∞ and the products are cooled and heat ΔQ is transferred to the surrounding.

$$(\Delta W + \Delta Q)_{T_\infty} = (H_p - H_R)_{T_\infty} = (\Delta H)_{T_\infty}$$

$$\therefore \quad (\Delta H)_{T_\infty} = (\Delta Q)_{T_\infty}$$

$(\Delta H)_{T_\infty}$ is called heat of combustion or enthalpy of reaction at T_∞ and p_∞. However, enthalpy is not much affected by the pressure, $(\Delta H)_{T_\infty}$ may be taken as enthalpy of reaction at T_∞.

The reactants and products are usually composed of several constituents each, the SSSF energy equation can be written as:

$$\sum_R (mh) - \Delta Q = \sum_P (mh) + \Delta W$$

where m is the mass and h the specific enthalpy of each constituent.

8.4.2 Internal Energy of Reaction in a Closed System

Combustion equations for fuels burning in a closed system, such as a cylinder of I.C. engine or bomb calorimeter, can be obtained from first law equation for the closed system.

$$U_R - \Delta Q = U_p + \Delta W$$

where, U_R = Internal energy of reactants

U_p = Internal energy of products

ΔW = Non-flow work

For gases,

$$H = U + pV$$

$$= U + n\overline{R}\,T$$

where, $\overline{R}$ = Universal gas constant

The combustion equation for closed system is:

$$\sum_R (nMh - n\overline{R}\,T) - \Delta Q = \sum_P (nMh - n\overline{R}\,T) + \Delta W$$

8.4.3 Enthalpy of Formation

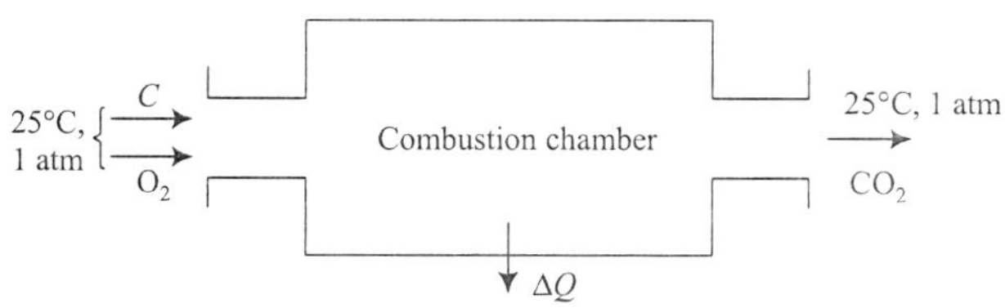

Fig. 8.2 Enthalpy of formation

A combustion chamber is shown in Fig. 8.2 for the combustion of carbon and oxygen to form CO_2. The reactants C and O_2 enter the chamber at 25°C and 1 atm, heat ΔQ is removed and product CO_2 is cooled and leaves the system at 25°C and 1 atm. ΔQ is measured as 393,552 kJ/kmol of CO_2 formed

$$C + O_2 = CO_2$$

$$H_R + \Delta Q = H_p$$

$$\sum_R n_i \overline{h}_i + \Delta Q = \sum_P n_e \overline{h}_e$$

The enthalpy of all elements at 25°C and 1 atm is assigned the value of zero.

$$\therefore \qquad H_R = 0$$

$$\Delta Q = H_p = -393{,}522 \text{ kJ/kmol}$$

$$= \bar{h}_f^o$$

$\bar{h}_f^o$ is called enthalpy of formation of CO_2 at 25°C and 1 atm.

$$(\bar{h}_f^o)_{CO_2} = -393{,}522 \text{ kJ/kmol}$$

The enthalpy of formation for different elements and chemical compounds is given in Table 8.3.

The change of enthalpy between standard state of 25°C, 1 atm and actual state can be found out as:

$$\bar{h}_{T,p} = \bar{h}_f^o + \Delta\bar{h}$$

where, $\Delta\bar{h}$ = Difference in enthalpy between a given state and enthalpy at 298.15 K, 1 atm

The value of $\Delta\bar{h}$ for various substances at different temperatures is given in Table 8.4.

Table 8.3 Enthalpy of formation h_f at 25°C and 1 atm

Substance	*Formula*	*M*	*State*	h_f kJ/kg
Carbon	C	12.011	solid	0
Oxygen	O_2	32.000	gas	0
Hydrogen	H_2	2.016	gas	0
Nitrogen	N_2	28.016	gas	0
Sulphur	S	32.060	solid	0
Carbon monoxide	CO	28.011	gas	–3948.3
Carbon dioxide	CO_2	44.011	gas	–8946.8
Water	H_2O	18.016	liquid	–15,875.5
			vapour	–13,430.8
Methane	CH_4	16.043	gas	–4669.8
Ethane	C_2H_6	30.070	gas	–2817.3
Propane	C_3H_8	44.097	gas	–2356.3
Butane	C_4H_{10}	58.124	gas	–2171.6
Octane	C_8H_{18}	114.230	liquid	–2189.5
			vapour	–1826.0
Nitric oxide	NO	30.008	gas	–3020.8
Nitrogen dioxide	NO_2	46.008	gas	–733.3
Sulphur dioxide	SO_2	64.060	gas	–4632.8

Table 8.4 Enthalpies of formation h_f at different temperatures and 1 atm pressure, kJ/kg

Temperature, K	CO_2	CO	H_2O	O_2	H_2	N_2
298	−8946.808	−3948.346	−13430.796	0	0	0
400	−8855.636	−3842.055	−13239.146	94.662	1468.535	122.804
500	−8757.718	−3736.462	−13046.567	190254	2850.317	317.477
1000	−8187.422	−3123.609	−11988.077	710.079	10266.723	766.364
1500	−7543.630	−2560.517	−10759.800	1269.909	17999.908	1374.106
2000	−6867.740	−1921.376	−9393.835	2161.403	26272.689	2005.107
2500	−6174.873	−1268.281	−7934.603	2450.738	34987.600	2654.160

8.4.4 Heating Value

The heating value of a fuel is same as the heat of combustion.

$$HV = \left[\sum_P (nMh_f) - \sum_R nMh_f \right]_{T_1}$$

where, T_1 = standard temperature.

This is heating value at constant pressure for steady flow open system.

The heating value at constant value for a closed system

$$HV_{nf} = \sum_P (nMR_f - n\bar{R}T) - \sum_R (nMh_f - n\bar{R}T)$$

$$= HV - (n_p - n_R)\,\bar{R}T_1$$

where, HV_{nf} = heating value for non-flow closed system.

HV = heating value for steady flow open system.

n_p = total number of moles of products.

n_R = total number of moles of reactants.

Two heating values are used for a fuel containing hydrogen atoms. If water contained in the product of combustion is condensed, the heating value is called higher heating value (HHV). If water exists in vapour state, the heating value is called lower heating value (LHV).

$$HHV = LHV + mh_{fg}$$

where, m = mass of water formed

h_{fg} = specific heat of vaporization at given temperature.

For non-flow closed system,

$$(HHV)_{nf} = (LHV)_{nf} + mu_{fg}$$

where u_{fg} = change of specific internal energy during vaporization at given temperature.

8.4.5 Combustion Temperatures

The heat of combustion increases the enthalpy of product gases (internal energy for a closed system) and hence its temperature. High temperature is used to transfer the heat to heat transfer surfaces (water tubes, superheaters, reheaters in a steam generator) by radiation and/or convection. The combustion temperature can be calculated by applying first law of thermodynamics to the combustion process.

$$\sum_P (nMh_f)_{T_2} = \sum_R (nMh_f) + \Delta Q$$

1. Calculate $\sum_R (nMh_f)$ from combustion equation for a given volumetric composition of fuel.
2. Calculate $\sum_P (nMh_f)_{T_2}$ for a given value of ΔQ.
3. Find out T_2 by trial and error.
   T_2 is called combustion temperature.

8.4.6 Adiabatic-Combustion Temperature

A fuel burning with no heat exchange with the surrounding and no work done will result in the adiabatic-combustion or adiabatic-flame temperature. It is greater when fuel is burnt in oxygen than in air because of dilution effect of nitrogen.

Maximum flame temperature is obtained for stoichiometric mixture.

A lean mixture has dilution effect.

A rich mixture results in incomplete combustion.

At adiabatic-flame temperature, $\Delta Q = 0$

$$\therefore \qquad \sum_P (nMh) = \sum_R (nMh)$$

8.5 HEAT BALANCE SHEET AND COMBUSTION EFFICIENCY OF A BOILER FURNACE

The mass and heat balance of boiler furnaces is shown in Fig. 8.3. The composition of fuel, ash from pit, flue gas, dry bulb and wet bulb temperature are measured for various estimates.

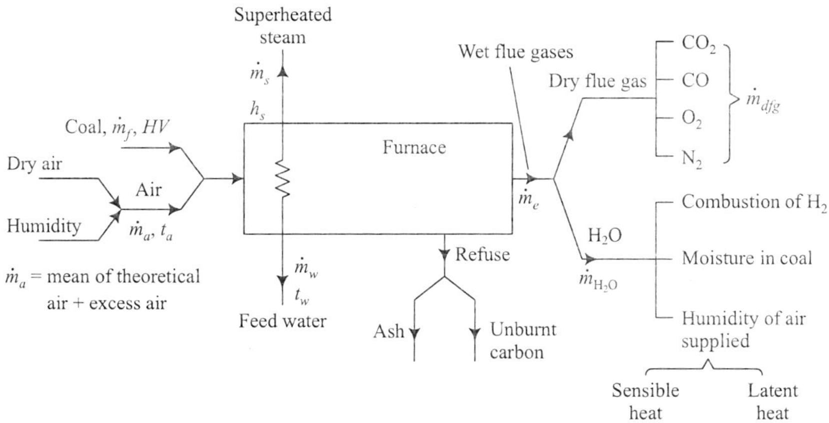

Fig. 8.3 Mass and energy balance of boiler furnace

1. **Coal Composition**

The various elements present in different types of coals used in power plants are shown in Fig. 8.4.

Pulverized coal is used in modern boilers to increase the surface area of 1 kg of coal to achieve 400-600 m^2 surface area.

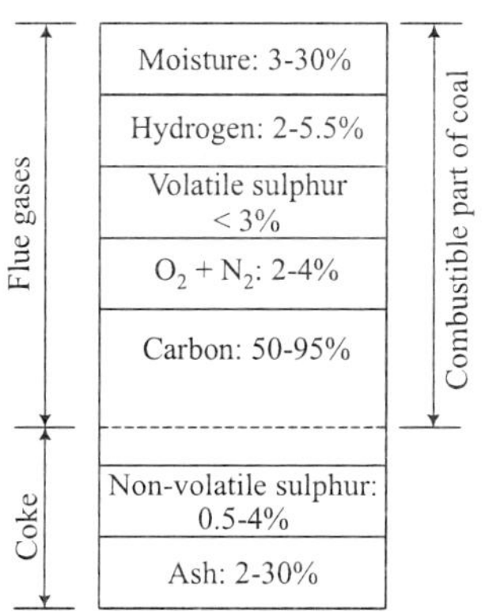

Fig. 8.4 Coal composition

2. Combustion Equations

$$\begin{array}{ccccccc} C & + & O_2 & = & CO_2 \\ (12)\text{ kg} & + & (2 \times 16)\text{ kg} & = & (44)\text{ kg} \\ 2H_2 & + & O_2 & = & 2H_2O \\ (2 \times 2)\text{ kg} & + & (2 \times 16)\text{ kg} & = & (36)\text{ kg} \\ S & + & O_2 & = & SO_2 \\ (32)\text{ kg} & + & (2 \times 16)\text{ kg} & = & (64)\text{ kg} \end{array}$$

3. Mass of air supplied

The mass of theoretical air can be calculated from the combustion equation and analysis of fuel. In order to ensure complete combustion of fuel, 15-20% excess air is supplied. The air is drawn from the atmosphere and preheated before supplying to the fuel burners to achieve better combustion.

Mass of theoretical air per kg of fuel,

$$\dot{m}_{a_{\text{stoic}}} = \frac{100}{23}\left[\frac{32}{12}C + \frac{32}{4}H + \frac{32}{32}S\right]$$

where, C = carbon content per kg of fuel

H = hydrogen content per kg of fuel.

S = sulphur content per kg of fuel.

The amount of air actually supplied can be calculated from Orsat Analysis of the flue gases. Masses of different constituents of flue gas will be proportional to their molecular masses.

$$\frac{\text{Mass of N}_2 \text{ in flue gas}}{\text{Mass of Carbon in flue gas}} = \frac{\text{Mass of N}_2 \text{ in air supplied per kg of fuel } (\dot{m}_{N_2})}{\text{Mass of carbon per kg of fuel}}$$

$$\text{Mass of air, } \dot{m}_a = \dot{m}_{N_2} \times \frac{100}{77}$$

$$\text{Excess air} = (\dot{m}_a) - (\dot{m}_{a_{\text{stoic}}})$$

4. Mass of flue gases

Mass of flue gases,

$$\dot{m}_e = (\dot{m}_f + \dot{m}_a)$$

Moisture content of flue gases,

$\dot{m}_{H_2O}$ = Moisture present in fuel + Humidity of supplied air + Moisture due to combustion of H_2 in fuel = $(H \times 9)$.

Mass of dry flue gas

$$\dot{m}_{d_{fg}} = \dot{m}_e - \dot{m}_{H_2O}$$

5. Heat Losses

(i) Heat carried away in dry flue gas

$$= \dot{m}_{d_{fg}} \times C_e (t_d - t_0)$$

where, C_e = specific heat of dry flue gas [kJ/kgK]

t_d = temperature of dry flue gas

t_0 = ambient temperature

(ii) Heat loss in water vapour discharged

$$= \dot{m}_{H_2O} \times (h_d - h_{f0})$$

where, h_d = enthalpy of superheated steam at partial pressure of moisture ($p_{H_2O} \approx 0.1$ bar) and temperature t_d from steam tables.

where, $$p_{H_2O} = \frac{n_{H_2O}}{n} \times p$$

where, n_{H_2O} = no. of moles of water vapour

n = total number of moles of flue gas

p = flue gas discharge pressure

h_{f0} = enthalpy of saturated water at room temperature from steam tables.

(iii) Heat loss due to incomplete combustion of carbon

$$= (\text{mass of unburnt carbon of fuel}) \times 33980 \text{ (kJ)}$$

where heat of combustion from C to CO_2 is 33980 kJ/kg of fuel.

(iv) Heat loss due to incomplete combustion of fuel

$$= 23640 \times \frac{\%(CO)}{\%(CO_2 + CO)} \times \text{mass of carbon burnt/kg of fuel}$$

where heat of combustion from CO to CO_2 is 23640 kJ/kg of fuel.

(v) Miscellaneous heat losses including radiation

$$= \text{HV} - \sum_{(i)}^{(iv)} \text{Losses} - \Delta Q$$

where ΔQ = Heat transferred to feed water

6. Heat Balance of Boiler

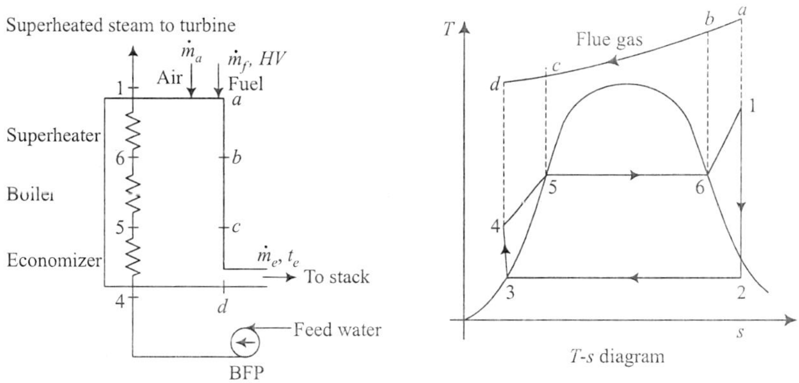

Fig. 8.5 Heat balance of a boiler

The heat balance of a boiler and T-s diagram are shown in Fig. 8.5.

$$\text{Energy input} = \text{Chemical energy of fuel } (\dot{m}_f \text{HV})$$
$$+ \text{Enthalpy of air supplied } (\dot{m}_a C_a t_a)$$
$$+ \text{Enthalpy of fuel } (\dot{m}_f C_f t_0)$$
$$\text{Energy output} = \text{Heat transfer to feedwater } [(\dot{Q}_{in} = \dot{m}_s (h_1 - h_4))]$$
$$+ \text{Enthalpy of products of combustion } (\dot{m}_e C_e t_e)$$
$$+ \text{Radiation losses } (\dot{Q}_r)$$
$$\text{Energy input} = \text{Energy output}$$
$$\dot{m}_f \text{HV} + (\dot{m}_f + m_a) C_o t_o = \dot{m}_s (h_1 - h_4) + \dot{m}_e C_e t_e + \dot{Q}_r$$

The combustion efficiency of boiler,

$$\eta_{\text{comb}} = \frac{\dot{m}_s (h_1 - h_4)}{\dot{m}_f HV}$$

Example 8.4 A boiler is fired with 1600 kg/hr of coal which has a heating value of 35,000 kJ/kg and composition by weight of carbon: 85.2%; Hydrogen: 4.8%, ash: 10% which as fired contains 1.8% moisture. Air is supplied to the furnace at a temperature of 25°C. Analysis of flue gases by volume: $CO_2 = 9.4\%$; $O_2 = 11.1\%$; $N_2 = 79.5\%$. The boiler evaporates 14000 kg/hr of steam at 12.8 bar and 249.5°C. The temperature of hot well from which the feed pump draws water is 55°C. The chimney exhaust temperature is 310°C. Draw a heat balance of boiler per kg of coal.

Solution

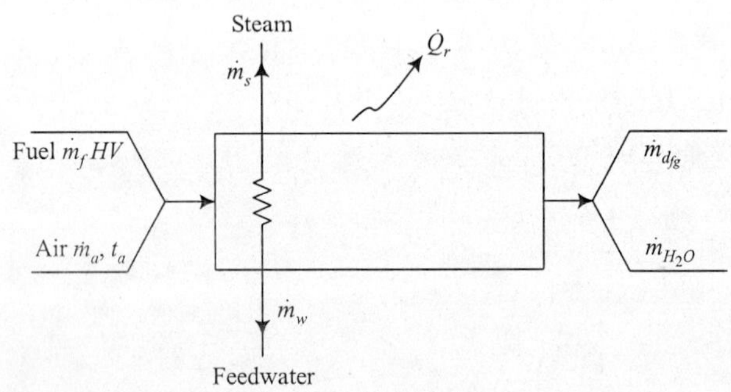

Fig. 8.6 Energy balance of boiler

Table 8.5 Heat balance sheet Basis: 1 kg of coal

Credit			*Debit*		
Details	*kJ/kg*	*% age*	*Details*	*kJ/kg*	*%age*
1. Heat supplied by coal $= \frac{\dot{m}_f}{\dot{m}_f} \cdot HV$	35,000	100	1. Heat transferred to water for steam generation $= \frac{\dot{m}_s}{\dot{m}_f}(h_s - h_w) = \frac{14000}{1600}(2935 - 230.2)$	23667	67.6
			2. Heat in dry flue gas	6349.8	18.14
			$= \frac{\dot{m}_{df_g}}{\dot{m}_f} C_e (t_e - t_a)$		
			$= \frac{22.28}{1} \times 1 \times (310 - 25)$	63649.8	18.14
			3. Heat in moisture of flue gas $= \frac{\dot{m}_{H_2}}{\dot{m}_f}(h_e - h_a)$		
			$= 6.45\ (3077 - 104.8)$	1337.5	3.8
			4. Radiation and unaccounted for (by difference)	3645.7	10.46
Σ	35000	100		35000	100

Supporting Calculations

1. *Mass of flue gas per kg of coal*

$$\frac{\text{Mass of Nitrogen in flue gas } (\dot{m}_{N_2})}{\text{Mass of carbon in flue gas}} = \frac{\text{Mass of } N_2 \text{ in air supplied per kg of coal}}{\text{Mass of carbon per kg of coal}}$$

$$\frac{\dot{m}_{N_2}}{0.852} = \frac{79.5 \times 28}{9.4 \times 44 \times \frac{12}{44}}$$

$$\dot{m}_{N_2} = 16.81 \text{ kg/kg of coal.}$$

$$\text{Mass of air supplied per kg of coal} = \frac{100}{77} \times \dot{m}_{N_2}$$

$$= \frac{100}{77} \times 16.81 = 21.83 \text{ kg}$$

Combustible matter in coal = 0.9 kg/kg of coal

∴ Total mass of flue gas/kg of coal = 21.83 + 0.9 = 22.73 kg

Mass of moisture in flue gas

$\dot{m}_{H_2O}$ = Moisture in coal + Moisture due to combustion of H_2

$$= 0.18 + 0.048 \times 9 = 0.45 \text{ kg}$$

Mass of dry flue gas, $\dot{m}_{d_{fg}} = 22.73 - 0.45 = 22.28$ kg

2. *Enthalpy of steam in flue gas*

The partial pressure of H_2O in flue gas

$$p_{H_2O} = \frac{\dot{m}_{H_2O}}{\dot{m}_{d_{fg}}} = \frac{0.45}{22.73} = 0.02 \text{ bar}$$

Enthalpy of steam in flue gas

$(h_e)_{H_2O}$ = Enthalpy of superheated steam at 0.02 bar and 310°C from steam table

= 3077 kJ/kg

3. *Enthalpy of moisture in atmospheric air*

h_a = Enthalpy of saturated water at atmospheric temperature from steam tables.

$= h_{f_{25°C}} = 104.8$ kJ/kg

8.6 HEAT BALANCE OF I.C. ENGINES

The energy flow diagram for I.C. engine is shown in Fig. 8.7.

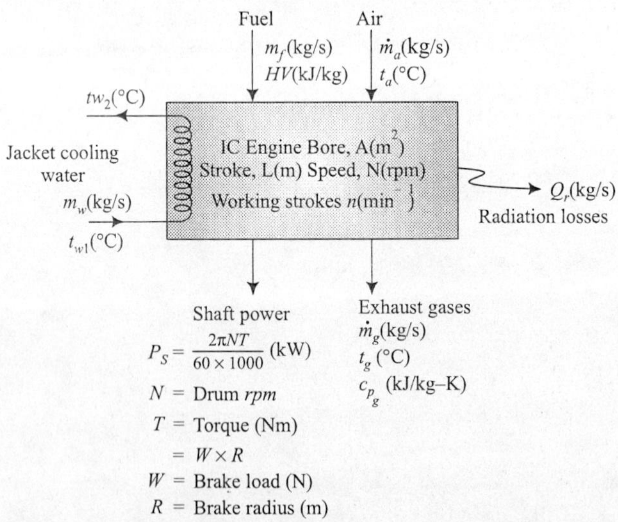

Fig. 8.7 Energy balance diagram of I.C. engine

The heat balance sheet is shown in Table 8.6. All the energy inflows are shown as credit and all energy outflows from the engines are shown as debit.

(a) Credit

Fuel prower, $Q_f = \dot{m}_f HV$ [kW]

where, $\dot{m}_f$ = mass flow rate of fuel [kg/s]

HV = heating value of fuel [kJ/kg]

(b) Debit

(i) Shaft power, $$P_s = \frac{2\pi NT}{60 \times 1000} \text{ [kW]}$$

Indicated power, $$P_i = \frac{p_m \times 10^5 \times L \times A \times n}{60 \times 1000} \text{ [kW]}$$

where, p_m = mean effective pressure (bar)

(ii) Friction power, $P_f = P_i - P_s$ [kW]

(iii) Energy lost in jacket water,

$$\dot{Q}_w = \dot{m}_w \, C_w \, (t_{w_2} - t_{w_1}) \text{ [kW]}$$

where, $\dot{m}_w$ = cooling water flow rate [kg/s]

C_w = specific heat of water

= 4.186 kJ/kgK.

t_{w_1} = Inlet temperature of cooling water (°C)

t_{w_2} = Outlet temperature of cooling water (°C)

(iv) Exhaust gas loss,

$$\dot{Q}_e = \dot{m}_e C_e \, (t_e - t_a) \text{ [kW]}$$

where, $\dot{m}_e$ = mass flow rate of exhaust gas [kg/s]

$= \dot{m}_f + \dot{m}_a$

C_e = specific heat of exhaust gases [kJ/kg-K]

≈ 1.005 kJ/kg-K

t_e = exhaust gas temperature (°C)

t_a = inlet air temperature (°C)

= t_0 = ambient air temperature

(v) Miscellaneous losses include radiation losses and other unacconted for losses.

$$\dot{Q}_r = \Sigma \text{Credit} - \Sigma \text{Debit}$$

Table 8.6 Heat balance sheet of I.C. engine

S.No.	Description	Credit		Debit	
		kW	% age	kW	% age
1.	Fuel power, $\dot{Q}_f$	√	√	–	–
2.	Shaft power, P_s	–	–	√	√
3.	Friction power, P_f,	–	–	√	√
4.	Energy lost in jacket water, $\dot{Q}_W$	–	–	√	√
5.	Energy lost in exhaust gas, $\dot{Q}_e$	–	–	√	√
6.	Unaccounted for losses, $\dot{Q}_r$			√	√
	Total	√	100	√	100

Example 8.5 A four-stroke gas engine with 27 cm bore and 45 cm stroke was subject to a test.

Effective brake diameter	= 1.62 m
Test duration	= 38 min. 30 sec
Total number of explosion	= 3230
Total number of revolutions	= 8080
Net load on brake	= 902.5 N
Mean effective pressure	= 5.75 bar
Gas consumption	= 7.7 m^3
Gas pressure	= 135 mm of H_2O above atmosphere
Gas temperature	= 27°C
Barometer reading	= 75 cm Hg
Calorific value of gas at NTP	= 18436 kJ/m^3
Jacket water	= 183 kg
Water temperature rise	= 47°C

Draw up a heat balance sheet.

Solution Gas pressure $= \left(750 + \dfrac{135}{13.6}\right) \times \dfrac{1}{735.56} = 1.03$ bar

Gas volume at NTP $= 7.7 \times \dfrac{1.03}{1.033} \times \dfrac{273}{300} = 6.98\ m^3$

Heat supplied in gas, $Q_f = \dfrac{6.98}{38.5} \times 18436 = 3342.42$ kJ/min

$$= \frac{3342.42}{60} = 55.7 \text{ kW}$$

Shaft power, $$P_s = \frac{2\pi NT}{60 \times 1000}$$

$$= \frac{2\pi \times 8080}{38.5 \times 60} \times \frac{902.5 \times 1.62}{1000 \times 2} = 16 \text{ kW}$$

Indicated power, $$P_i = \frac{p_m LAN}{60 \times 1000}$$

$$= 5.75 \times 10^5 \times 0.45 \times \frac{\pi}{4}(0.27)^2 \times$$

$$\frac{3230}{38.5 \times 60} \times \frac{1}{1000} = 20.7 \text{ kW}$$

Friction power, $P_f = P_i - P_s = 20.7 - 16 = 4.7$ kW

Energy lost to jacket water,

$$Q_{CW} = \frac{183}{38.5 \times 60} \times 47 \times 4.186 = 15.6 \text{ kW}$$

Table 8.7 Heat balance sheet

S.No.	Energy density	Debit (kW)	Credit (kW)	% age
1.	Fuel power, Q_f	–	55.7	100.00
2.	Shaft power, P_s	16.0	–	28.7
3.	Friction power, P_f	4.7	–	8.435
4.	Energy lost in jacket water, Q_{CW}	15.6	–	28.0
5.	Energy lost in exhaust gases and unaccounted for (by difference)	19.4	–	34.865
	Σ	65.7	55.7	

Brake thermal efficiency,

$$\eta_{th_b} = \frac{P_s}{Q_f} = \frac{16.0}{55.7} \times 100 = 28.7\%$$

Indicated thermal efficiency,

$$\eta_{th_i} = \frac{P_i}{Q_f} = \frac{20.7}{55.7} = 37.16\%$$

QUESTION BANK NO. 8

1. What is the importance of analysis of products of combustion?
2. What are different types of combustion equations? Illustrate with examples.

3. What is the significance of air-fuel ratio? How is it calculated?
4. Explain the procedure for mutual conversion of volumetric and gravimetric analysis of products of combustion?
5. Explain heat of combustion for an open system and a closed system.
6. What is enthalpy of formation?
7. Write notes on:
   (a) Heating value
   (b) Combustion temperature
8. Illustrate mass and energy balance for a boiler on a line diagram. How are different losses calculated?
9. Explain the procedure for drawing up a heat balance sheet for a boiler furnace.
10. Draw an energy balance diagram for an I.C. engine. How are various inputs and outflows of energy estimated?
11. How is heat balance sheet prepared for an I.C. engine?

PART III

THERMODYNAMIC CYCLES

CHAPTER 9

Air Standard Cycles

9.1 INTRODUCTION

A thermodynamic device which can produce mechanical work from heat continuously or transport heat from low temperature to higher temperature works on a cyclic process. There is a working substance which continuously flows through the engine or reversed engine to ensure continuous cyclic operation. Air and water are most common working substances used in thermodynamic cycles because of their availability in abundance and free for the taking. However, there are other gases and vapours used in thermodynamic cycles which have favourable thermodynamic and thermophysical properties for the operation of the devices.

The thermodynamic cycles may be closed cycle in which the same working substance flows continuously in a closed loop. There may be open cycles where the working substance may be thrown into environment and a fresh charge enters the cycle again. Open cycles are more possible when air is the working substance. Most of the internal combustion engines and gas turbines work on open cycle where cycle exhausts are emitted into atmosphere and cycle starts again with a fresh change of air. Most of steam power cycles and refrigeration cycles are closed type.

Air is heated in a regenerator in hot air cycles working as closed cycles. A fuel is injected into air directly and combusted to generate heat by chemical reaction in engines working on open thermodynamic cycles.

A thermodynamic cycle consists of a series of processes carried out sequentially. The working substance is heated and cooled isobarically or isochorically. The expansion and compression processes are carried out adiabatically or isothermally, the former being more common.

9.2 PROPERTIES OF AIR

Air is most important working medium for thermodynamic cycles. It is used in hot air engines, internal combustion engines, open-cycle gas turbines, air refrigeration and air-conditioning system. The main properties of air as working medium are given in gas tables. Some of them are listed below:

1. Air is treated as ideal gas satisfying the equation of state:

$$pv = RT$$

The value of R is 0.287 kJ/kg-K.

2. The composition of air is taken based on a mixture of nitrogen and oxygen.

Air Composition

Elements:	By volume	By mass
N_2:	79%	77%
O_2:	21%	23%

The molecular mass, $M = 28.967 \approx 29$.

3. Thermodynamic properties of air are given in tables prepared by Keenan and Keys for various values of temperature. The value of datum temperature is taken as OK. The values are tabulated for the following properties:

(i) Enthalpy, $h = \int C_p \, dT$

where, $C_p = \psi(T)$

(ii) Internal energy,

$$u = h - pv$$
$$= h - RT$$

(iii) The change of entropy for any process:

$$s_2 - s_1 = C_p \frac{dT}{T} - R\frac{dp}{p}$$

(iv) The change of entropy for the constant pressure process,

$$s_2 - s_1 = \int_{T_1}^{T_2} C_p \frac{dT}{T}$$

$$= \int_{T_0}^{T_1} C_p \frac{dT}{T} - \int_{T_0}^{T_2} C_p \frac{dT}{T} = \phi_2 - \phi_1$$

(v) Pressure ratio for an *isotropic* process.

For an ideal gas,

$$ds = C_p \frac{dT}{T} - R\frac{dp}{P}$$

$$ds = 0 \text{ for isotropic process}$$

$$\therefore \quad C_p \frac{dT}{T} = R\frac{dp}{p}$$

$$\ln \frac{p}{p_o} = \frac{1}{R}\int C_p \frac{dT}{T}$$

The relative pressure,

$$p_r = \frac{p}{p_o} = \exp\left\{\frac{1}{R}\int C_p \frac{dT}{T}\right\}$$

The pressure ratio,

$$\frac{p_2}{p_1} = \frac{(p_r)_2}{(p_r)_1} \quad \text{(isotropic process)}$$

(vi) Volume ratio

$$ds = 0 = C_V \frac{dT}{T} + R\frac{dv}{v}$$

$$\therefore \quad \frac{dv}{v} = -\frac{1}{R}\left(C_V \frac{dT}{T}\right)$$

$$\ln\frac{v}{v_0} = -\frac{1}{R}\int_{T_0}^{T} C_V \frac{dT}{T} = \ln v_r$$

where volume ratio, $v_r = \dfrac{v}{v_0}$.

4. Thermodynamic properties of air are not strongly pressure dependent and may be used as a function of temperature alone. These properties are given in air tables. Important properties at ambient conditions are:

$$C_p = 1.005 \text{ kJ/kg-K}$$
$$C_V = 0.718 \text{ kJ/kg-K}$$
$$\gamma = 1.4$$
$$\rho = 1.1774 \text{ kg/m}^3$$
$$\mu = 1.983 \times 10^5 \text{ kg/ms}$$
$$\nu = 15.68 \times 10^5 \text{ m}^2\text{/s}$$
$$K = 0.02624 \text{ W/mK}$$
$$\alpha = 0.2216 \times 10^5 \text{ m}^3\text{/s}$$
$$P_r = 0.708$$

5. Psychometric chart and tables are available for psychometric properties of air for design and analyses of air-conditioning plants and cooling towers, etc.

6. Thermodynamic properties and thermophysical properties of air and p-h chart are used for design and analysis of air refrigeration plants and aircraft cooling system.

7. The products of combustion of fuel and air in I.C. engines and gas turbines are assumed to have same thermodynamic and thermophysical properties as air in absence of accurate data.
8. Dry air is assumed to be inert to all metals.

9.3 AIR STANDARD CYCLE

The operating cycle of an internal combustion engine consists of the following separate processes:

1. Intake or suction process
2. Compression process
3. Combustion process
4. Expansion process
5. Exhaust process

It is necessary to devise an engine cycle based on these processes which can be used to assess the performance of the engine. The mass of fuel in the air-fuel mixture is very small as compared to mass of air, the mixture can be assumed to obey the perfect gas laws. The engines can be assumed to be working on thermodynamic cycles. For simplification air can be taken as working substance inside the engine cylinder. The ideal engine cycle employing air as working fluid is called *air standard cycle*.

9.3.1 Assumptions

The following assumptions are usually made in the analysis of air standard cycles.

1. Air considered an ideal gas is the working medium. The composition of working medium remains constant throughout the cycle.
2. The compression and expansion processes are reversible and adiabatic. These processes are internally reversible and there is no heat loss or friction.
3. The combustion process is replaced by a reversible heat transfer process from a heat reservoir.
4. The exhaust process is replaced by a constant volume heat rejection process to a heat reservoir.
5. The specific heats are assumed constant at the ambient temperature,

$$C_p = 1.005 \text{ kJ/kg-K}$$

$$C_v = 0.718 \text{ kJ/kg-K}$$

$$\gamma = 1.4$$

6. The charging and discharging operations are omitted. Therefore, a constant mass of the medium is carried through the entire cycle.

In practice, there are following deviations from the ideal conditions assumed:

1. There are internal and external irreversibilities like fluid friction, mechanical friction, combustion, etc.
2. There are heat losses to the cooling medium and by radiation.
3. The properties of the working medium change throughout the cycle.
4. The specific heat of gases changes with temperature.

However, the air standard cycles are useful and are important tool for comparison of cycle performance.

9.3.2 Air Standard Efficiency

The efficiency of the theoretical cycle with working medium as air is called air standard efficiency or ideal efficiency. It is the ratio of ideal work done and heat supplied.

$$\text{Air standard efficiency} = \frac{\text{Ideal work done}}{\text{Heat supplied}}$$

$$\text{Air standard efficiency} = \frac{\text{Heat supplied} - \text{Heat rejected}}{\text{Heat supplied}}$$

The *thermal efficiency* is the ratio of actual work done and heat supplied.

$$\text{Thermal efficiency} = \frac{\text{Actual work done}}{\text{Heat supplied}}$$

Relative efficiency is the ratio of thermal efficiency and air standard efficiency.

$$\text{Relative efficiency} = \frac{\text{Thermal efficiency}}{\text{Air standard efficiency}}$$

$$= \frac{\text{Actual work done}}{\text{Ideal work done}}$$

9.3.3 Power Cycles

Following power cycles are used depending upon type of heat addition process:

1. Otto cycle or constant volume heat addition process.
2. Diesel cycle or constant pressure heat addition process
3. Dual cycle in which heat is added partly at constant volume and partly at constant pressure.

9.4 OTTO CYCLE

The cycle was first proposed by a French engineer Beau de Rochas but is named after a German engineer Nicholas Otto who successfully built the engine working on this cycle. All petrol engines and other spark ignition engines work on Otto cycle. The air standard Otto cycle is composed of four internally reversible processes. The p-V and T-s diagrams are shown:

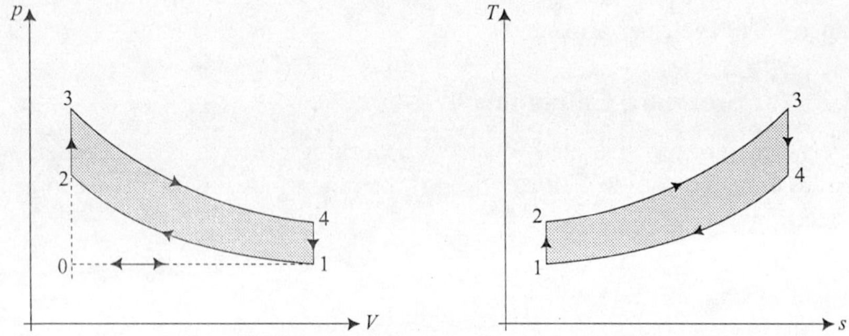

Fig. 9.1 Otto cycle

The Otto cycle consists of:

1. Process 1-2 : Isentropic compression
2. Process 2-3 : Constant volume heat addition
3. Process 3-4 : Isentropic expansion process
4. Process 4-1 : Constant volume heat rejection process

In the theoretical Otto cycle, the following processes are eliminated:

5. Process 0-1 : Suction process
6. Process 1-0 : Exhaust process.

Let engine cylinder contain m kg of air at point 1 where air parameters are p_1, V_1, and T_1.

1. *Process 1-2: Reversible adiabatic or isentropic compression*

   No heat is added or rejected during the process,

   $$Q_{1-2} = 0$$

   The law of process is $pV^\gamma = C$.

2. *Process 2-3: Constant volume heating process*

   Heat is added to air from a heat reservoir at constant volume and temperature of air changes from T_2 to T_3,

   Heat added, $Q_{2-3} = mCv(T_3 - T_2)$

3. *Process 3-4: Reversible adiabatic or isentropic expansion*

   No heat is added or rejected during the process,

   $$Q_{3-4} = 0$$

   The law of process is $pV^{\gamma} = C$.

4. *Process 4-1: Constant volume heat rejection*

   Heat is rejected to a heat reservoir at constant volume and temperature of air changes from T_4 to T_1,

   Heat rejected, $\quad Q_{4-1} = m\,Cv\,(T_4 - T_1)$

9.4.1 Cycle Analysis

Work done per cycle = Heat absorbed – Heat rejected

$$W = m\,Cv\,(T_3 - T_2) - m\,Cv\,(T_4 - T_1)$$

Air standard efficiency,

$$\eta = \frac{\text{Work done}}{\text{Heat absorbed}} = \frac{W}{m\,Cv(T_3 - T_2)}$$

$$= \frac{m\,Cv(T_3 - T_2) - m\,Cv\,(T_3 - T_1)}{m\,Cv(T_3 - T_2)}$$

$$= 1 - \frac{[T_4 - T_1]}{[T_3 - T_2]}$$

$$= 1 - \frac{T_1\left[\dfrac{T_4}{T_1} - T_1\right]}{T_2\left[\dfrac{T_3}{T_2} - 1\right]}$$

Compression ratio, $\quad r_c = \dfrac{V_1}{V_2}$

Expansion ratio, $\quad r_e = \dfrac{V_4}{V_3}$

For Otto cycle, $\quad r_c = r_e \qquad [\because \; V_1 = V_4 \;$ and $\; V_2 = V_3]$

For reversible adiabatic compression process 1-2,

$$\frac{T_2}{T_1} = \left(\frac{V_1}{V_2}\right)^{\gamma - 1} = (r_c)^{\gamma - 1} \qquad \text{...(1)}$$

For reversible adiabatic expansion process 3-4,

$$\frac{T_3}{T_4} = \left(\frac{V_4}{V_3}\right)^{\gamma-1} = (r_e)^{\gamma-1} = (r_c)^{\gamma-1} \qquad ...(2)$$

From equations (1) and (2),

$$\frac{T_2}{T_1} = \frac{T_3}{T_4} = (r_c)^{\gamma-1}$$

$$\therefore \qquad \frac{T_1}{T_4} = \frac{T_2}{T_3}$$

or

$$\frac{T_4}{T_1} = \frac{T_3}{T_2}$$

$$\therefore \qquad \eta = 1 - \frac{T_1}{T_2} = 1 - \frac{1}{(r_c)^{\gamma-1}}$$

9.4.2 Observations

Efficiency of Otto cycle depends on compression ratio, r_c only. As r_c increases, efficiency increases. But r_c cannot be increased beyond a value of 10 due to the following reasons:

(i) The cost of engine increases with increase of r_c as the engine weight and size increase. Therefore, a compromise has to be made between increase of efficiency and cost of engine.

(ii) There is a limit on r_c due to auto-ignition of gasoline. As r_c increases, there is tendency of pre-ignition and detonation of fuel. In practice, a compression ratio of 6 to 10 is used for petrol engines.

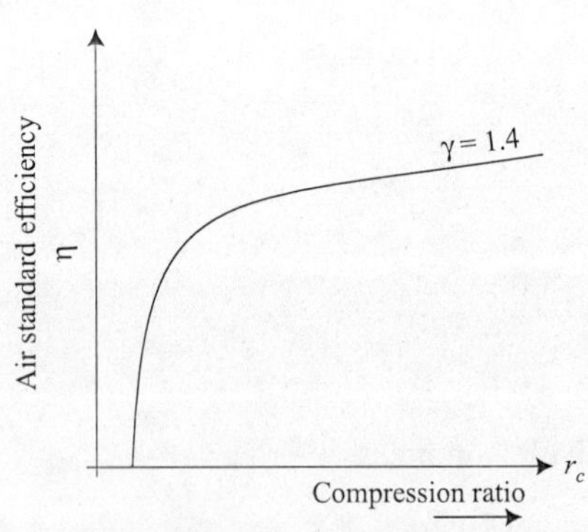

Fig. 9.2 Efficiency v/s compression ratio

9.4.3 Mean Effective Pressure

Mean effective pressure is the ratio of work done and swept volume,

$$p_m = \frac{\text{Work done}}{\text{Swept volume}} = \frac{W}{V_s} = \frac{W}{V_1 - V_2}$$

Example 9.1 A petrol engine works on an air standard Otto cycle. The pressure and temperature of air at the beginning of the cycle are 1 bar and 15°C. The compression ratio is 8 and the heat supplied per kg of air is limited to 2000 kJ. Find:

(i) The air standard efficiency of the cycle.

(ii) The mean effective pressure.

Solution

(i) The efficiency of air standard Otto cycle.

$$\eta = 1 - \frac{1}{(r_c)^{\gamma-1}} = 1 - \frac{1}{(8)^{1.4-1}}$$

$$= 0.5647 = 56.47\%$$

(ii) Heat supplied, $Q_{2-3} = 2000$ kJ

Work done, $W = Q_{2-3} \times \eta = 2000 \times 0.5647$

$$= 1229.4 \text{ kJ}$$

Now, $p_1 v_1 = m\,RT_1$

$$\therefore \quad V_1 = \frac{m\,RT_1}{p_1} = \frac{1 \times 287 \times 288}{1 \times 10^5} = 0.826 \text{ m}^3$$

[$\because R = 287$ J/kg-K for air]

$$V_2 = \frac{V_1}{r_c} = \frac{0.826}{8} = 0.10325 \text{ m}^3$$

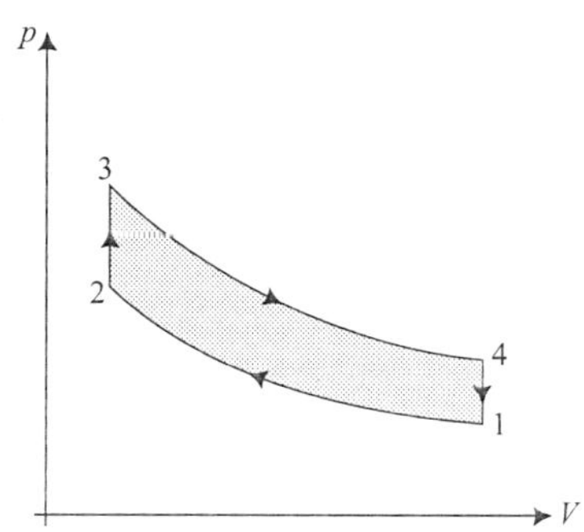

Mean effective pressure

$$p_m = \frac{W}{V_s} = \frac{1229.4}{V_1 - V_2} = \frac{1229.4}{0.826 - 0.10325}$$

$$= 1562.64 \text{ kN/m}^2$$

$$= \mathbf{15.626 \text{ bar}} \quad \mathbf{Ans.}$$

Example 9.2 2 kg of air at a pressure of 1 bar and 300 K is taken through an Otto cycle of compression ratio 8 involving 5000 kJ of heat addition. Determine the thermal efficiency, work done and mean effective pressure, if $\gamma = 1.4$ and $R = 287$ J/kg-K.

Solution

$$m = 2 \text{ kg}$$

$$p_1 = 1 \text{ bar}$$

$$T_1 = 300 \text{ K}$$

$$r_c = 8$$

$$Q_{2-3} = 5000 \text{ kJ}$$

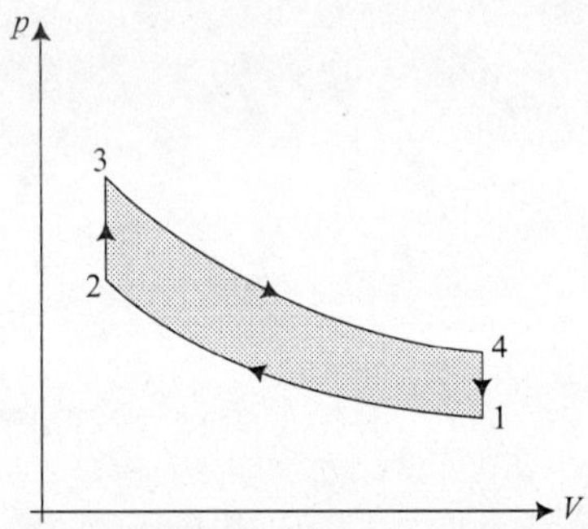

(i) Thermal efficiency,

$$\eta = 1 - \frac{1}{(r_c)^{\gamma-1}} = 1 - \frac{1}{(8)^{0.4}} = 0.5647 = 56.47\%$$

$$\text{Work done} = Q_{2-3} \times \eta = 0.5647 \times 5000$$

$$= 2823.5 \text{ kJ.}$$

$$p_1 V_1 = m R T_1$$

$$V_1 = \frac{m R T_1}{p_1} = \frac{2 \times 287 \times 300}{1 \times 10^5} = 1.1722 \text{ m}^3$$

$$V_2 = \frac{V_1}{r_c} = \frac{1.722}{8} = 0.215 \text{ m}^3.$$

Mean effective pressure,

$$p_m = \frac{W}{V_1 - V_2} = \frac{2823.5}{1.722 - 0.215} = 1873.6 \text{ kN/m}^3$$

$$= \mathbf{18.74 \text{ bar}} \quad \textbf{Ans.}$$

9.5 DIESEL CYCLE

Diesel cycle is named after a German engineer Rudolf Diesel who invented compression ignition engine. The ideal Diesel cycle consists of two reversible adiabatic or isentropic, a constant pressure and a constant volume processes. Slow speed diesel engines work on Diesel cycle, however, modern high speed diesel engines work on dual cycle. The air standard Diesel cycle is composed of four internally reversible processes. The *p-V* and *T-s* diagrams are shown:

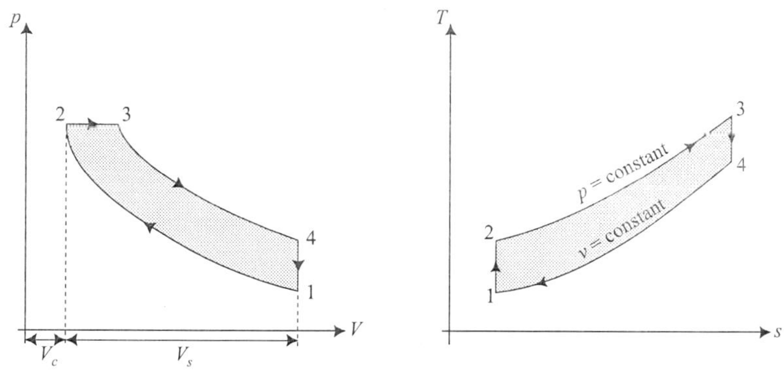

Fig. 9.3 Diesel cycle

The Diesel cycle consists of the following processes.

(i) Process 1-2 : Isentropic compression

(ii) Process 2-3 : Constant pressure heat addition

(iii) Process 3-4 : Isentropic expansion

(iv) Process 4-1 : Constant volume heat rejection

Let engine cylinder contain m kg of air at state 1 where air parameters are p_1, V_1, T_1.

1. *Process 1-2: Reversible adiabatic or isentropic compression*

No heat is added or rejected through the process.

$$Q_{1-2} = 0$$

The law of process is $pV^{\gamma} = C$.

2. *Process 2-3: Constant pressure heat absorption*

   Heat is absorbed by air from a heat reservoir at constant pressure and temperature of air changes T_2 to T_3,

   Heat absorbed, $\quad Q_{2-3} = m\,Cp\,(T_3 - T_2)$

3. *Process 3-4: Reversible adiabatic or isentropic expansion*

   No heat is added or rejected during the process,

$$Q_{3-4} = 0$$

   The law of process is $pV^{\gamma} = C$.

4. *Process 4-1: Constant volume heat rejection*

   Heat is rejected by air to a heat sink at constant volume and the temperature of air changes for T_4 to T_1.

   Heat rejected, $\quad Q_{4-1} = m\,Cv\,(T_4 - T_1)$

9.5.1 Cycle Analysis

Work done by the cycle = Heat absorbed – Heat rejected

$$W = m\,Cp\,(T_3 - T_2) - m\,Cv\,(T_4 - T_1)$$

∴ Air standard efficiency,

$$\eta = \frac{\text{Work done}}{\text{Heat absorbed}} = \frac{m\,Cp(T_3 - T_2) - m\,Cv(T_4 - T_1)}{m\,Cp(T_3 - T_2)}$$

$$= 1 - \frac{Cv}{Cp}\left[\frac{T_4 - T_1}{T_3 - T_2}\right] = 1 - \frac{1}{\gamma}\left[\frac{T_4 - T_1}{T_3 - T_2}\right]$$

The following ratios may be defined,

Compression ratio, $\quad r_c = \dfrac{V_1}{V_2}$

Expansion ratio, $\quad r_e = \dfrac{V_4}{V_3}$

Cut-off ratio, $\quad \rho = \dfrac{V_3}{V_2}$

Now expansion ratio, $$r_e = \frac{V_4}{V_3} = \frac{V_4}{V_2} \times \frac{V_2}{V_3}$$

$$= r_c \times \frac{1}{\rho} = \frac{r_c}{\rho}$$

For process 2-3 (constant pressure)

$$\frac{V_2}{T_2} = \frac{V_3}{T_3} \qquad (\because \; p_2 = p_3)$$

$$\therefore \quad T_3 = T_2 . \frac{V_3}{V_2} = T_1 \times \rho \qquad \text{...(1)}$$

For process 3-4 (isentropic expansion)

$$\frac{T_4}{T_3} = \left(\frac{V_3}{V_4}\right)^{\gamma-1} = \left(\frac{1}{r_e}\right)^{\gamma-1} = \left(\frac{\rho}{r_c}\right)^{\gamma-1}$$

$$\therefore \quad T_4 = T_3 \left(\frac{\rho}{r_c}\right)^{\gamma-1} = T_1 \times \rho \left(\frac{\rho}{r_c}\right)^{\gamma-1} \qquad \text{...(2)}$$

For process 1-2 (isentropic compression)

$$\frac{T_2}{T_1} = \left(\frac{V_1}{V_2}\right)^{\gamma-1} = (r_c)^{\gamma-1}$$

$$\therefore \quad T_2 = T_1 \, (r_c)^{\gamma-1} \qquad \text{...(3)}$$

Substituting the value of T_2 from equation (3) into equations (1) and (2)

$$\therefore \quad T_3 = T_1 \, (r_c)^{\gamma-1} \times \rho$$

$$T_4 = T_1 \, (r_c)^{\gamma-1} \times \rho \left(\frac{\rho}{r_c}\right)^{\gamma-1} = T_1 \, \rho^{\gamma}$$

Substituting the values of T_2, T_3 and T_4 into expression for efficiency,

$$\eta = 1 - \frac{1}{r}\left[\frac{T_1 \rho^{\gamma} - T_1}{T_1 (r_c)^{\gamma-1} \rho - T_1 (r_c)^{\gamma-1}}\right]$$

$$\therefore \quad \eta = 1 - \frac{1}{(r_c)^{\gamma-1}}\left[\frac{\rho^{\gamma} - 1}{\gamma(\rho - 1)}\right]$$

9.5.2 Observations

1. The Diesel cycle efficiency depends upon compression ratio, r_c, γ and cut-off ratio ρ.
2. The air standard efficiency of Diesel cycle is lower than Otto cycle for same compression ratio. However, the compression ratio used in diesel engine is more than petrol engine. Overall, the thermal efficiency of diesel engine is more than petrol engine.
3. The Diesel cycle efficiency increases with decrease in cut-off ratio and maximum value can be equal to Otto cycle efficiency.
4. The cut-off ratio in diesel engines is proportional to the load on the engine, for this reason, the efficiency of the Diesel cycle decreases with the load on the engine.
5. The Diesel cycle efficiency differs from Otto cycle by the bracketed term $\left[\dfrac{\rho^\gamma - 1}{\gamma(\rho - 1)}\right]$ which is always greater than unity.

Example 9.3 Air enters at 1 bar and 230°C in an engine running on Diesel cycle whose compression ratio is 18. Maximum temperature of the cycle is limited to 1500°C. Calculate:

(i) Cut-off ratio
(ii) Heat supplied per kg of air
(iii) Cycle efficiency.

Solution

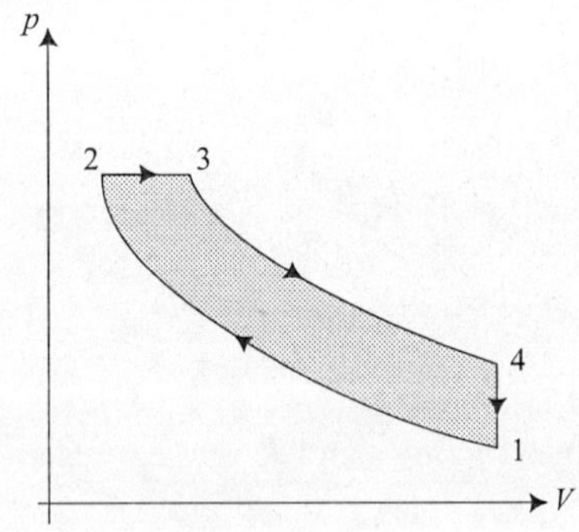

The p-v diagram of Diesel cycle is drawn.

$$p_1 = 1 \text{ bar}$$

$$T_1 = 230°\text{C} + 273 = 503 \text{ K}$$

$$T_3 = 1500°\text{C} + 273 = 1773 \text{ K}$$

$r_c = 18.$

(i) For compression process 1–2,

$$\frac{T_2}{T_1} = \left(\frac{V_1}{V_2}\right)^{\gamma-1}$$

$\therefore \qquad T_2 = T_1\,(r_c)^{\gamma-1} = 503 \times (18)^{0.4} = 1598 \text{ K}$

For constant pressure process 2-3,

Cut-off ratio, $\quad \rho = \dfrac{V_3}{V_2} = \dfrac{T_3}{T_2} = \dfrac{1773}{1598} \qquad (\because \; p_2 = p_3)$

$= \mathbf{1.1}$ **Ans.**

(ii) Heat supplied per kg of air,

$$Q_{2-3} = Cp\,(T_3 - T_2) = 1.005\,(1773 - 1598)$$

$= \mathbf{175.875\ kJ/kg}$ **Ans.**

(iii) Cycle efficiency,

$$\eta = 1 - \frac{1}{(r_c)^{\gamma-1}}\left[\frac{\rho^{\gamma}-1}{\gamma(\rho-1)}\right]$$

$$= 1 - \frac{1}{(18)^{0.4}}\left[\frac{(1.1)^{1.4}-1}{1.4(1.1-1)}\right] = 1 - \frac{1}{3.17767}\left[\frac{0.142746}{0.14}\right] = 0.679$$

$= \mathbf{67.9\%}$ **Ans.**

9.6 SUPERCHARGED OTTO CYCLE

The power produced by an engine is approximately proportional to the quantity of fuel burnt and the latter is controlled by the air supplied to the engine. The quantity of air in naturally aspirated engine is proportional to the volume swept by the piston. An external source of air supply has to be provided in order to increase the air quantity. This is called supercharging. Supercharging has been applied to both S.I. Engines as well as C.I. Engines.

An ideal supercharged Otto cycle is shown on *p-v* diagram. The power loop 1–2–3–4–1 has higher pressures on all the salient points. The pumping loop 1–5–6–7–1 is positive and gives additional power. But the higher pressure p_4 results in loss of power as high pressure gas is exhausted.

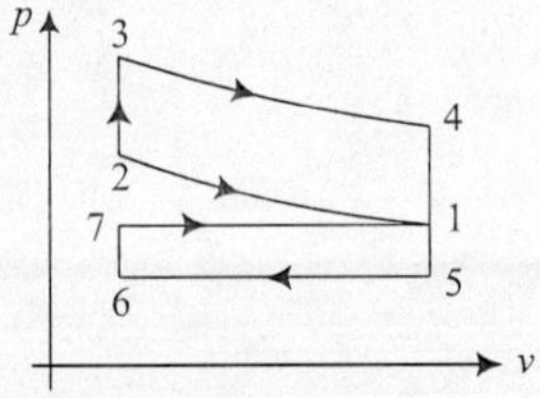

Fig. 9.4 Supercharged Otto cycle

The indicated power = Area of power loop 1–2–3–4–1
+ Area of pumping loop 1–5–6–7–1.

There is power consumption if the blower is directly connected to the engine.

$$\text{Shaft power} = (\text{Indicated power} \times \eta_m) - (\text{power consumed by supercharger})$$

Observations

1. The increase in supercharging pressure increases its tendency to detonate. Therefore, compression ratio has to be decreased resulting in loss of efficiency. Supercharging is employed in S.I. engines only for aircraft and racing cars.
2. Supercharging of petrol engines, because of its poor fuel economy, is not very popular and is used only when more power is needed to compensate altitude loss.

9.7 SUPERCHARGED DIESEL CYCLE

An ideal supercharged Diesel cycle is shown in Fig. 9.5.

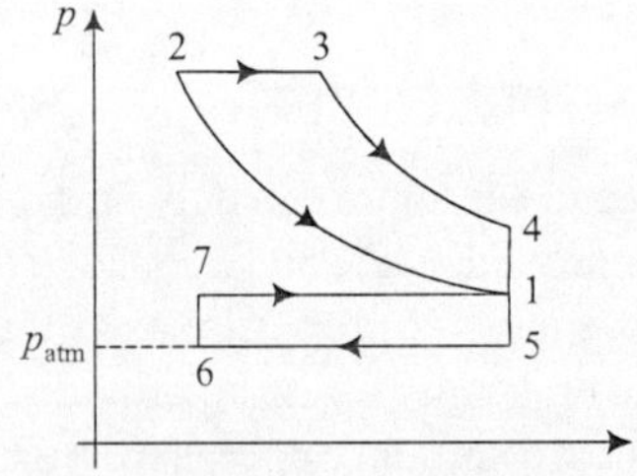

Fig. 9.5 Supercharged Diesel cycle

1. There is improvement of combustion process due to better mixing of air and fuel as a result of turbulence effect created by supercharger.
2. Increase in pressure and temperature of intake air reduces the delay significantly. The rate of higher pressure rise results in better, quieter and smooth combustion.

3. The operation of diesel engine is not sensitive to poor quality of fuel used.
4. Increase in air density results in higher volumetric efficiency and power output.
5. The reliability and durability of a supercharged engine increase over naturally aspirated engine due to smooth combustion and lower exhaust temperature.
6. The power output of a supercharged engine is higher than naturally aspirated engine due to combined effect of higher volumetric, combustion and scavenging efficiencies.
7. Although there is additional power required to drive the mechanical supercharger, however, the mechanical efficiency of a supercharged engine is better than naturally aspirated engine.
8. The supercharged diesel engines are widely used for the above advantages.

9.8 DUAL COMBUSTION CYCLE

High speed diesel engines work on dual combustion cycle where heat is absorbed partly at constant volume and partly at constant pressure. Therefore, this cycle is a combination of Otto cycle and Diesel cycle. The dual engine cycle is represented on *p-v* diagram and T-s diagram in Fig. 9.6.

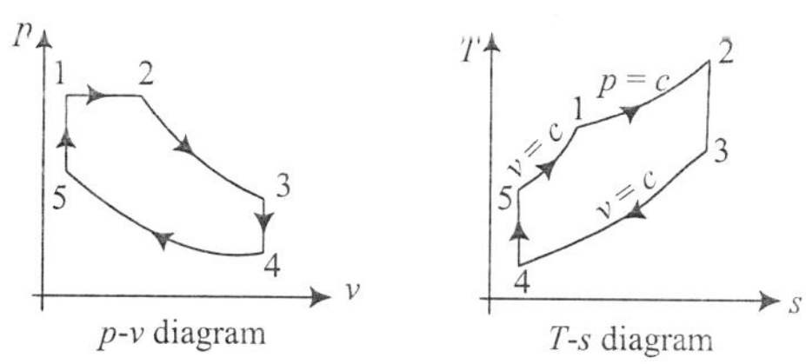

Fig. 9.6 Dual cycle

Let engine cylinder contain m kg of air at point 1 where air parameters are p_1, T_1 and v_1.

1. *Process 1-2: Constant Pressure Heating*

   Heat is absorbed by air getting heated from temperature T_1 to T_2.

   $$Q_{1-2} = m\, C_p(T_2 - T_1)$$

2. *Process 2-3: Reversible Adiabatic or Isotropic expansion*

   No heat is absorbed or rejected by air.

   $$Q_{2-3} = 0$$

3. *Process 3-4: Constant volume cooling*

   The air is cooled from temperature T_3 to T_4.

   $$Q_{3-4} = mCv(T_3 - T_4)$$

4. *Process 4-5: Reversible Adiabatic or Isotropic Compression*

   No heat is absorbed or rejected.

$$Q_{4-5} = 0$$

5. *Process 5-1: Constant Volume Heating*

   The air is heated from temperature T_1 to T_5.

$$Q_{5-1} = mC_V(T_1 - T_5)$$

9.8.1 Cycle Analysis

$$\text{Work done} = [\text{Heat absorbed}] - [\text{Heat rejected}]$$

$$W = [mC_p(T_2 - T_1) + mC_V(T_1 - T_5)] - [m\,C_V(T_3 - T_4)]$$

Air standard efficiency,

$$\eta = \frac{\text{Work done}}{\text{Heat absorbed}}$$

$$= \frac{[mC_p\,(T_2 - T_1) + mC_V\,(T_1 - T_2)] - [mC_V\,(T_3 - T_4)]}{[mC_p(T_2 - T_1) + mC_V\,(T_1 - T_5)]}$$

$$= 1 - \frac{[C_V\,(T_3 - T_4)]}{[C_P(T_2 - T_1) + C_V\,(T_1 - T_5)]}$$

$$= 1 - \frac{(T_3 - T_4)}{[\gamma(T_2 - T_1) + (T_1 - T_5)]} \qquad \text{...(i)}$$

Comparison ratio, $r_C = \dfrac{v_4}{v_3} = \dfrac{v_3}{v_1}$ $[\therefore v_3 = v_4 \text{ and } v_5 = v_1]$

Cut-off ratio, $\rho = \dfrac{v_2}{v_1} = \dfrac{v_2}{v_5}$ $[\therefore v_1 = v_5]$

Pressure ratio, $\alpha = \dfrac{p_1}{p_5}$

For process 1-2 (Constant pressure),

$$\frac{v_1}{T_1} = \frac{v_2}{T_2} \; (\because \; p_1 = p_2)$$

$$T_2 = T_1 \times \frac{v_2}{v_1} = T_1 \times \rho \qquad \text{...(ii)}$$

For process 2-3 (Isotropic expansion)

$$\frac{T_3}{T_2} = \left(\frac{v_2}{v_3}\right)^{\gamma-1}$$

$$\frac{T_3}{T_2} = \left(\frac{v_2}{v_1} \times \frac{v_1}{v_3}\right)^{\gamma-1} = \left(\frac{\rho}{r_C}\right)^{\gamma-1}$$

$$T_3 = T_2\left(\frac{\rho}{r_C}\right)^{\gamma-1}$$

$$T_3 = T_1\rho\left(\frac{\rho}{r_C}\right)^{\gamma-1} \quad \text{...(iii)}$$

For process 4-5 (Isotropic compression)

$$\frac{T_5}{T_4} = \left(\frac{v_4}{v_5}\right)^{r-1} = (r_C)^{\gamma-1}$$

$$T_5 = T_4(r_C)^{\gamma-1} \quad \text{...(iv)}$$

For process 5-1 (Constant volume)

$$\frac{p_5}{T_5} = \frac{p_1}{T_1} \quad (\because v_5 = v_1)$$

$$\therefore \quad T_1 = T_5 \times \frac{p_1}{p_5} = T_5 \times \alpha$$

$$T_1 = T_4(r_C)^{\gamma-1}\alpha \quad \text{...(v)}$$

Substituting the value of T_1 from equation (v) into equations (ii) and (iii)

$$T_2 = T_4(r_C)^{\gamma-1}\,\alpha\,\rho$$

$$T_3 = T_4(r_C)^{\gamma-1}\,\alpha\,\rho\left(\frac{\rho}{r_C}\right)^{\gamma-1} = T_4\,\alpha\,\rho^{\gamma}$$

Substituting the value of T_1, T_2, T_3 and T_5 in equation (i)

$$\eta = 1 - \frac{(T_4\,\alpha\,\rho^{\gamma}) - (T_4)}{\gamma[T_4\{[(r_C)^{\gamma-1}\,\alpha\,\rho] - T_4(r_C)^{\gamma-1}\alpha]\} + [T_4(r_C)^{\gamma-1}\alpha - T_4(r_C)^{\gamma-1}]}$$

$$= 1 - \frac{T_4(\alpha\rho^{\gamma} - 1)}{T_4(r_C)^{\gamma-1}[\gamma\,\alpha\,(\rho - 1) - (\alpha - 1)]}$$

$$\eta = 1 - \frac{1}{(r_C)^{\gamma-1}}\left[\frac{\alpha\rho^{\gamma} - 1}{(\alpha - 1) + \gamma\,\alpha\,(\rho - 1)}\right] \quad \text{...(vi)}$$

9.8.2 Observation

(i) For Otto cycle, $\rho = 1$

From equation (vi)

$$\eta_{otto} = 1 - \frac{1}{(r_C)^{\gamma-1}}$$

(ii) For Diesel cycle, $\alpha = 1$

From equation (vi)

$$\eta_{diesel} = 1 - \frac{1}{(r_C)^{\gamma-1}}\left[\frac{\rho^{\gamma}-1}{\gamma(\rho-1)}\right]$$

(iii) The efficiency of dual cycle is more than Diesel cycle and less than Otto cycle for the same compression ratio.

Example 9.4 The pressure at the intake of dual combustion cycle is 1 bar and maximum pressure is 40 bar. The compression ratio is 10 and cut off takes place at $\frac{1}{10}$ of the stroke. Determine the air standard efficiency of the cycle.

Solution

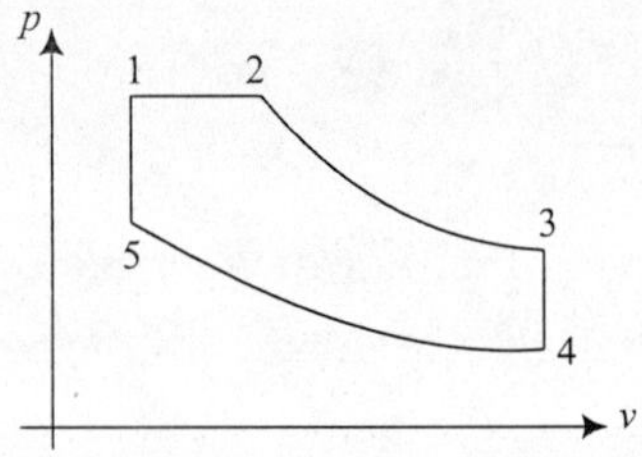

Fig. 9.7

$$p_4 = 1 \text{ bar}$$

$$p_1 = p_2 = 40 \text{ bar.}$$

$$r_C = 10$$

$$\rho = \frac{1}{10}$$

$$\gamma = 1.4$$

The volume at cut-off,

$$v_2 = v_1 + \frac{1}{10}(v_4 - v_1)$$

$$= v_5 + 0.1\,(v_4 - v_5)$$

∴ Cut-off ratio,

$$\rho = \frac{v_2}{v_5} = 1 + 0.1\left(\frac{v_4}{v_5} - 1\right)$$

$$= 1 + 0.1\,(10 - 1) = 1.9$$

For compression process 4-5,

$$p_4 v_4^{\gamma} = p_5 v_5^{\gamma}$$

$$p_5 = p_4\left(\frac{v_4}{v_5}\right)^{\gamma} = 1\,(10)^{1.4} = 25.12 \text{ bar}$$

∴ Pressure ratio,

$$\alpha = \frac{p_1}{p_5} = \frac{40}{25.12} = 1.59$$

Air standard cycle efficiency,

$$\eta = 1 - \frac{1}{(r_c)^{\gamma-1}}\left[\frac{(\alpha\rho^{\gamma} - 1)}{(\alpha - 1) + \gamma\alpha(\rho - 1)}\right]$$

$$= 1 - \frac{1}{(10)^{1.4-1}}\left[\frac{1.59(1.9)^{1.4} - 1}{(1.59 - 1) + 1.4 \times 1.59(1.9 - 1)}\right]$$

$$= 1 - 0.44 = 0.554 \text{ or } \mathbf{55.4\%} \quad \textbf{Ans.}$$

9.9 COMPARISON OF OTTO, DIESEL AND DUAL CYCLES

The three cycles can be compared for different design and operating parameters.

9.9.1 For same Compression Ratio and Heat Input

The *p-v* diagram and *T-s* diagram of the three cycles are superimposed.

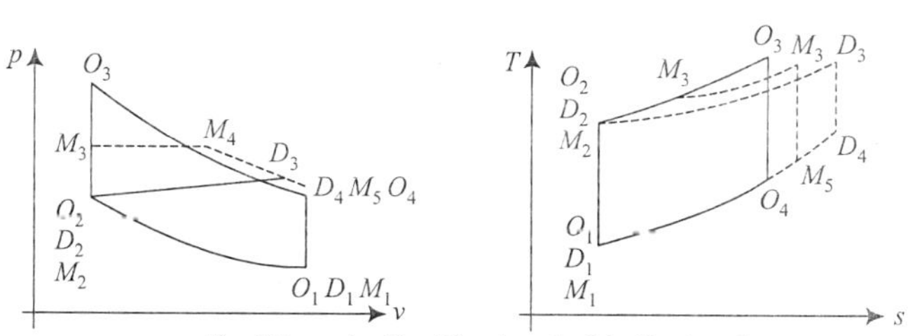

O = Otto cycle; *D* = Diesel cycle, *M* = Dual cycle

Fig. 9.8 Comparison of air standard cycle

On *p-v* diagram, points 1 and 2 are same for same compression ratio.

On *T-s* diagram, area under curves $O_2 - O_3$, $M_2 - M_3 - M_4$ and $D_2 - D_3$ are same for same heat input.

$$\eta_O = 1 - \frac{1}{(r_C)^{\gamma-1}}$$

$$\eta_D = 1 - \frac{1}{(r_C)^{\gamma-1}}\left[\frac{(\rho^\gamma - 1)}{\gamma(\rho - 1)}\right]$$

$$\eta_D = 1 - \frac{1}{(r_C)^{\gamma-1}}[K]$$

where, $$K = \left[\frac{(\rho^\gamma - 1)}{\gamma(\rho - 1)}\right]$$

If $\gamma = 1.4$

ρ	1.5	2.0	2.5	3.0
K	1.092	1.17	1.24	1.31

The value of *K* is always greater than unity.

$$\eta_O > \eta_D$$

or $$\eta_O > \eta_M > \eta_D$$

Also, $$\eta = 1 - \frac{Q_2}{Q_1}$$

where, Q_1 = Heat absorbed

Q_2 = Heat rejected

From *T-s*, diagram,

Q_2 = Area under the curve $O_4 - O_1$ for Otto cycle

= Area under the curve $M_5 - M_1$ for dual cycle

= Area under the curve $D_4 - D_1$ for diesel cycle

$$[Q_2]_D > [Q_2]_M > [Q_2]_O$$

∴ $$\eta_D < \eta_M < \eta_O$$

9.9.2 For same Maximum Pressure and Temperature

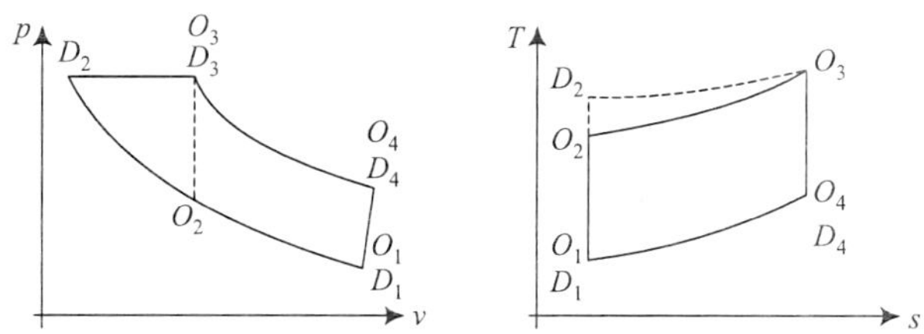

O = Otto cycle; D = Diesel cycle

Fig. 9.9 Comparison of air standard cycles

For same heat rejection,

$$[Q_2]_D = [Q_2]_O = \text{Area under the curve 4-1 on T-s diagram}$$

$$[Q_1]_D > [Q_1]_O = \text{Area under 2-3}$$

Now, $$\eta = 1 - \frac{Q_2}{Q_1}$$

$\therefore$ $$\eta_D > \eta_O$$

Hence, $$\eta_D > \eta_M > \eta_O$$

9.9.3 For same Maximum Pressure and Heat Input

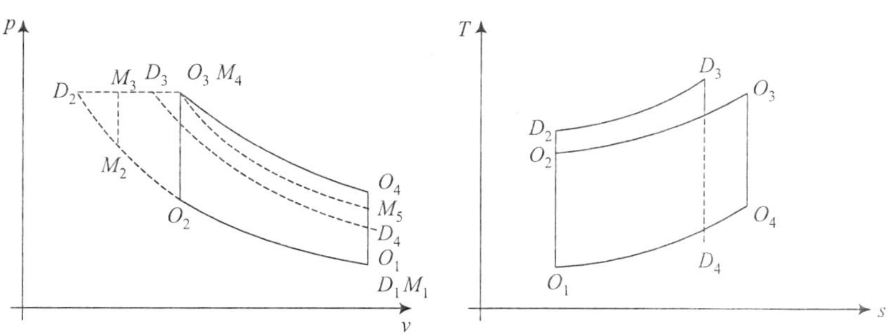

O = Otto cycle; D = Diesel cycle, M = Dual cycle

Fig. 9.10 Comparison of air standard cycles

$$[Q_1]_O = [Q_1]_M = [Q_1]_D \quad \text{(Given)}$$

Maximum pressure,

$$p_{O_3} = p_{D_2} = p_{D_3} = p_{M_2} = p_{M_4}$$

$$[Q_2]_D < [Q_2]_M < [Q_2]_O \quad \text{[Area under curve 4-1]}$$

$$\eta = 1 - \frac{Q_2}{Q_1}$$

$$\therefore \qquad \eta_D > \eta_M > \eta_O$$

9.10 HOT AIR ENGINE CYCLES

Hot air engines using air as the working fluid operate on a closed cycle. Internal combustion engines where the combustion of fuel takes place inside the engine cylinder are non-cyclic heat engines working on open cycle. In hot air engine, air is compressed outside the engine, heated and expanded in the engine cylinder.

9.10.1 Carnot Cycle

The *p-v* diagram and *T-s* diagram for a Carnot cycle are shown in Fig. 9.11. Assume 1 kg of air at state 1 which has parameters p_1, v_1 and T_1.

1. *Process 1-2 Isothermal Heating Process*

$$Q_{1-2} = RT_1 \ln \frac{v_2}{v_1}$$

$$W_{1-2} = RT_1 \ln \frac{v_2}{v_1}$$

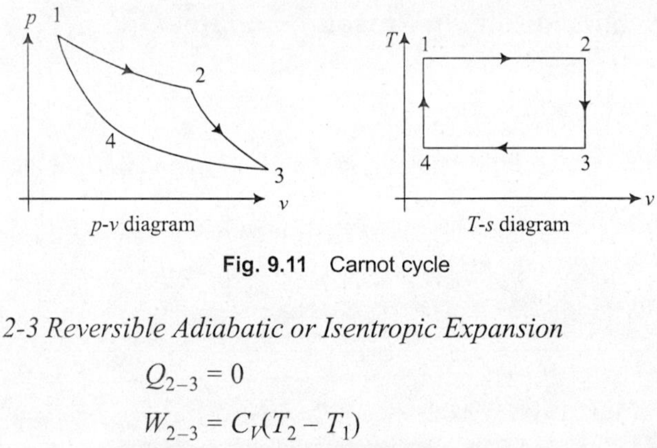

Fig. 9.11 Carnot cycle

2. *Process 2-3 Reversible Adiabatic or Isentropic Expansion*

$$Q_{2-3} = 0$$

$$W_{2-3} = C_V(T_2 - T_1)$$

3. *Process 3-4 Isothermal Compression Process*

$$Q_{3-4} = RT_2 \ln \frac{v_4}{v_3}$$

$$W_{3-4} = RT_2 \ln \frac{v_4}{v_3}$$

4. *Process 4-1 Reversible adiabatic or Isentropic Compression*

$$Q_{4-1} = 0$$

$$W_{4-1} = -C_V(T_1 - T_4)$$

(a) *Analysis*

$$\sum_{\text{cycle}} \delta Q = \sum_{\text{cycle}} \delta W$$

Now,

$$\frac{v_2}{v_3} = \left(\frac{T_2}{T_1}\right)^{\frac{1}{\gamma-1}}$$

and

$$\frac{v_1}{v_4} = \left(\frac{T_2}{T_1}\right)^{\frac{1}{\gamma-1}}$$

∴

$$\frac{v_2}{v_3} = \frac{v_1}{v_4}$$

or

$$\frac{v_2}{v_1} = \frac{v_3}{v_4}$$

Heat absorbed,

$$Q_1 = RT_1 \ln\frac{v_2}{v_1}$$

Work done

$$W = Q_1 - Q_2 = R \ln\frac{v_2}{v_1}(T_1 - T_2)$$

$$\eta_{\text{cycle}} = \frac{W}{Q_1} = \frac{T_1 - T_2}{T_1}$$

(b) *Observation*

The compression work is large which is a big drawback with Carnot cycle.

9.10.2 Stirling Cycle

The Stirling cycle is shown on p-v diagram and T-s diagram in Fig. 9.12.

Assume 1 kg of air at state 1 which has parameters p_1, v_1, T_1.

1. *Process 1-2: Isothermal Expansion Process*

$$Q_{1-2} = W_{1-2} = RT_1 \ln\frac{v_2}{v_1}$$

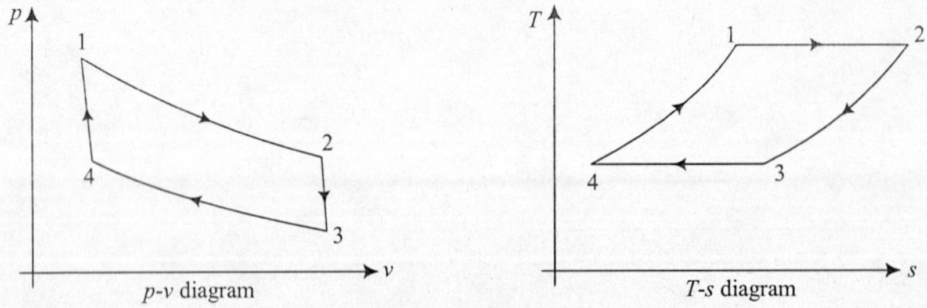

Fig. 9.12 Stirling cycle

2. *Process 2-3: Constant Volume Cooling Process*

$$Q_{2-3} = C_V(T_2 - T_3) = C_V(T_1 - T_3)$$
$$W_{2-3} = 0$$

3. *Process 3-4: Isothermal Comparison Process*

$$Q_{3-4} = W_{3-4} = -RT_2 \ln\frac{v_3}{v_4}$$

4. *Process 4-1: Constant Volume Heating Process*

$$Q_{4-1} = C_V(T_1 - T_2)$$
$$W_{4-1} = 0$$

(a) *Analysis*

The heat addition and rejection take place at constant volume, the efficiency of Stirling cycle is less than Carnot cycle. However, if a regenerator is used so that $Q_{2-3} = Q_{4-1}$, the cycle efficiency.

$$\eta = \frac{RT_1 \ln\frac{v_2}{v_1} - RT_2 \ln\frac{v_3}{v_4}}{RT_1 \ln\frac{v_2}{v_1}} = \frac{T_1 - T_2}{T_1}$$

(b) *Observation*

A regenerative Stirling cycle has same efficiency as Carnot cycle.

9.10.3 Ericsson Cycle

The *p-v* diagram and *T-s* diagram of Ericsson cycle are shown in Fig. 9.13.

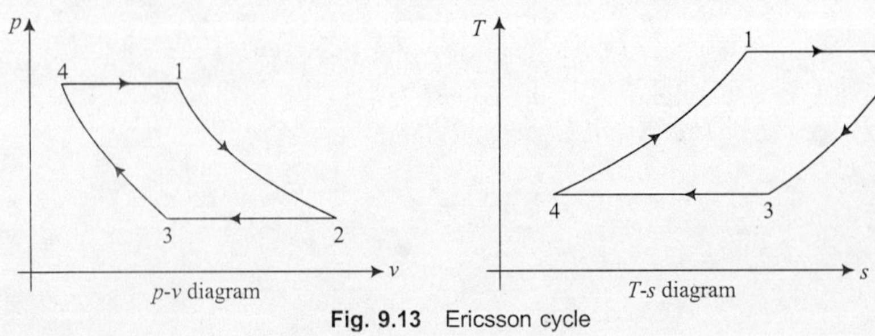

Fig. 9.13 Ericsson cycle

Assume 1 kg of air at state 1 which has parameters p_1, v_1, T_1,

1. *Process 1-2: Isothermal Expansion Process*

$$Q_{1-2} = W_{1-2} = RT_1 \ln \frac{v_2}{v_1}$$

2. *Process 2-3: Isobaric Cooling Process*

$$Q_{2-3} = C_p(T_2 - T_3)$$
$$W_{2-3} = p_2(v_3 - v_2) = R(T_2 - T_1)$$

3. *Process 3-4: Isothermal Compression Process*

$$Q_{3-4} = W_{3-4} = -RT_2 \ln \frac{p_1}{p_2}$$

4. *Process 4-1: Isobaric Heating Process*

$$Q_{4-1} = C_p(T_1 - T_4)$$
$$W_{4-1} = p_1(v_1 - v_4) = R(T_2 - T_1)$$

(a) *Analysis*

$$\eta = 1 - \frac{Q_2}{Q_1} = 1 - \frac{RT_2 \ln \dfrac{p_1}{p_2}}{RT_1 \ln \dfrac{p_1}{p_2}}$$

$$= 1 - \frac{T_2}{T_1}$$

(b) *Observations:*

1. Part of heat is transferred at constant pressure and part at constant temperature, efficiency of Ericsson cycle is less than Carnot cycle.
2. With ideal regeneration $Q_{2-3} = Q_{4-1}$, i.e., all heat is added from external source at temperature T_1 and all heat is rejected to external sink at temperature T_2, the cycle efficiency is same as that for Carnot cycle.
3. In Ericsson cycle, the work of compression is big. Also, it is very difficult to transfer heat to air at high rate due to low thermal conductivity of air film, much progress has not been made.
4. The cost of fuels used in I.C. engines is increasing fast, hot air engine may find application in near future.

Example 9.5 A regenerative Ericsson hot air engine works between temperature limits of 318 K and 503 K. The expansion ratio is 2. Calculate

(i) Work done per kg of air.

(ii) Cycle efficiency.

Solution

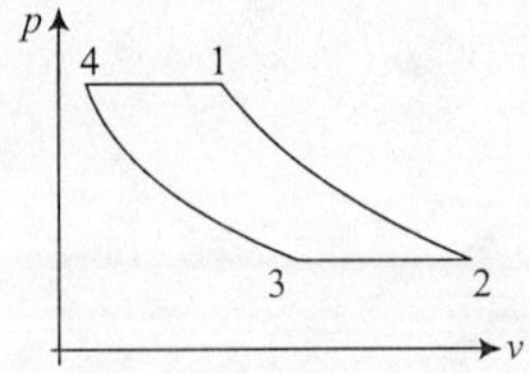

Fig. 9.14 Ericsson cycle

$$T_1 = T_4 = 318 \text{ K}$$
$$T_2 = T_3 = 503 \text{ K}$$
$$r = 2$$
$$R = 0.287 \text{ kJ/kgK (for air)}$$

(i) Work done per kg of air,

$$W = (T_2 - T_3)R \ln r$$
$$= (503 - 318) \times 0.287 \ln 2 = \mathbf{36.76\ kJ/kg} \quad \textbf{Ans.}$$

(ii)
$$\eta = 1 - \frac{T_2}{T_1} = 1 - \frac{318}{503} = 0.368 = \mathbf{36.8\%} \quad \textbf{Ans.}$$

QUESTION BANK NO. 9

1. Explain the properties of air for use as working fluid in thermodynamic cycles.
2. What are the assumptions made in the analysis of air standard cycles? Define air standard efficiency.
3. Stating the assumptions made, describe air standard Otto cycle.
4. Show Otto cycle and Diesel cycles on *p-v* and *T-s* diagram.
5. Describe a Diesel cycle with the help of *p-v* diagram and *T-s* diagram.
6. Compare supercharged Otto cycle and supercharged Diesel cycle. Why supercharged Diesel cycles are widely used?
7. What is dual combustion cycle? How can the efficiency of a dual cycle be compared with an Otto cycle and Diesel cycle?
8. What is hot air engine? What are important hot air engine cycles? What are their limitations?

CHAPTER 10

Gas Turbine Cycles

10.1 INTRODUCTION

Gas turbines are used for a very wide range of services.

1. Most notable use of gas turbines is for powering aircraft of all types. In *jet-aircraft engine*, the gas turbine supplies work of compression only. The exhaust gases from the turbine are expanded in the exhaust cone to a high velocity, the resulting reaction force propelling the aircraft. The turbine and compressor are mounted on a single shaft. It has favourable performance at high-altitude and high-speed flights.

   In *prop-jet aircraft*, the turbine utilizes nearly all the energy of the gas and drives the compressor and a conventional propeller. The turbine exhaust gases are expanded in an exhaust cone, thereby assisting propulsion. It is used in commercial aircraft for more extensive range, shorter length of take-off required, large cargo capacity, and suitability to moderate speeds and altitudes.
2. Gas turbines are used in industrial plants for driving mechanical equipment such as pumps, compressors and small generators. They offer flexibility in supplying process needs, such as compressed air, in addition to electric power. These can be run on a range of liquid and gaseous fuels including synthetic fuels and low heat waste gases. They are subject to fewer environmental restrictions than other prime movers.
3. Gas turbines are used in electrical utilities for producing electric power for peak loads as well as for intermediate and some base load duties. However, these turbines have one major disadvantage that prevents utilities from using them as major base-load prime mover. They are incompatible with solid fuels and have low efficiencies. Gas turbines have many advantages. Compared with steam plants they, and their total systems, are small in size, mass and initial cost per unit output. They are available with relatively short delivery times and are quick to install and put to use. They are quick-starting, often by remote control, and are smooth-running. The

combination of low capital cost and low efficiency have determined their use primarily as power-peaking units where they are not expected to be on line for more than 1000 or 2000 h/year. Under these conditions, a large steam plant designed to meet peak loads would operate at an uneconomical load factor during most of the year.

4. There is growing interest in using gas turbines in combined-cycle plants. These plants use combinations of gas and steam turbines in a variety of configurations of turbines, heat-recovery boilers, and regenerators. The use of gas turbines in combined cycle is one scheme to overcome their present low cycle efficiency for utility base-load use, while at the same time offering the utilities the gas turbine advantages of quick starting and flexible operation over a wide range of loads.
5. Gas turbines designed for aircraft propulsion are sometimes modified and used for industrial service.

10.2 CLASSIFICATION OF GAS TURBINE CYCLES

Gas turbines are available in one-or two-shaft models. Single-shaft turbines have compressor, turbine and electrical generator on one shaft running at constant speed. This configuration is used to drive small generators as well as large generators for utility use. Gas turbines with two shafts that rotate at different speeds one shaft has the compressor and a turbine that drives it, the other has the power turbine connected to the electrical generator.

The hot gas emerging from a combustor or gas-cooled reactor can be used directly as the primary working fluid, i.e., by expanding through a gas turbine, or indirectly by heating a secondary fluid acting as the working fluid. For each of these cases, i.e., the direct or indirect cycle, there can be an open or a closed cycle.

Following are the possible combinations:

1. Direct open cycle
2. Indirect open cycle
3. Direct closed cycle
4. Indirect closed cycle

10.2.1 Direct Open Cycle

The direct open cycle is shown schematically in Fig. 10.1.

The air enters the compressor at point 1, where it is compressed to point 2. The air then enters the combustion chamber or reactor where it receives heat at constant pressure (ideally) and emerges hot at point 3. From there it expands through the turbine to point 4. The hot exhaust mixes with the atmosphere outside the cycle, and a fresh cool supply is taken in at point 1. The turbine supplies power to the compressor and surplus power is used to drive an electric generator.

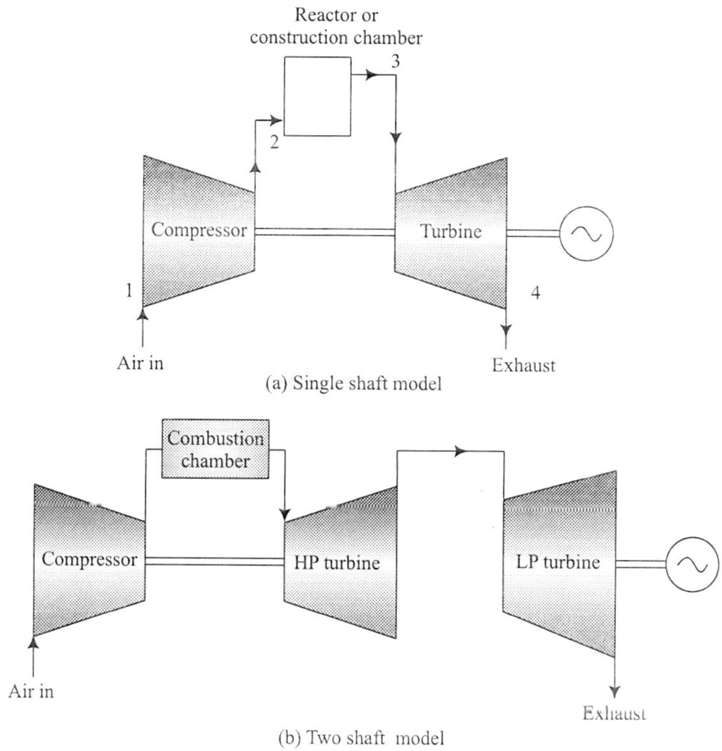

Fig. 10.1 Schematic of direct open cycle

10.2.2 Indirect Open Cycle

An indirect open gas turbine cycle is shown in Fig. 10.2.

The air is a secondary fluid that receives its heat from a primary coolant in a heat exchanger.

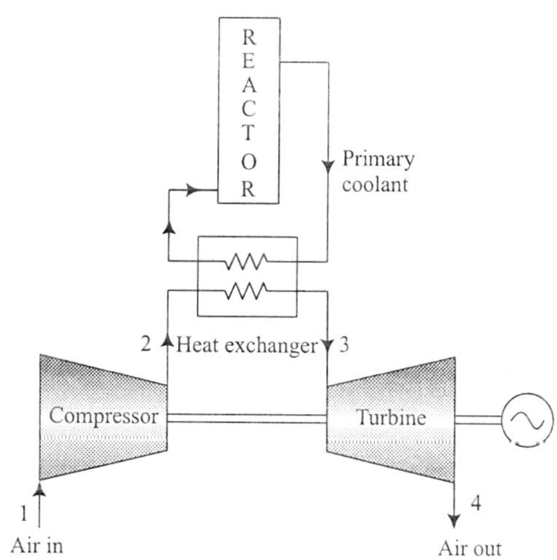

Fig. 10.2 Schematic of indirect open gas turbine cycle

10.2.3 Direct Closed Cycle

Direct closed gas turbine cycle is shown in Fig. 10.3.

The gas coolant is heated in the reactor, expanded through the

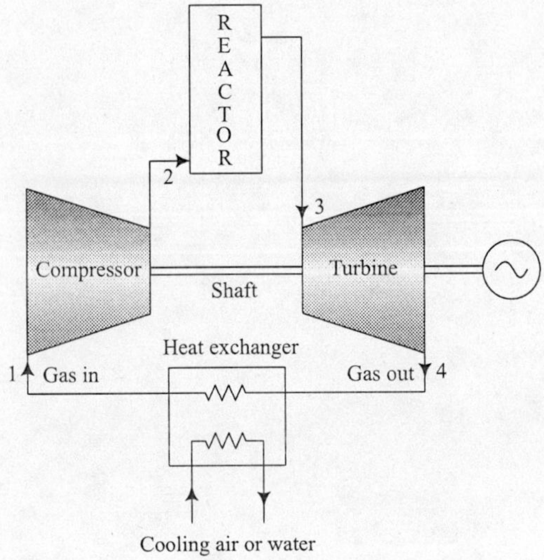

Fig. 10.3 Schematic of direct closed gas turbine cycle

turbine, cooled in a heat exchanger, and compressed back to the reactor. Air or a suitable gas may be used in this cycle. The most suitable working fluid is helium which is pressurized to reduce the size of rotating machinery. No effluent of radioactive gases passes into the atmosphere under normal operating conditions.

10.2.4 Indirect Closed Cycle

Indirect closed gas turbine cycle is shown in Fig. 10.4.

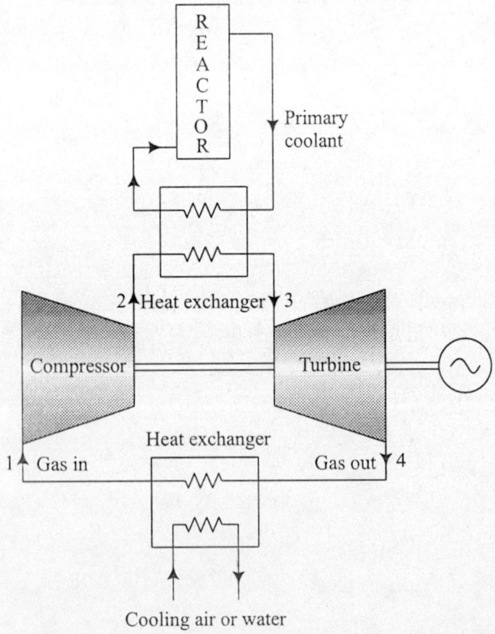

Fig. 10.4 Schematic of indirect closed gas turbine cycle

It combines the indirect open cycle and direct closed cycle in that the reactor is separated from the working fluid by a heat exchanger, whereas the working gas rejects heat to the atmosphere via a heat exchanger. The primary coolant may be water or liquid metal, or a gas such as helium.

10.3 THE IDEAL BRAYTON CYCLE

Open cycle, closed cycle and semiclosed gas turbine plants, operate on Brayton cycle. The following assumptions are made for ideal Brayton cycle.

1. The compression and expansion processes are reversible adiabatic or isentropic.
2. The specific heat of working fluid remains constant throughout the cycle.
3. The pressure losses in the cycle are neglected.
4. The heat losses due to various factors are neglected.
5. The kinetic energy of the working fluid, while entering the compressor and leaving the turbine is equal.

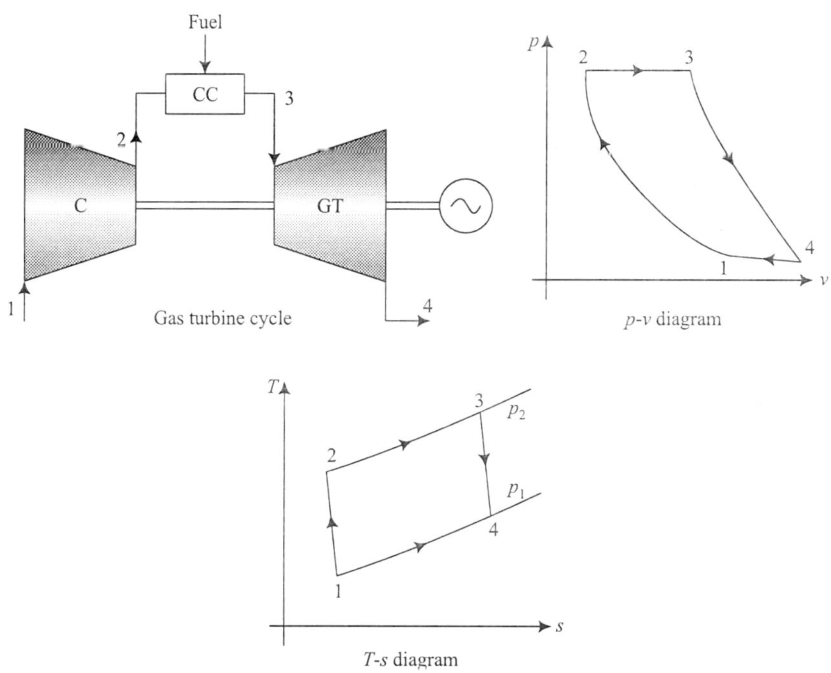

Fig. 10.5 Brayton cycle

The Brayton cycle is shown on p–v diagram and T-s diagram in Fig. 10.5. It consists of the following processes:

1-2: Isentropic compression (Compressor)

2-3: Constant pressure heating (Combustion chamber)

3-4: Isentropic expansion (Gas turbine)

4-1: Constant pressure cooling

Heat supplied to the cycle per unit mass,

$$Q_s = Q_{2-3} = C_p\,(T_3 - T_2)$$

Work done in the turbine,

$$W_T = W_{3-4} = C_p\,(T_3 - T_4)$$

Work consumed in the compressor,

$$W_C = W_{1-2} = C_p\,(T_2 - T_1)$$

Thermal efficiency or cycle efficiency,

$$\eta = \frac{\text{Net work done}}{\text{Heat supplied}} = \frac{W_T - W_C}{Q_S}$$

$$= \frac{C_p(T_3 - T_4) - C_p(T_2 - T_1)}{C_p(T_3 - T_2)}$$

$$= \frac{(T_3 - T_2) - (T_4 - T_1)}{(T_3 - T_2)}$$

$$= 1 - \frac{(T_4 - T_1)}{(T_3 - T_2)}$$

Assuming that expansion and compression ratios are equal to R.

$$\frac{T_2}{T_1} = \frac{T_3}{T_4} = \left(\frac{p_2}{p_1}\right)^{\frac{\gamma-1}{\gamma}} = R^{\left(\frac{\gamma-1}{\gamma}\right)}$$

$$\eta = 1 - \left(\frac{1}{R}\right)^{\frac{\gamma-1}{\gamma}}$$

The cycle efficiency is independent of temperature and depends upon pressure ratio only.

Gas turbines based on Brayton cycle are widely used generating electricity, driving locomotives and aeroplane propellers, and jet propulsion of aircraft.

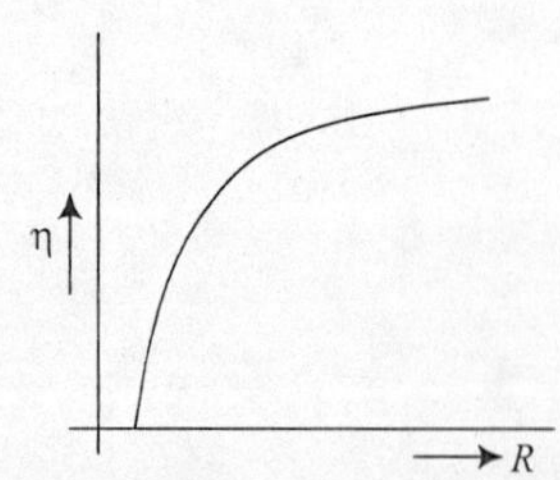

Fig. 10.6 Cycle efficiency

10.4 MAXIMUM WORK OUTPUT

The work done by gas turbine plant per unit mass,

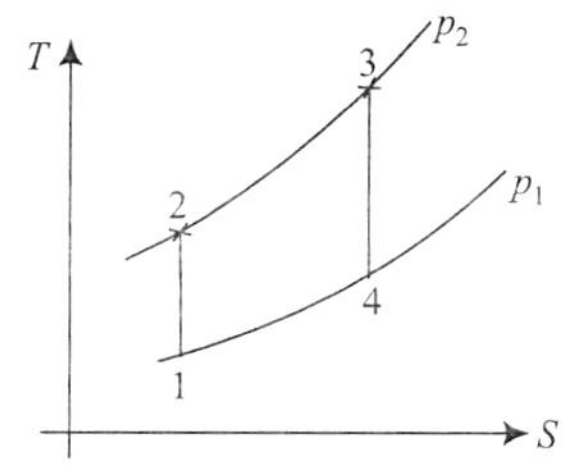

Fig. 10.7 Max. work output

$$W_N = W_T - W_C$$

$$= C_p\,[(T_3 - T_4) - (T_2 - T_1)]$$

$$= C_p\,T_1\left[\left(\frac{T_3}{T_1} - \frac{T_4}{T_1}\right) - \left(\frac{T_2}{T_1} - 1\right)\right]$$

Assume $\quad \dfrac{T_3}{T_1} = \dfrac{T_{max}}{T_{min}} = t = \text{fixed}$

and $\quad \dfrac{T_2}{T_1} = C = \text{Constant for a cycle}$

Now, $\quad \dfrac{T_4}{T_1} = \dfrac{T_4}{T_3} \times \dfrac{T_3}{T_1}$

Now, $\quad \dfrac{T_4}{T_3} = \dfrac{T_1}{T_2}$

(Assuming expansion and compression ratios are same).

$\therefore \quad \dfrac{T_4}{T_1} = \dfrac{T_1}{T_2} \times \dfrac{T_3}{T_1} = \dfrac{t}{C}$

$$W_N = C_p\,T_1\left[\left(t - \frac{t}{C}\right) - (C-1)\right]$$

$$= C_p\,T_1\left[t\left(1 - \frac{1}{C}\right) - (C-1)\right]$$

T_1 = Constant being atmospheric temperature.

For maximization of work done,

$$\frac{dW}{dC} = 0 = C_p\,T_1\left[\frac{t}{C^2} - 1\right]$$

or
$$\frac{1}{C^2} - 1 = 0$$
$$C^2 = t$$
$$C = \sqrt{t}$$

But,
$$C = \frac{T_2}{T_1} = \frac{T_3}{T_4}$$

$\therefore$
$$\frac{T_2}{T_1} \times \frac{T_3}{T_4} = t = \frac{T_3}{T_1}$$

$\therefore$
$$T_2 = T_4$$

Hence, for maximum work output T_2 should be raised to T_4.

10.5 SIMPLE CYCLE WITH HEAT EXCHANGER

The heat of exhaust gases is a waste and should be utilized by modifying the cycle as per Fig. 10.8.

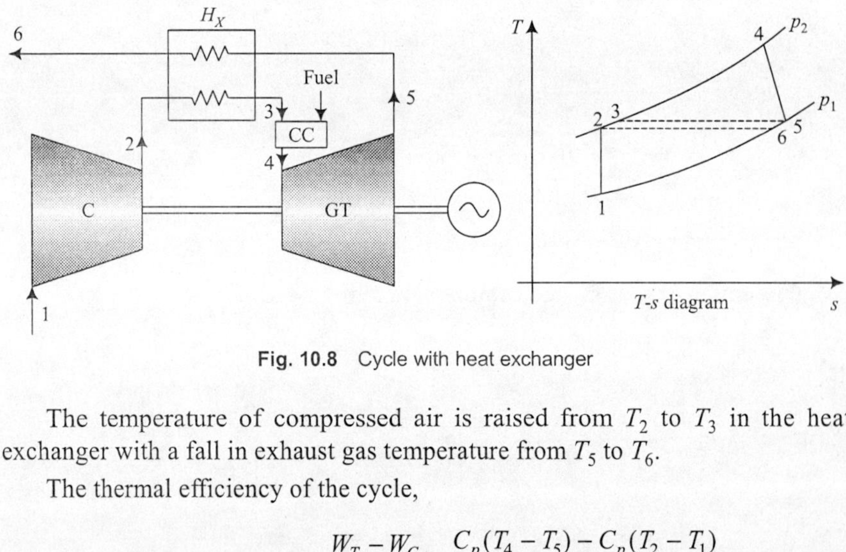

Fig. 10.8 Cycle with heat exchanger

The temperature of compressed air is raised from T_2 to T_3 in the heat exchanger with a fall in exhaust gas temperature from T_5 to T_6.

The thermal efficiency of the cycle,

$$\eta_{Th} = \frac{W_T - W_C}{Q_s} = \frac{C_p(T_4 - T_5) - C_p(T_2 - T_1)}{C_p(T_4 - T_3)}$$

For ideal case, $T_5 = T_3$ (refer # 10.4)

$\therefore$
$$\eta_{Th} = \frac{(T_4 - T_3) - (T_2 - T_1)}{(T_4 - T_3)} = 1 - \frac{(T_2 - T_1)}{(T_4 - T_3)}$$

But, $$\frac{T_2}{T_1} = \frac{T_4}{T_5} = R^{\frac{\gamma-1}{\gamma}}$$

or, $$\frac{T_1}{T_2} = \frac{1}{R^{\frac{\gamma-1}{\gamma}}}$$

$$1 - \frac{T_1}{T_2} = 1 - \frac{1}{R^{\frac{\gamma-1}{\gamma}}}$$

$\therefore$ $$(T_2 - T_1) = T_2\left[1 - \frac{1}{R^{\frac{\gamma-1}{\gamma}}}\right]$$

Also, $$(T_4 - T_5) = T_4\left[1 - \frac{1}{R^{\frac{\gamma-1}{\gamma}}}\right] \quad \left[\because \frac{T_4}{T_5} = \frac{T_2}{T_1}\right]$$

$\therefore$ $$\eta_{Th} = 1 - \frac{T_2}{T_4}$$

$$= 1 - \frac{T_2}{T_1} \cdot \frac{T_1}{T_4} = 1 - \frac{C}{t}$$

$$W_N = W_T - W_C = C_p\,[(T_4 - T_5) - (T_2 - T_1)]$$

The work output remains unchanged.

Under ideal conditions, effort of adding a heat exchanger is to raise the thermal efficiency of the cycle and work output remains unchanged.

10.6 SIMPLE CYCLE WITH REHEAT

In order to improve the performance of gas turbine cycle, two gas turbines are used with a provision for additional combustion chamber. The schematic diagram is shown in Fig. 10.9.

The net work output of the reheat cycle per unit mass of air.

$$W_N = W_{HPGT} + W_{LPGT} - W_C$$

where, W_{HPGT} = work output of high pressure gas turbine

W_{LPGT} = work output of low pressure gas turbine

W_C = compressor work input

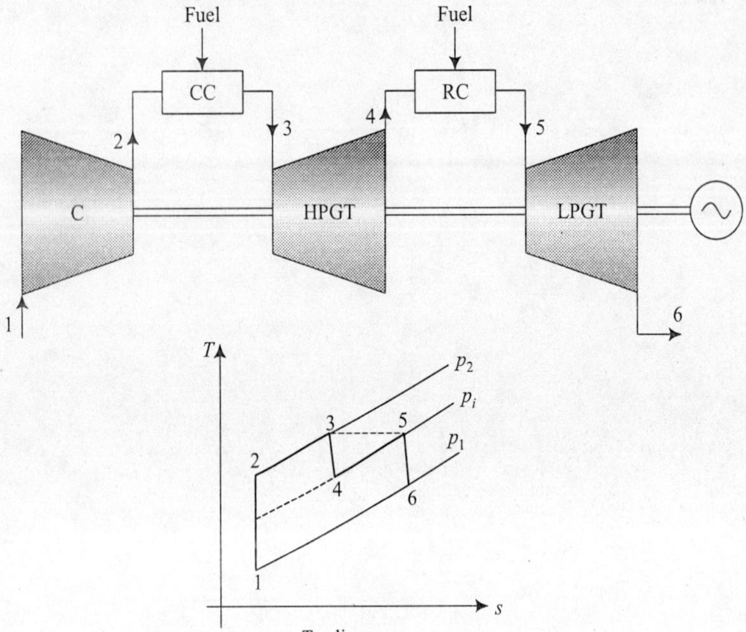

T-s diagram

Fig. 10.9 Reheat cycle

$$W_N = C_p[(T_3 - T_4) + (T_5 - T_6) - (T_2 - T_1)]$$

Let $\dfrac{T_2}{T_1} = \mathrm{C}$

$$\frac{T_3}{T_4} = C_1$$

$$\frac{T_5}{T_6} = C_2 \text{ and } T_3 = T_5$$

$\therefore$ $$C = \frac{T_2}{T_1} = \left(\frac{p_2}{p_1}\right)^{\frac{\gamma-1}{\gamma}}$$

$$C_1 = \frac{T_3}{T_4} = \left(\frac{p_2}{p_i}\right)^{\frac{\gamma-1}{\gamma}}$$

$$C_2 = \frac{T_5}{T_6} = \left(\frac{p_i}{p_1}\right)^{\frac{\gamma-1}{\gamma}}$$

Let $R = R_1 \times R_2$

$\therefore$ $$C_2 = \frac{C}{C_1}$$

and $C = C_1 \cdot C_2$

$\therefore$ $$W_N = C_p[(T_3 - T_4) + (T_3 - T_6) - (T_2 - T_1)]$$

$$= C_p\left[2T_3 - T_3.\frac{T_4}{T_3} - T_6\frac{T_4}{T_3} - T_2 + T_1\right]$$

$$= C_p\, T_1\left[2\frac{T_3}{T_1} - \frac{T_3}{T_1}.\frac{T_4}{T_3} - \frac{T_3}{T_1}.\frac{T_6}{T_5} - \frac{T_2}{T_1} - 1\right] \quad (\because T_3 = T_5)$$

$$W_N = C_p\, T_1\left[2t - \frac{t}{C_1} - t\frac{C_1}{C} - C + 1\right]$$

Assume t and C are constant for a given cycle.

$$\frac{dW_N}{dC_1} = C_p\, T_1\left[\frac{t}{C_2^1} - \frac{t}{C}\right] = 0$$

$$(C_1)^2 = C$$

or $$C_1 = \sqrt{C}$$

and $$R_1 = R_2 = \sqrt{R}$$

For maximum work output, expansion in the two stages should be divided equally.

10.7 SIMPLE CYCLE WITH INTERCOOLER

The work of compression can be reduced by the use of an intercooler between two stages of compression. The schematic diagram with intercooler is shown in Fig. 10.10 along with T-s diagram.

For ideal condition, $T_1 = T_3$ (Complete intercooling)

$$R_1 = R_2 = \sqrt{R}$$

For minimum compressor work input, the compression ratio should be divided equally

$$\frac{p_i}{p_1} = \frac{p_2}{p_i}$$

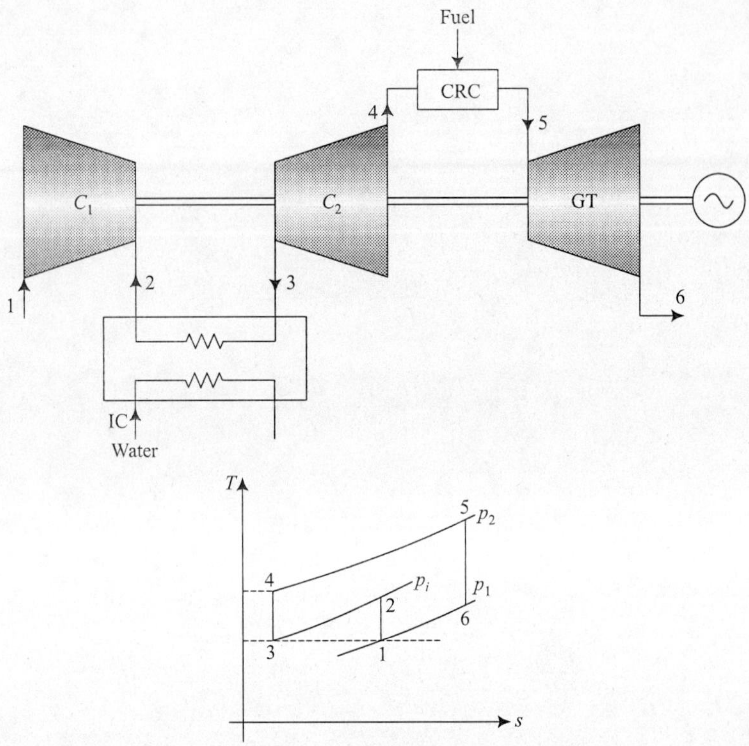

T-s diagram

Fig. 10.10 Cycle with intercooler

The cycle efficiency will be reduced due to heat loss to cooling water in the intercooler.

10.8 COMPLEX GAS TURBINE CYCLES

Complex cycles using intercooling, reheating and waste heat recovery (HX) are simultaneously incorporated to improve the plant performance. The quantitative effects of these modifications are given in Table 10.1.

Table 10.1 Complex cycles

S.No.	*Addition to simple cycle*	*Effect on cycle efficiency,* η_{TH}	*Effect on power output,* W_N
1.	Heat exchanger (HX)	+50%	No change
2.	Intercooling	–6.5%	+10.2%
3.	Reheating	–10.4%	+24.5%
4.	Reheat + HX	+66.7%	+24.5%
5.	Intercooling + HX	+68%	+10.2%
6.	Reheat + Intercooling	–18.2%	+34.7%
7.	Reheat + Intercooling + HX	+80%	+34.7%

10.9 BRAYTON CYCLE PERFORMANCE

The performance of gas turbine cycle is affected by various parameters.

1. Pressure ratio and turbine inlet temperature
2. Component efficiencies
3. Compressor inlet temperature
4. Pressure ratio and intercooling
5. Pressure ratio, intercooling and reheating
6. Working media other than air

Effect of some of the parameters are discussed in subsequent sections. The above effects have been shown graphically.

10.9.1 Effect of Pressure Ratio and Turbine Inlet Temperature

The cycle thermal efficiency is a function of pressure ratio and increases rapidly as the turbine inlet temperature increases.

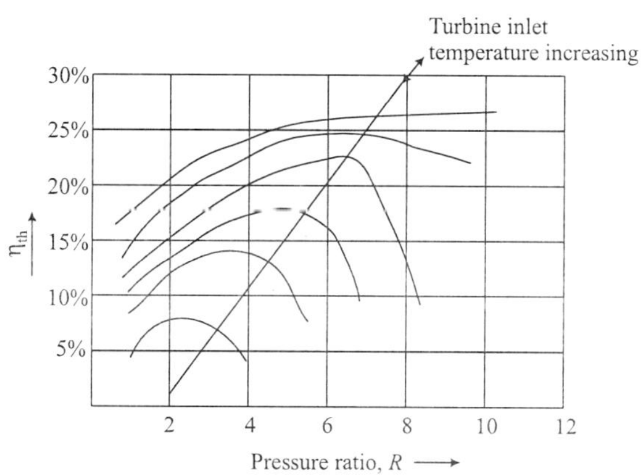

Fig. 10.11 Effect of turbine inlet temperature

10.9.2 Effect of Component Efficiencies

Turbine and compressor efficiencies have effect on the overall thermal efficiency of the gas turbine plant, see Fig. 10.12.

10.9.3 Effect of Compressor Inlet Temperature

The thermal efficiency of gas turbine plant decreases with increase of compressor inlet temperature whereas efficiency increases with turbine inlet temperature, see Fig. 10.13.

10.9.4 Effect of Intercooler

The thermal efficiency of gas turbine plant decreases with intercooler at low pressure ratios but the same improves with higher pressure ratio, see Fig. 10.14.

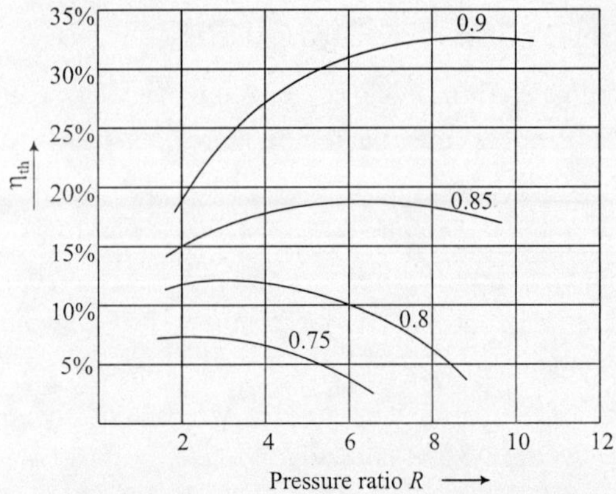

Fig. 10.12 Effect of component efficiencies

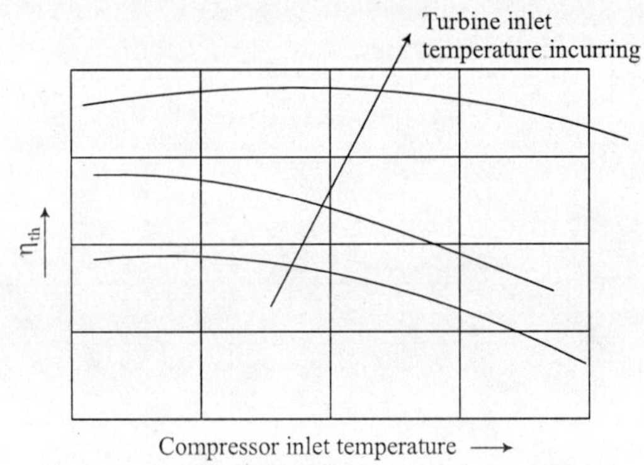

Fig. 10.13 Effect of compressor inlet temperature

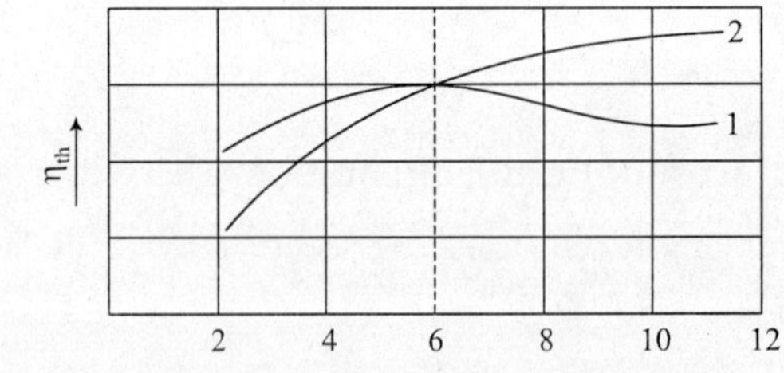

Fig. 10.14 Effect of intercooling

Curve (1): Without intercooling

Curve (2): With one intercooler

10.9.5 Effect of Intercooling and Reheat

The effect of various combinations of intercooler, reheaters and regenerators are compared in Fig. 10.15.

Curve (1): Simple Brayton Cycle

Curve (2): One intercooler and one reheater

Curve (3): One intercooler and two reheaters

Curve (4): One intercooler, one reheater and one regenerator

Curve (5): Two intercoolers, two reheaters and one regenerator

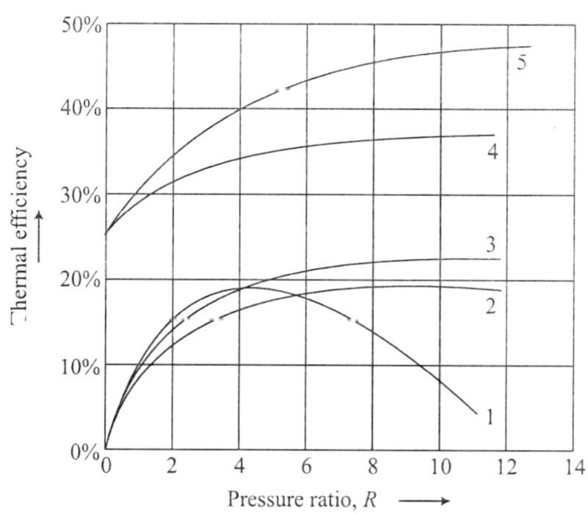

Fig. 10.15 Effect of intercooling and reheating

10.9.6 Effect of Operating Media other than Air

The operating medium for gas turbine plant should have the following characteristics:

1. Stability against high temperature of operation.
2. Cheapness and availability.
3. Non-inflammability.
4. Non-explosiveness.
5. Non-toxicity.
6. High specific heat (C_p)
7. High specific heat ratio $\left(\gamma = \frac{C_p}{C_v}\right)$

8. High thermal conductivity (U)
9. Non-corrosion.
10. Low gas constant, (R).

$$W_C = mC_pT_2\left(1 = \frac{1}{R^{\frac{\gamma-1}{\gamma}}}\right)$$

$$W_T = mC_pT_3\left(1 = \frac{1}{R^{\frac{\gamma-1}{\gamma}}}\right)$$

1. Low values of C_p and γ are required to reduce compressor work input, W_C.
2. High values of C_p and γ are required to increase turbine work output, W_T.
3. High values of C_p and γ also help to reduce mass flow rate for a given capacity of the plant. However, low value of R is needed for low specific volume ($pv = RT$), affecting the size of the plant. When the high values of C_p and γ are also accompanied by high values of R, there is counter effect on the size of the machines.
4. High value of thermal conductivity (U) affects the saving in the size of regenerators and intercoolers. Therefore, reduction in gas turbine plant size can be realized in heat exchangers by using gas with high thermal conductivity.

10.10 PRACTICAL GAS TURBINE CYCLE

There are some derivations in practical cycles from ideal cycles.

1. Compression and expansion processes are not isentropic.
2. There are pressure losses in flow through combustion chamber, heat exchangers and piping.
3. The effectiveness of heat exchangers is less than 100%.
4. There are mechanical losses in the turbine and compressor.
5. Specific heats vary due to temperature and gas mixtures.

All the above derivations need to be accounted for considering the performance of practical cycles.

10.10.1 Compressor and Turbine Efficiency

The net effect of irreversible processes in compressor and turbine is heating of gases due to friction.

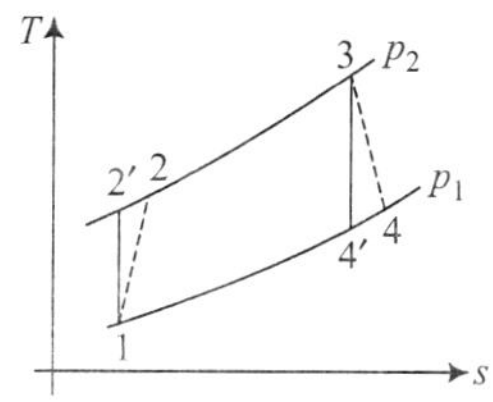

Fig. 10.16 Turbine and compressor efficiency

In the compressor, 1-2′ is isentropic process and 1-2 is irreversible process with friction. The compressor efficiency,

$$\eta_C = \frac{\text{Ideal work input}}{\text{Real work input}}$$

$$= \frac{mC_p(T_2' - T_1)}{mC_p(T_2 - T_1)}$$

$$= \frac{T_2' - T_1}{T_2 - T_1}$$

Now,

$$\frac{T_2'}{T_1} = R^{\frac{\gamma-1}{\gamma}}$$

∴

$$\eta_C = \frac{T_1\left(R^{\frac{\gamma-1}{\gamma}} - 1\right)}{(T_2 - T_1)}$$

$$(T_2 - T_1) = \frac{T_1}{\eta_C}\left(R^{\frac{\gamma-1}{\gamma}} - 1\right)$$

The value of η_C is approximately 0.87.

In the gas turbine, 3-4′ is isentropic expansion process and 3-4 is the irreversible process with friction. The turbine efficiency

$$\eta_T = \frac{\text{Actual work output}}{\text{Ideal work output}}$$

$$= \frac{\dot{m}C_p(T_3 - T_4)}{\dot{m}C_p(T_3 - T_4')} = \frac{T_3 - T_4}{T_3 - T_4'}$$

$$(T_3 - T_4) = \eta_T T_3\left(1 - \frac{1}{R^{\frac{\gamma-1}{\gamma}}}\right)$$

The value of η_T is approximately 0.85.

10.10.2 Pressure Losses

Due to pressure drop in the ducting, heat exchanger and combustion chamber, the pressure at turbine inlet will be slightly less than that at compressor outlet. Similarly, the exhaust pressure of turbine will be more to account for pressure drop in the regenerator. Both the effects reduce the expansion ratio in the turbine and hence the available turbine work output.

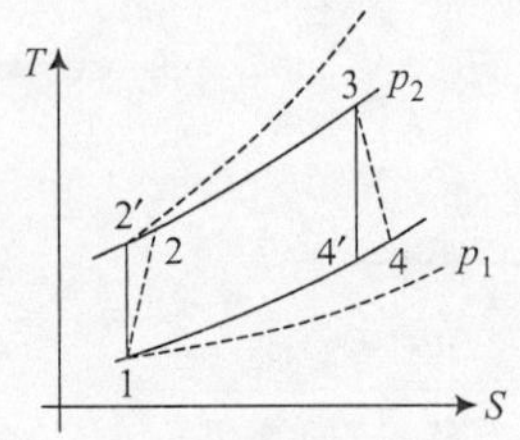

Fig. 10.17 Pressure losses

10.10.3 Effectiveness of Heat Exchangers

The effectiveness of a heat exchanger is the ratio of actual heat transfer to ideal heat transfer, i.e., actual heat picked up to maximum possible.

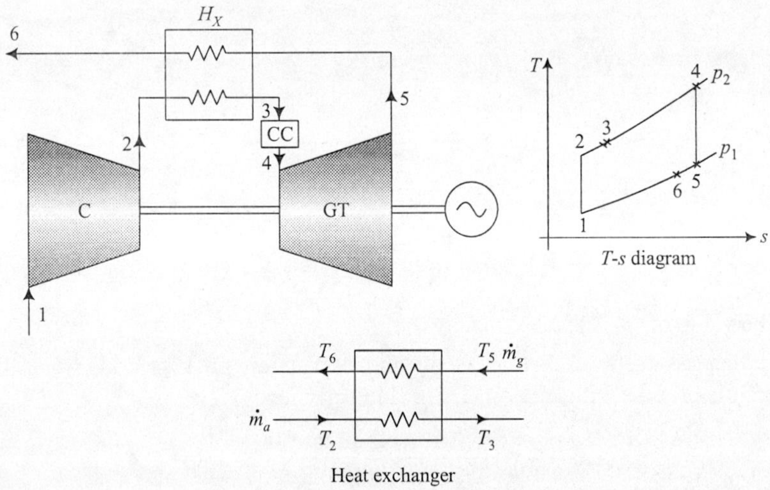

Fig. 10.18 Effectiveness of heat exchanger

The thermal ratio or effectiveness of regenerator of a gas turbine plant.

$$R_e = \frac{\dot{m}_a Cp_a (T_3 - T_2)}{\dot{m}_g Cp_g (T_5 - T_6)}$$

where, $\dot{m}_a Cp_a$ = water equivalent of air stream

$\dot{m}_g Cp_g$ = water equivalent of gas stream

Assume $\dot{m}_a Cp_a = \dot{m}_g Cp_g$

and $T_2 = T_6$ for ideal condition of heat exchanger.

$$\therefore \quad R_e = \frac{T_3 - T_2}{T_5 - T_2}$$

$$= \frac{\text{Actual heat possible up}}{\text{Maximum possible}}$$

The effectiveness of regenerator may be of the order of 0.75.

10.10.4 Mechanical Losses

There are mechanical losses in compressor and turbine due to friction in the bearings and windage losses due to friction between rotor and surrounding air or gas.

For a compressor, mechanical efficiency,

$$\eta_{m_C} = \frac{W_C}{mC_p(T_2 - T_1)} \approx 0.99$$

For a gas turbine, mechanical efficiency,

$$\eta_{m_T} = \frac{W_T}{mC_p(T_4 - T_5)} \approx 0.99.$$

10.10.5 Variation of Specific Heats

The value of specific heat (C_p) changes both with temperature as well as fuel-air ratio as shown in Fig. 10.19.

In order to take care of the above effects, the work done can be integrated.

$$W_C = \dot{m}\int_{T_1}^{T_2} C_p(t)\, dT$$

The value of $C_p(t)$ should be taken from the curves. By neglecting the above effects,

$$W_C = \dot{m}\, C_p(T_2 - T_1)$$

The following average values can be used.

For compression of air, $Cp_a = 1.005$ kJ/kgK, $\gamma = 1.4$

For expansion and heating of gases $Cp_g = 1.156$ kJ/kgK, $\gamma = 1.33$

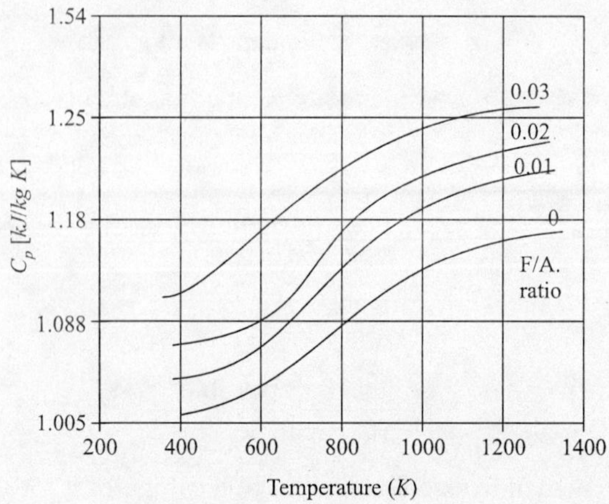

Fig. 10.19 Variation of specific heat

Example 10.1 The following data refer to a gas turbine plant working on Brayton cycle with intercooling and reheat.

$$\text{Pressure ratio} = 4$$
$$t_{min} = 27°C$$
$$t_{max} = 600°C$$
$$\eta_T = 0.8$$
$$\eta_C = 0.8$$

Determine: (a) Cycle thermal efficiency,

(b) Cycle work ratio,

(c) Cycle air rate.

Assume the pressure ratio is same in both the stages. Intake pressure = 1 bar and C_p = 1.005 kJ/kgK.

Solution The line diagram of the plant along with *T-s* diagram are shown in Fig. 10.20.

$$T_1 = T_2 = 27°C + 273 = 300 \text{ K}$$

$$T_5 = T_7 = 600°C + 273 = 873 \text{ K}$$

$$p_1 = p_8 = 1 \text{ bar}$$

$$p_4 = p_5 = 4 \text{ bar}$$

$$p_2 = p_3 = p_6 = p_7 = \sqrt{4} = 2 \text{ bar}$$

$$T'_2 = T_1\left(\frac{p_i}{P_1}\right)^{\frac{\gamma-1}{\gamma}} = 300(2)^{\frac{1.4-1}{1.4}} = 366 \text{ K}$$

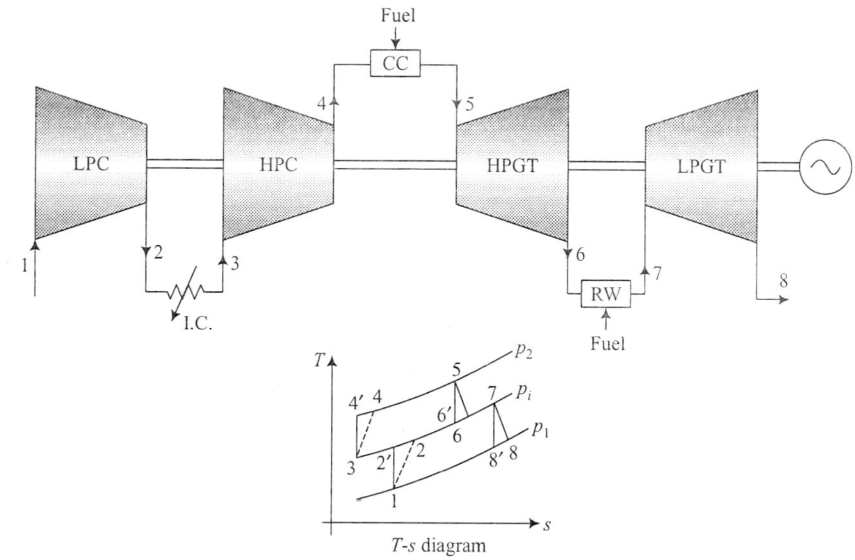

Fig. 10.20 Gas turbine cycle

$$\eta_C = \frac{T_2' - T_1}{T_2 - T_1} = 0.8 = \frac{366 - 300}{T_2 \quad 300}$$

$\therefore$ $T_2 = 382.5$ K

$T'_4 = T'_2 = 366$ K

$T_4 = T_2 = 382.5$ K

$$T'_6 = T_5\left(\frac{p_2}{p_i}\right)^{\frac{\gamma}{\gamma-1}} = 718 \text{ K}$$

$$\eta_T = \frac{T_5 - T_6}{T_5 - T_6'} = 0.8 = \frac{873 - T_6}{873 - 718}$$

$\therefore$ $T_6 = 749$ K

$T'_8 = T'_6 = 718$ K

$T_6 = T_8 = 749$ K

The turbine work per unit mass flow rate,

$$W_T = C_p(T_5 - T_6) + C_p(T_7 - T_8)$$
$$= 1.005\,(873 - 749) + 1.005\,(873 - 749)$$
$$= 249.24 \text{ kJ/kg of air,}$$

The compressor work

$$W_C = C_p(T_2 - T_1) + C_p(T_4 - T_3)$$
$$= 2 \times 1.005\ (382.5 - 300)$$
$$= 165.825 \text{ kJ/kg of air}$$

Heat supplied,

$$Q_s = C_p\ (T_5 - T_4) + C_p(T_7 - T_6)$$
$$= 1.005\ (873 - 382.5) + 1.005\ (873 - 749)$$
$$= 613.2 \text{ kJ/kg of air}$$

(a) $\eta_{\text{cycle}} = \dfrac{W_T - W_C}{Q_S} = \dfrac{249.24 - 165.825}{613.2} = 0.136$

$= \mathbf{13.6\%}$ **Ans.**

(b) Cycle work ratio $= \dfrac{W_T - W_C}{W_T} = \dfrac{249.24 - 165.825}{249.24}$

$= \mathbf{0.3346}$ **Ans.**

(c) Cycle air rate, $\dot{m} \times W_N = 1$ kWh

$$\therefore \quad \dot{m} = \frac{1 \times 3600}{(249.24 - 165.825)}$$

$= \mathbf{43.16\ kg/kWh}$ **Ans.**

Example 10.2 Helium ($C_p = 5.23$ kJ/kgK, $\gamma = 1.64$) enters the compressor of an ideal Brayton cycle at 20 bar and 27°C and leaves at 60 bar. The gas is heated in heat exchanger to 850°C and expanded through a turbine and then passed through a precooler. Determine

(a) The temperature of helium at end of expansion and compression processes.
(b) The heat supplied, heat rejected and net work per kg of Helium.
(c) The thermal efficiency of the cycle and work ratio.

Solution The line diagram of the cycle and *T-s* diagram are shown in Fig. 10.21.

$$p_1 = 20 \text{ bar}$$
$$T_1 = 27°\text{C} + 273 = 300 \text{ K}$$
$$p_2 = 60 \text{ bar}$$

$$\frac{T_2}{T_1} = \left(\frac{p_2}{p_1}\right)^{\frac{\gamma-1}{\gamma}}$$

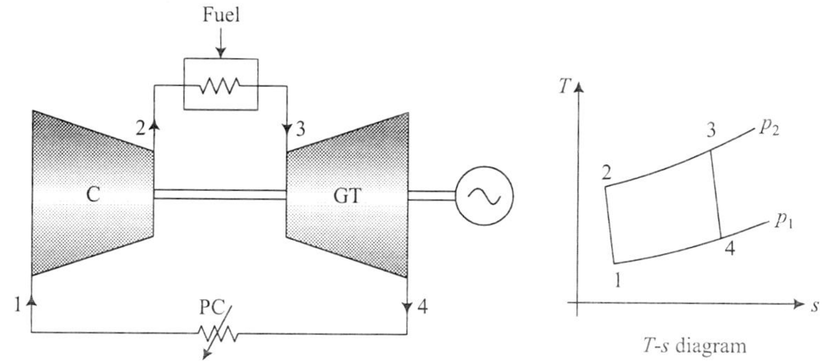

Fig. 10.21 Closed cycle plant

$$T_2 = 300\,(3)^{\frac{1.64-1}{1.64}} = 460.6 \text{ K}$$

$$T_3 = 850°\text{C} + 273 = 1123 \text{ K}$$

$$T_4 = T_3\left(\frac{p_4}{p_3}\right)^{\frac{\gamma-1}{\gamma}} = 1123\left(\frac{20}{60}\right)^{\frac{1.64-1}{1.64}} = 731.5 \text{ K}$$

(a) Temperature of Helium at the end of compression,

$$T_2 = 460.6 \text{ K}$$

Temperature of Helium at the end of expansion,

$$T_4 = 731.45 \text{ K}$$

(b) Heat supplied per kg of gas flow,

$$Q_s = C_p\,(T_3 - T_2) = 5.23\,(1123 - 460.6)$$
$$= 3464.35 \text{ kJ/kg}$$

Heat rejected in the precooler,

$$Q_R = C_p\,(T_4 - T_1) = 5.23\,(731.45 - 300)$$
$$= 225648 \text{ kJ/kg}$$

Work done,

$$W = Q_S - P_R = (3464.35 - 225648)$$
$$= 1207.87 \text{ kJ/kg}$$

(c) Cycle thermal efficiency,

$$\eta_{Th} = \frac{W}{Q_S} = \frac{1207.87}{3464.35} = 0.3486$$
$$= 34.86\%$$

Work ratio,

$$WR = \frac{W_T - W_C}{W_T}$$

$$= \frac{C_p(T_3 - T_4) - C_p(T_2 - T_1)}{C_p(T_3 - T_4)}$$

$$= \frac{(1123 - 731.45) - (460.6 - 300)}{(1123 - 731.45)}$$

$$= \mathbf{0.59} \quad \textbf{Ans.}$$

Example 10.3 A 500 kW gas turbine generating set operates with two compressor stages; the overall pressure ratio is 9 : 1. A high pressure turbine drives the generator. The gas temperature at entry to HP turbine is 650°C and the gases are reheated to 650°C after expansion in the first turbine. The exhaust gases leaving the low pressure turbine are passed through a heat exchanger to heat air leaving the HP compressor. The compressors have equal pressure ratios and intercooling is complete between the stages. The air inlet temperature to the unit is 150°C. The isentropic efficiency of each turbine stage is 0.85 and that of each compressor stage is 0.80. The heat exchanger effectiveness is 0.75. A mechanical efficiency of 98% can be assumed for both the power shaft and the compressor turbine shaft. Neglecting all pressure losses and changes in kinetic energy, calculate the thermal efficiency and work ratio of the plant and mass flow rate in kg/s.

For air, take C_p = 1.005 kJ/kgK, γ= 1.4 and for gases in the combustion chamber and in the turbine and heat exchanger, take C_p = 1.15 kJ/kgK and γ= 1.333. Neglect the mass of the fuel.

Solution The line diagram along with *T-s* diagram are drawn as Fig. 10.22.

Pressure $\quad p_1 = p_{10} = p'_9 = p_9 = 1$ bar

$$p_4 = p_5 = p_6 = 9 \text{ bar}$$

$$p_2 = p_3 = \sqrt{9} = 3 \text{ bar}$$

Temperatures $\quad T_1 = 15°\text{C} + 273 = 288$ K

$$\frac{T_2'}{T_1} = \left(\frac{p_2}{p_1}\right)^{\frac{\gamma-1}{\gamma}}$$

$$\therefore \quad T'_2 = 288\ (3)^{\frac{1.4-1}{1.4}} = 392 \text{ K}$$

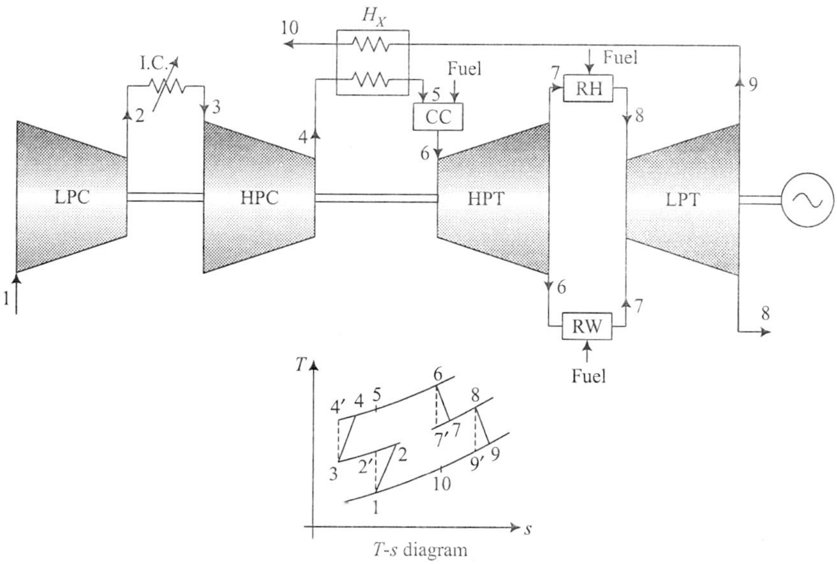

Fig. 10.22 Double shaft gas turbine plant

$$\eta_C = \frac{T_2' - T_1}{T_2 - T_1} = 0.8 = \frac{392 - 288}{T_2 - 288}$$

$\therefore$ $$T_2 = 418 \text{ K}$$

$$T_3 = T_1 = 288 \text{ K}$$

$\therefore$ $$T_4 = T_2 = 418 \text{ K}$$

$$T_8 = T_6 = 650°\text{C} + 273 = 923 \text{ K}$$

For 1 kg/s of air flow rate,

$$\dot{W}_C = [Cp_a (T_2 - T_1) + (T_4 - T_3)]$$

$$= 1.005\ (418 - 288) \times 2 = 261.3 \text{ kW}$$

$$\dot{W}_{HPT} = \frac{\dot{W}_C}{\eta_m} = \frac{261.3}{0.98} = 266.63 \text{ kW}$$

But, $$\dot{W}_{HPT} = Cp_g (T_6 - T_7) \times \eta_m = 266.63$$

$\therefore$ $$(T_6 - T_7) = \frac{266.63}{1.15 \times 0.98}$$

$$T_7 = 923 - \frac{266.63}{1.15 \times 0.98} = 686.4 \text{ K}$$

$$\eta_T = \frac{T_6 - T_7'}{T_6 - T_7}$$

$$T'_7 = T_6 - (T_6 - T_7)\,\eta_T$$

$$= 923 - (923 - 684.4) \times 0.85 = 721.9 \text{ K}$$

$$\frac{T_6}{T_7'} = \left(\frac{p_6}{p_7}\right)^{\frac{\gamma-1}{\gamma}}$$

$$\therefore \quad \frac{p_6}{p_7} = \left(\frac{T_6}{T_7'}\right)^{\frac{\gamma}{\gamma-1}}$$

$$p_7 = \frac{p_6}{\left(\frac{T_6}{T_7'}\right)^{\frac{\gamma}{\gamma-1}}} = \frac{9}{\left(\frac{923}{721.0}\right)^{\frac{1.333}{0.333}}} = 3.365 \text{ bar}$$

$$\frac{T_8}{T_9'} = \left(\frac{p_8}{p_9}\right)^{\frac{\gamma-1}{\gamma}}$$

$$T'_9 = \frac{T_8}{\left(\frac{p_6}{p_9}\right)^{\frac{\gamma-1}{\gamma}}} = \frac{923}{\left(\frac{3.365}{1}\right)^{\frac{0.333}{1.333}}} = 681.64 \text{ K}$$

$$\eta_T = \frac{T_8 - T_9}{T_8 - T_9'}$$

$$T_9 = T_8 - (T_8 - T'_9)\,\eta_T$$

$$= 923 - (923 - 681.64) \times 0.85 = 717.84 \text{ K}$$

$$\dot{m}_g\, Cp_g(T_8 - T_9) \times \eta_m = 5000$$

$$\dot{m}_g = \frac{5000}{1.15(923 - 717.84) \times 0.98} = 21.625 \text{ kg/s}$$

$$\varepsilon_{HX} = \frac{T_5 - T_4}{T_9 - T_4}$$

$$T_5 = \varepsilon_{HX}(T_9 - T_4) + T_4$$

$$T_5 = 0.75\,(717.84 - 418) + 418 = 642.88\text{ K}$$

$$\dot{Q}_{in} = (\dot{m}_g\, Cp_g\, T_6 - \dot{m}_a\, Cp_a\, T_5) + \dot{m}_g\, Cp_g\, (T_8 - T_7)$$

$$= 21.62\,[1.15 \times 923 - 1.0059 \times 642.88] + 1.15\,[923 - 684.4]$$

$$= 14866.106\text{ kW}$$

$$\eta_{TH} = \frac{5000}{14866.106} = 0.3363 \text{ or } \mathbf{33.63\%} \quad \textbf{Ans.}$$

$$\dot{W}_T = \dot{m}_g\, Cp_g\, [(T_6 - T_7) + (T_8 - T_9)]$$

$$= 21.62 \times 1.15\,[(923 - 686.4) + (923 - 717.84)]$$

$$= 10983.5\text{ kW}$$

$$WR = \frac{\dot{W}_N}{\dot{W}_T} = \frac{5000}{10983.5} = \mathbf{0.455} \textbf{ Ans.}$$

10.11 GAS TURBINE CYCLE FOR JET PROPULSION

In a turbojet engine, the gas turbine supplies work of compression only. The exhaust gases from the turbine are expanded in an exhaust cone to a high velocity, the resulting reaction force propelling the aircraft. The turbine and compressor are mounted on a single shaft. It has favourable performance at high altitude and high-speed flights.

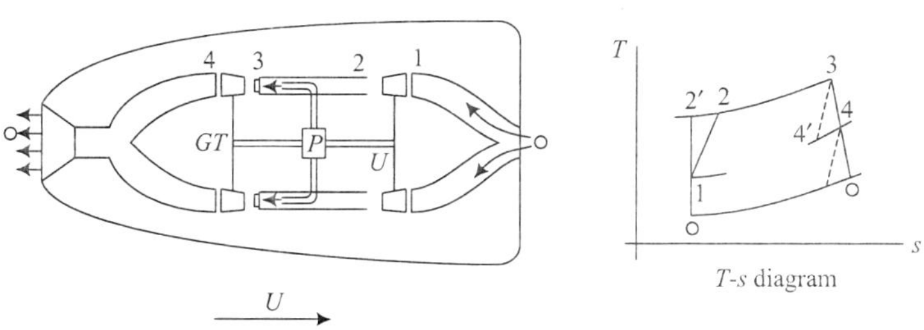

Fig. 10.23 Turbojet engine

The line diagram of turbojet engine along with *T-s* diagram is shown in Fig. 10.23.

0 = Orifice

0-1 = Diffuser or ram jet

1-2 = Rotary air compressor (centrifugal or axial flow)

2-3 = Annular combustion chamber (ring of fuel nozzle 8 to 10. Combustion chambers arranged circumferentially)

3-4 = Gas turbine expansion

4-0 = Expansion in the nozzle

GT = Gas turbine

C = Compressor

P = Oil pump

The turbojet works on Brayton cycle as shown in *T-s* diagram.

The power balance gives

$$W_T = W_C + W_p$$

where,
W_T = Turbine power developed
W_C = Compressor work consumed
W_p = Fuel pump power consumed

$$\eta = 1 - \left(\frac{1}{R}\right)^{\gamma-1}$$

where compression ratio, $R \approx 3.5$

Overall efficiency,

$$\eta_O = \eta_C \times h_T \times h_N$$

where,
η_T = Turbine efficiency (18 – 20%)
η_C = Compressor efficiency (75 – 80%)
η_N = Nozzle efficiency (85 – 90%)

$$\eta_O = \frac{\text{Work done by jet per hour}}{\text{Energy supplied by fuel per hour}}$$

$$= \frac{F_p \times U \times 3600}{\dot{m}_f \times CV}$$

where,
F_p = Propulsion force
U = Velocity of aircraft.

Example 10.4 In a turbojet engine, a compressor of pressure ratio 10 : 1 and isentropic efficiency of 85% is driven by a turbine of isentropic efficiency of 90%. The temperature at entry to the compressor and turbine are 300 K and 1600 K respectively. The specific heat is given as:

For air, $C_p = 0.00015\,T + 0.96$ [kJ/kg-K]

For gas, $C_p = 0.00014\,T + 1.06$ [kJ/kg-K]

Assuming that the isentropic law $pv^\gamma = C$ can be used with mean value of $\gamma = 1.39$ for air and $\gamma = 1.26$ for the gas, determine the ratio of turbine exit pressure to the compressor inlet pressure.

Solution

Pressure $p_1 = 1$ bar

$p_2 = p_3 = 10$ bar

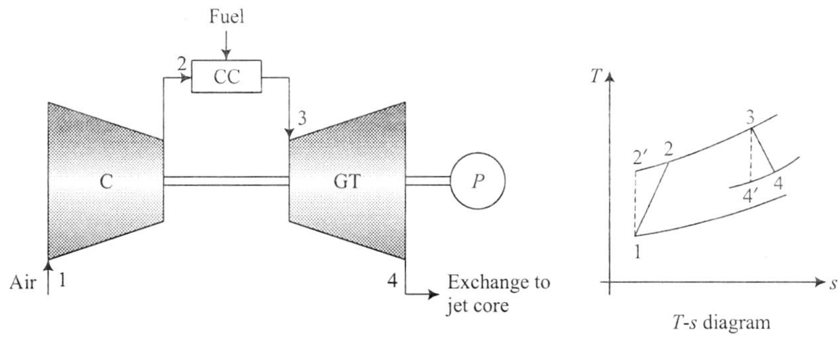

Fig. 10.24 Turbojet cycle

$$p_4 = ?$$

Temperatures $\quad T_1 = 300 \text{ K}$

$$\frac{T_2'}{T_1} = \left(\frac{p_2}{p_1}\right)^{\frac{\gamma-1}{\gamma}}$$

$$T'_2 = 300\ (10)^{\frac{1.39-1}{1.39}} = 572 \text{ K}$$

$$\eta_C = \frac{T_2' - T_1}{T_2 - T_1}$$

$$T_2 = T_1 + \frac{T_2' - T_1}{\eta_C} = 300 + \frac{572 - 300}{0.85}$$

$$= 620 \text{ K}$$

$$T_3 = 1600 \text{ K}$$

$$T_4 = ?$$

Neglecting oil pump work,

$$\dot{W}_T = \dot{W}_C = \dot{m}_a Cp_a(T_2 - T_1)$$

Assume $\quad \dot{m}_a = 1 \text{ kg/s}$

$$W_C = \int_{T_1}^{T_2} (0.00015\ T + 0.96)\ dT$$

$$\dot{W}_C = \left|0.000075T^2 + 0.96T\right|_{300}^{620} = 329.28 \text{ kJ/kg}$$

$$\dot{W}_T = \int_{T_4}^{1600} (0.00014 \text{ T} + 1.06) \text{ dT}$$

$$\dot{W}_T = \left|0.00007T^2 + 1.06T\right|_{T_4}^{1600}$$

$$0.0000T_4^2 + 1.064\ T_4 - 1545.92 = 0$$

$$\therefore \quad T_4 = 1340\ \text{K}$$

$$\eta_T = \frac{T_3 - T_4}{T_3 - T_4'}$$

$$T'_4 = 1600 - \frac{1600 - 1340}{0.9} = 1311\ \text{K}$$

$$\frac{p_3}{p_4} = \left(\frac{T_3}{T_4}\right)^{\frac{\gamma}{\gamma-1}} = \left(\frac{1600}{1300}\right)^{\frac{1.26}{0.26}} = 2.626$$

$$p_4 = \frac{10}{2.626} = 3.81\ \text{bar}$$

$$\therefore \quad \frac{p_4}{p_1} = 3.81 \quad \textbf{Ans.}$$

QUESTION BANK NO. 10

1. Show that the efficiency of an ideal Brayton cycle depends only on pressure ratio.
2. Discuss the main fields of application of gas turbines.
3. Classify gas turbine cycles.
4. How is regeneration useful in gas turbines?
5. What are the different methods for improving the efficiency of a gas turbine plant?
6. Discuss the effect of various parameters on the performance of Brayton cycle.
7. Describe a gas turbine cycle for jet propulsion.
8. Discuss the various deviations of practical gas turbine cycle from ideal cycle.

CHAPTER 11

Steam Power Cycles

11.1 INTRODUCTION

Steam power plants form the largest contribution for electricity generation. Water is heated in a boiler with the help of a fuel or steam generator in a nuclear reactor to form superheated steam. This high pressure and high temperature steam rotates a steam turbine to produce mechanical work and then electrical power. The steam power plants can operate as:

1. Thermal power plants where chemical energy of a fossil fuel (coal, oil, gas) is released into heat by combustion to raise steam.
2. Nuclear power plants where nuclear reactions of fission or fusion release heat to generate steam.
3. Geothermal power plants where springs of hot water or steam from the interior earth can be directly used to turn a turbine.
4. Solar power plants where solar rays are focused and collected to raise steam.

In ocean thermal power plants, the hot water at the water surface of the sea can be used to run a power plant.

Coal is most abundantly available in India and other countries and is the main fuel for steam power plants.

11.2 STEAM POWER PLANT

Steam power plants are used to produce bulk electrical power in a thermal power plant and a nuclear power plant.

1. The heat energy of combustion of a fuel (coal, oil or gas) or heat of fusion of nuclear fuels (uranium, plutonium or thorium) is converted into mechanical work which in turn is used to generate electricity.
2. The plant works on thermodynamic cycle.
3. The working fluid is water/steam.
4. 1st law of thermodynamics can be applied to the plant.

The total energy flowing into the fluid during the cycle must be equal to that flowing out of the cycle.

$$\dot{Q}_{in} + \dot{W}_{in} = \dot{Q}_{out} + \dot{W}_{out}$$

where, $\dot{Q}$ = rate of heat transfer

$\dot{W}$ = rate of work transfer, i.e., power.

5. The boiler or nuclear reactor works as the heat source and condenser works on heat sink.
6. The performance of the plant is measured by thermal efficiency.

$$\eta_{th} = \frac{\dot{W}_{out} - \dot{W}_{in}}{\dot{Q}_{in}} = \frac{\dot{Q}_{in} - \dot{Q}_{out}}{\dot{Q}_{in}}$$

$$= 1 - \frac{\dot{Q}_{out}}{\dot{Q}_{in}}$$

The components of a steam power plant

The line diagram of a steam power plant is shown in Fig. 11.1.

There are four main components of a steam power plant:

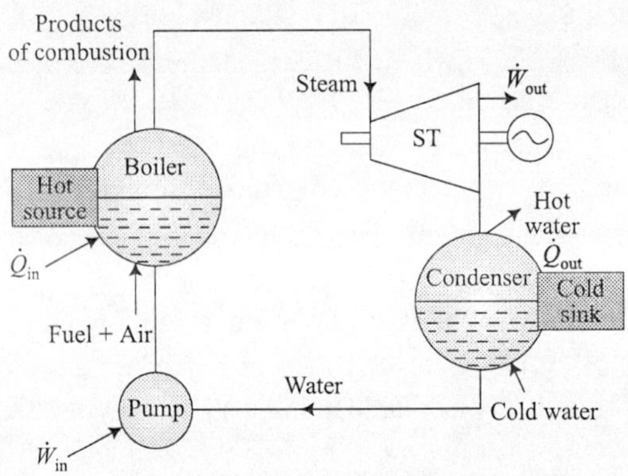

Fig. 11.1 Steam power plant

1. Steam boiler or generator: It is the hot source reservoir. The fuel (coal, oil or gas) is burnt in the presence of air. The heat energy of combustion gases raise steam by heating feed water.

$$\dot{Q}_{in} = \dot{m}_f CV \text{ [kJ/s]}$$

$$= \dot{m}\,(h_{out} - h_{in})$$

where, $\dot{m}_f$ = mass rate of fuel burnt, (kg/s)

Cv = calorific value of fuel, (kJ/kg)

$\dot{m}$ = steam generation rate, (kg/s)

h_{out} and h_{in} are enthalpies of working fluid at outlet and inlet of boiler (kJ/kg).

2. Steam turbine: The steam generated in the boiler is expanded in a steam turbine adiabatically to a low pressure to produce mechanical work $\dot{W}_{out}$ which is further converted into electrical energy by the generator.

$$\dot{W}_{out} = \dot{m}\,(h_1 - h_2),\ [\text{kW}]$$

where, $\dot{m}$ = flow rate of steam [kg/s]

h_1 = specific enthalpy of steam at turbine inlet (kJ/kg)

h_2 = specific enthalpy of steam at turbine outlet (kJ/kg)

3. Condenser: Condenser acts as cold sink into which heat is rejected. It is a tubular heat exchanger in which exhaust and low pressure steam from the turbine is cooled with the help of cooling water and condensed into water called condensate.

$$\dot{Q}_{out} = \dot{m}_w\, C_w\, (t_1 - t_2)\ (\text{kJ/s})$$

where, $\dot{m}_w$ = flow rate of cooling water, (kg/s)

C_w = specific heat of water.

= 4.186 kJ-kg-K

t_1 = inlet temperature of cooling water (°C)

t_2 = outlet temperature of cooling water (°C)

Also, $$s\,\dot{Q}_{out} = \dot{m}\,(h_{in} - h_{out})$$

$\dot{m}$ = flow rate of steam into the condenser, (kg/s)

h_{in} = enthalpy of incoming steam (kJ/kg)

h_{out} = enthalpy of outgoing condensate (kj/kg)

4. Pump: Condensate extraction pump or boiler feed pump or both return the condensate back to the boiler. This pump draws its power from the grid.

$$\dot{W}_{in} = \dot{m}\,(h_{out} - h_{in})$$

where, $\dot{m}$ = flow rate of feed water, (kg/s)

h_{out} = enthalpy of water at pump outlet (kJ/kg)

h_{in} = enthalpy of feed water at pump inlet (kJ/kg)

$\dot{W}_{in}$ = work done by the pump

$$= \dot{m} v_f (p_{out} - p_{in}) \times 10^2,\ (\text{kW})$$

where, v_f = specific volume of feed water which is at saturated condition at pump inlet

p_{out} = pump exit pressure, (bar)
= boiler pressure

p_{in} = pump inlet pressure, (bar)
= condenser pressure.

Note: Requirement of utilities

1. Fuel required in boiler

$$\dot{m}_f = \frac{\dot{Q}_{in}}{CV} = \frac{\dot{m}(h_{out} - h_{in})}{CV}$$

2. Cooling water required in condenser $\dot{m}_w = \dfrac{\dot{Q}_{out}}{Cw(t_1 - t_2)} = \dfrac{\dot{m}(h_{in} - h_{out})}{Cw(t_1 - t_2)}$

11.3 THE CARNOT VAPOUR CYCLE

It is more convenient to analyse the performance of steam power plant by means of idealised cycles which are theoretical approximations of the real cycles. The Carnot cycle is an ideal cycle, but non-practical cycle, giving the maximum possible thermal efficiency for a cycle operating on selected maximum and minimum temperature range. The line diagram of Carnot cycle along with its presentation on *T-s* diagram are shown in Fig. 11.2.

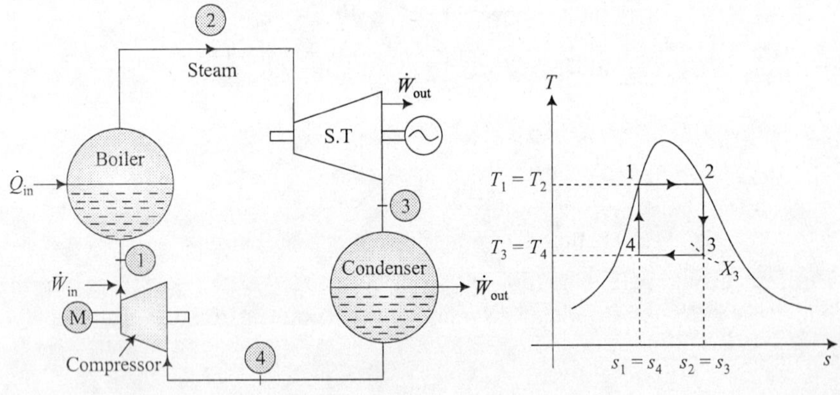

Fig. 11.2 The Carnot vapour cycle

The Carnot cycle is made up of four ideal processes.

1. Isothermal Process 1-2

Saturated water at state 1 (T_1) is fed into the boiler and evaporated into saturated steam at state 2 (T_2).

$$T_1 = T_2$$

The heat supplied in the boiler

$$\dot{Q}_{in} = \dot{m}\ T_1\ (s_2 - s_1),\ (kJ)$$

where, $\dot{m}$ = Flow rate of working fluid, (kg/s)

T_1 = Temperature of evaporation, (K)

s_1 = Entropy of saturated water at boiler inlet, (kJ/kg-K)

s_2 = Entropy of saturated steam at boiler outlet, (kJ/kg-K)

2. Isentropic Process 2-3

The saturated steam from state $2(T_2, s_2)$ is expanded isentropically in the turbine upto state 3 (T_3, s_3, X_3)

$$s_2 = s_3$$

$$\dot{W}_{out} = \dot{m}\ (h_2 - h_3)$$

where, $\dot{m}$ = Flow rate of steam through turbine, (kg/s)

h_2 = Enthalpy of saturated steam at turbine inlet, (kJ/kg).

h_3 = Enthalpy of wet steam at turbine outlet, (kJ/kg).

3. Isothermal Process 3-4

The wet steam from turbine exhaust is partially condensed to state 4 (T_4, s_4, X_4) losing heat $\dot{Q}_{out}$ to cooling water

$$T_3 = T_4$$

$$\dot{Q}_{out} = \dot{m}\ T_3\ (s_3 - s_4)\ (kJ)$$

where, $\dot{m}$ = Flow rate of working fluid (kg/s) (wet steam)

T_3 = Temperature of condensation, (K)

s_3 = Entropy of wet steam at condenser inlet [kJ/kg-K]

s_4 = Entropy of wet steam at condenser outlet [kJ/kg-K]

4. Isentropic Process 4-1

The wet steam at state 4 enters the compressor where it is compressed to saturated water 1 $(T_1, s_1,)$ to complete the cycle. The compressor draws power for its rotation.

$$s_1 = s_4$$

$$W_{in} = \dot{m}\ (h_1 - h_4)$$

where, $\dot{m}$ = Flow rate of working fluid through the compressor (kg/s)

h_1 = Enthalpy of working fluid at compressor outlet (kJ/s)

h_4 = Enthalpy of wet steam at compressor inlet (kJ/s)

11.3.1 Cycle Analysis

1. Efficiency: The thermodynamic efficiency or thermal efficiency is the ratio of net work output of steam power plant and heat supplied.

$$\eta_{th} = \frac{\dot{W}_{out} - \dot{W}_{in}}{\dot{Q}_{in}}$$

But, $\dot{W}_{out} - \dot{W}_{in} = \dot{Q}_{in} - \dot{Q}_{out}$ [First law of thermodynamics]

$$\therefore \quad \eta_{th} = \frac{\dot{Q}_{in} - \dot{Q}_{out}}{\dot{Q}_{in}} = 1 - \frac{\dot{Q}_{out}}{\dot{Q}_{in}}$$

$$\dot{Q}_{in} = \dot{m}\, T_1\, (s_2 - s_1)$$

$$\dot{Q}_{out} = \dot{m}\, T_3\, (s_3 = s_4)$$

But, $(s_2 - s_1) = (s_3 - s_4)$

$$\therefore \quad \eta_{th} = 1 - \frac{\dot{m} T_3 (s_2 - s_1)}{\dot{m} T_1 (s_2 - s_1)}$$

$$= 1 - \frac{T_3}{T_1}$$

2. Specific Steam Consumption (S.S.C): Specific steam consumption or steam rate is the amount of steam required to produce 1 kWh of work.

$$s.s.c = \frac{\dot{m}}{\dot{W}} \frac{[\text{kg/s}]}{[\text{kW}]} = \frac{\dot{m}}{\dot{W}} \frac{kg}{kWs} = \frac{3600\,\dot{m}}{\dot{W}} \frac{\text{kg}}{\text{kWh}}$$

$$= \frac{3600}{\dot{W}/\dot{m}} \left[\frac{\text{kg}}{\text{kWh}}\right]$$

3. Work Ratio (W.R.): Work ratio is the ratio of net plant output to gross (turbine) output.

$$W.R. = \frac{\dot{W}_{out} - \dot{W}_{in}}{\dot{W}_{out}} = \frac{\dot{W}_T - \dot{W}_C}{\dot{W}_T}$$

11.3.2 Limitations of Carnot Cycle

The Carnot cycle is not practical due to the following limitations:

1. Carnot cycle is an ideal cycle consisting of four reversible processes. It is not possible to achieve reversible process in practice.

2. Carnot limitation: The Carnot efficiency depends upon boiler temperature T_1 and condenser temperature T_3.

The *condenser temperature* cannot be reduced below 300 due to the following reasons.

(i) Condensation of steam requires a continuous bulk supply of cooling water. Natural water supply below atmospheric temperature of about 15°C is unavailable.

(ii) If condenser is to be of a reasonable cost and size, the temperature difference between the condensing steam and cooling water must be at least 10°C.

The *boiler temperature* cannot be more than 900 K due to non-availability of materials of construction of boiler tubes and turbine blades to withstand higher temperatures.

The material of construction should be of high reliability when operating at high temperature and the cost should also be reasonable. This is called *metallurgical limitation.*

In fact, the steam in Carnot cycle has a maximum cycle temperature of well below the metallurgical limitation owing to the properties of steam as it is limited by *critical temperature* of 374°C (647 K) otherwise special supercritical steam boilers will be required which suffer from low plant life.

Under best circumstances, the Carnot efficiency will be:

$$\eta_{\text{Carnot}} = 1 - \frac{T_3}{T_1} = 1 - \frac{300}{900} = 66.7°\text{C}$$

Therefore, maximum cycle efficiency would be well below 100%.

3. Compression of wet steam: Compressing a very wet steam mixture would require a compressor of size and cost comparable with the turbine. Such a compressor would absorb large amount of work produced by turbine and work ratio of power plant would be very low. Such a compressor would also have a short life because of blade erosion and cavitation problems.

4. High steam rate: Due to low steam parameter used in a Carnot cycle as discussed above, steam rate will be very high to produce certain amount of power. The plant would be very large in size and very expensive.

5. Cost of power generation: Cost of power generation has two components:

(i) *Fixed charges consisting interest* and depreciation of plant cost.

(ii) *Running cost* of fuel.

The plant operating on Carnot cycle will produce power at high cost as fixed charges are high due to large size and cost of plant and running cost is high due to Carnot limitations of thermal efficiency.

6. Low plant life: A Carnot cycle works on saturated steam and the wetness of steam increases during its expansion in the turbine. The turbine blades get eroded and its efficiency decreases. It is not possible to supperheat steam at constant temperature process of Carnot cycle.

11.3.3 Uses of Carnot Cycle

The Carnot cycle is not practical due to its limitations discussed above. Still it has certain uses.

1. Design of practical cycle: The Carnot cycle is useful in helping us appreciate the desirable factors for design of practical plant, i.e.,

(i) Maximum possible temperature range of cycle.

(ii) Maximum possible heat addition into the cycle at this maximum cycle temperature.

(iii) A maximum possible temperature of heat rejection.

(iv) A maximum possible work input into the cycle.

2. Thermodynamic constraints: The Carnot cycle also helps to understand the thermodynamic constraints on the design of cycles.

(i) Carnot limitation.

The cycle efficiency will be low due to Carnot limitation.

(ii) Use of superheated steam helps not only to improve cycle efficiency but also helps to control the erosion of turbine blades and hence improve the life of plant.

11.4 THE IDEAL RANKINE CYCLE

In order to overcome most of the limitations of Carnot cycle, the Rankine cycle is used as the basic cycle for steam power plants. An idealized cycle for a simple power plant is shown in Fig. 11.3. The Rankine cycle with superheated steam is 1-2-3-4-1 and that with dry and saturated steam is 1-2′-3′-4-1.

It is made up of four practical processes:

1. Isobaric heating process 1-2: Heat is added to feedwater in three steps:

(i) The water is heated upto its saturated value (process 1-A) in the economizer at constant pressure.

(ii) The saturated water is evaporated at constant temperature and pressure (process A-2) in the boiler.

(iii) The saturated steam is superheated at constant pressure (process 2′-2) in the superheater.

Applying 1st law of thermodynamics to boiler,

$$\dot{Q}_{in} = \dot{m}\,(h_2 - h_1)$$

where, $\dot{m}$ = mass flow rate of working fluid, (kg/s)

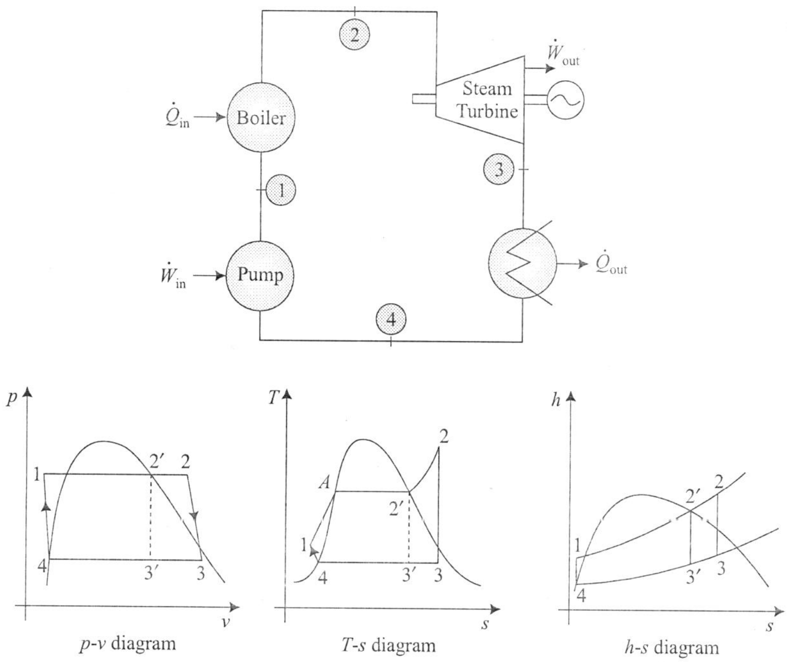

Fig. 11.3 Ideal Rankine cycle

2. Isentropic expansion process 2-3: The high pressure superheated steam is expanded to a low pressure in a steam turbine isentropically in an ideal Rankine cycle. However, in actual plant, friction of steam flow results in the expansion with increasing entropy.

Applying first law to this process

$$\dot{Q}_{out} = \dot{W}_T = \dot{m}\,(h_2 - h_3)$$

where, $\dot{W}_T$ = Turbine work [kW]

3. Isobaric cooling process 3-4: The low pressure wet steam from the steam turbine is *completely* condensed at constant condenser pressure back into saturated water. The latent heat of condensation is thereby rejected to the condenser cooling water, which, in turn, rejects this heat to the atmosphere.

Applying 1st law of thermodynamics to this process,

$$\dot{Q}_{out} = \dot{m}(h_3 \quad h_4)$$

4. Isentropic compression process 4-1: The low pressure saturated water is pumped back to the boiler pressure and, in doing so, it becomes sub-saturated. The water then enters the boiler and a new cycle begins.

Applying 1st law to this process,

$$\dot{W}_{in} = \dot{m}(h_1 - h_4)$$

However, $\dot{W}_{in}$ can be neglected with reasonable accuracy and it can be assumed that $h_1 = h_4$. The thermal efficiency of the cycle is given by:

$$\eta_{th} = \frac{\dot{W}_{out} - \dot{W}_{in}}{\dot{Q}_{in}} = \frac{\dot{m}(h_2 - h_3) - \dot{m}(h_1 - h_4)}{\dot{m}(h_2 - h_1)}$$

$$= \frac{(h_2 - h_3) - (h_1 - h_4)}{(h_2 - h_1)}$$

If $\dot{W}_{in}$ can be neglected,

$$\eta_{th} = \frac{\dot{W}_{out}}{\dot{Q}_{in}} = \frac{(h_2 - h_3)}{(h_2 - h_1)}$$

Specific steam consumption or steam rate is given by:

$$\text{s.s.c.} = \frac{3600}{\dot{W}/\dot{m}}\left[\frac{\text{kg}}{\text{kwh}}\right] = \frac{3600}{h_2 - h_3} \text{ kg/kWh.}$$

Work ratio, $$WR = \frac{\dot{W}_{out} - \dot{W}_{in}}{\dot{W}_{out}}$$

$$= \frac{(h_2 - h_3) - (h_4 - h_1)}{h_2 - h_3}$$

11.5 COMPARISON OF A CARNOT CYCLE AND A RANKINE CYCLE

The advantages of a Rankine cycle over a Carnot cycle are listed in Table 11.1. The Table also explains how the limitations of Carnot cycle are overcome in a Rankine cycle. (Points 4, 5, 9, 10 of Table 11.1).

Table 11.1 Comparison of Carnot and Rankine cycle

Carnot cycle	*Rankine cycle*
1. It consists of two isothermals and two isentropic processes.	1. It consists of two isobaric and two isentropic processes.
2. It is an ideal cycle and Carnot be achieved in practice.	2. All steam power plants are desigred on Rankine cycle.
3. The steam at turbine input is saturated steam.	3. The steam at turbine input is superheated steam in modern power plants.
4. Steam is partially condensed in the condenser.	4. Steam is fully condensed in the condenser.

Contd.

Carnot cycle	Rankine cycle
5. A compressor is used to compress two-phase wet steam. The size and power consumption are very large.	5. A pump is used to raise the pressure of water to boiler pressure. The size and power consumption are very small.
6. The thermal efficiency of a Carnot cycle is higher than an equivalent Rankine cycle working on saturated steam.	6. The thermal efficiency of a Rankine cycle can be improved by the use of high pressure superheated steam.
7. Superheated steam cannot be used due to difficulty of superheating at constant temperature simultaneously expanding the steam.	7. Superheated steam is used in Rankine cycle as superheating can be easily carried out at constant pressure.
8. Specific steam consumption is high.	8. Specific steam consumption is low.
9. Work ratio is very low as lot of power is required in the compression of wet steam.	9. Work ratio is very high as pump power consumption is very low.
10. Plant life is low due to operation of steam turbine and compressor with two-phase mixture of steam and water.	10. Plant life is high as steam turbine works mainly with superheated steam, or dryness fraction is very high. Pump operation is simple and reliable.

Example 11.1 A steam power plant working on Rankine cycle has steam parameters at turbine inlet as 100 bar and 550°C and condenser pressure of 0.05 bar. Determine the cycle efficiency, specific steam consumption and work ratio if all processes are reversible.

Solution

(i) Steam turbine

From superheated steam tables, at $p_1 = 100$ bar and $t_1 = 550$°C,

$$h_1 = 3500.9 \text{ kJ/kg}$$

$$s_1 = 6.756 \text{ kJ/kg}$$

Process 1-2 is reversible and isentropic

$$\therefore \quad s_1 = s_2$$

At turbine exit, from saturated steam tables,

$$p_2 = 0.05 \text{ bar}$$

$$s_{f2} = 0.476 \text{ kJ/kg}$$

$$s_{fg2} = 7.920 \text{ kJ/kg}$$

$$h_{f2} = 137.80 \text{ kJ/kg}$$

$$h_{fg2} = 2423.7 \text{ kJ/kg}$$

$$s_1 = s_{f2} + x_2 s_{fg2}$$

$$\therefore \quad X_2 = \frac{s_1 - s_{f2}}{s_{fg2}} = \frac{6.756 - 0.476}{7.920} = 0.80.$$

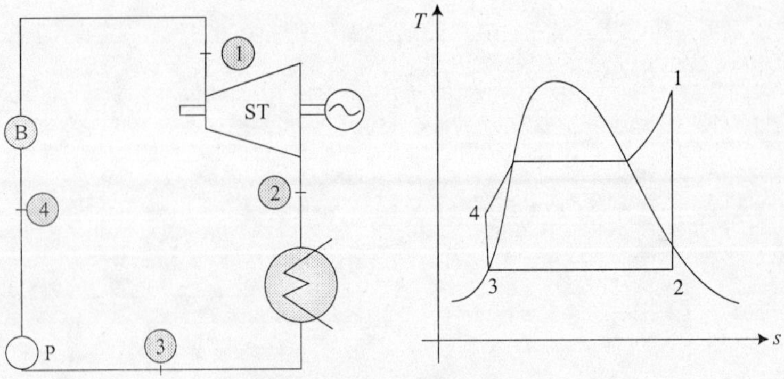

$\therefore \quad h_2 = h_{f2} + X_2 \, h_{fg2} = 137.80 + 0.8\,(2423.7)$

$= 2076.76$ kJ/kg.

$\therefore$ Turbine work,

$$W_T = h_1 - h_2 = 3500.9 - 2076.76 = 1424.14 \text{ kJ/kg}$$

(ii) Condenser

At inlet, $h_2 = 2076.76$ kJ/kg

At outlet $h_3 = h_{f_3} = 137.8$ kJ/bg

Heat removed $= h_2 - h_3 = 2076.76 - 137.8 = 1938.96$ kJ/kg

(iii) Pump

At pump inlet, $h_3 = 137.80$ kJ/kg

At pump outlet, $h_4 = h_3 + W_p$

Assuming water as incompressible,

Pump work, $W_p = v_{f3}\,(p_4 - p_3) \times 10^2$

$W_p = 0.001005\,(100 - 0.05) \times 10^2$

$= 10$ kJ/kg

$\therefore \quad h_4 = 137.80 + 10 = 147.8$ kJ/kg

(iv) Boiler

At inlet of boiler, $h_4 = 147.8$ kJ/kg

At outlet of boiler, $h_1 = 3500.9$ kJ/kg

Heat supplied, $Q_{1-4} = h_1 - h_4 = 3500.9 - 147.8$

$= 3357.1$ kJ/kg

(v) Steam power plant

Cycle efficiency,

$$\eta = \frac{\text{Net work done}}{\text{Heat supplied}}$$

$$= \frac{W_T - W_P}{Q_{1-4}} = \frac{1424.14 - 10}{3353.1} = 0.43$$

$= \mathbf{43\%}$ **Ans.**

Specific steam consumptions (s.s.c),

$$\text{s.s.c.} = \frac{3600}{W_T - W_p} = \frac{3600}{1414.14} = 2.55 \text{ kg/kWh}$$

Work Ratio (WR),

$$WR = \frac{W_T - W_p}{W_T} = \frac{1414.14}{1424.14} = \mathbf{0.99} \textbf{ Ans.}$$

11.6 THE NON-IDEAL RANKINE CYCLE

There are certain deviations from the assumptions made in the analysis of ideal Rankine cycle.

11.6.1 Thermal Irreversibility in Steam Generation

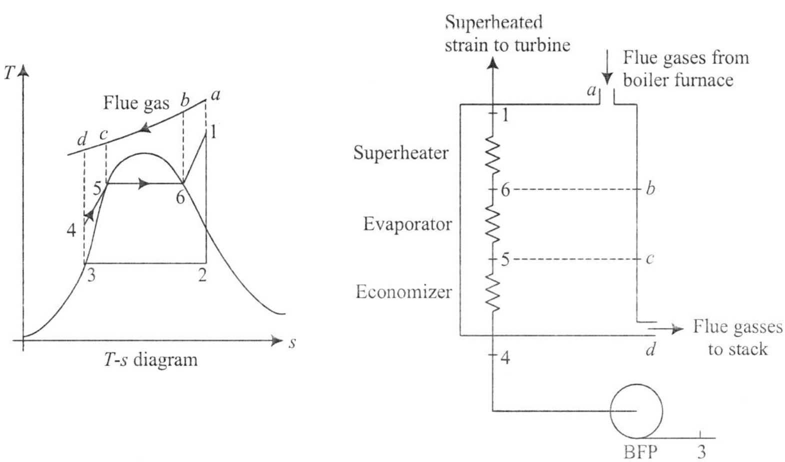

Fig. 11.4 Heat balance of a steam generator

Heat is added to feed water in three stages.

1. *Economizer:* The water is heated upto saturation temperature (process 4-5) in the economizer at constant pressure. For unit mass flow rate,

$$Q_{ECO} = h_5 - h_4$$

2. *Evaporator:* The saturated water vapour is evaporated to dry saturated steam (process 5-6) in the boiler at constant pressure and temperature

$$Q_{EVA} = h_6 - h_5 = h_{fg}$$

3. *Superheater:* The saturated steam is superheated at constant pressure (process 6-1) in the superheater.

$$Q_{SH} = (h_1 - h_6)$$

(a) *Mean temperature of heat supply:* The total heat supplied in the steam generator,

$$\begin{aligned} Q_{in} &= Q_{ECO} + Q_{EVA} + Q_{SH} \\ &= (h_5 - h_4) + (h_6 - h_5) + (h_1 - h_6) \\ &= (h_1 - h_4) \\ &= T_m (s_1 - s_4) \end{aligned}$$

where T_m = mean temperature of heat addition

$$T_m = \frac{(h_1 - h_4)}{(s_1 - s_4)}$$

The Rankine cycle efficiency,

$$\eta_{Rankine} = 1 - \frac{Q_{out}}{Q_{in}} = 1 - \frac{T_2(s_1 - s_4)}{T_m(s_1 - s_4)}$$

$$= 1 - \frac{T_2}{T_m}$$

where, T_2 = Temperature of heat rejection.

T_2 depends upon condenser pressure but cannot be less than the temperature of surrounding T_∞.

$\therefore$ $\eta_{Rankine} = f(T_m)$ only.

The higher the mean temperature of heat addition, the higher will be the cycle efficiency.

(b) *Pinch points:* The working fluid is heated from point 4 to 1 by flue gases getting cooled from *a* to *d* in counterflow heat exchangers. The minimum temperature differences are *c*-5 and 1-*a* between the two fluids. These points are called *pinch points*. A small pinch point causes increase in surface area and expensive steam generator whereas large pinch point results in lower plant efficiency due to large thermal irreversibility. The most economical pinch-point temperature difference is obtained by optimization to ensure minimum cost of steam generator and minimum operating costs.

11.6.2 Thermal Irreversibility in Condenser

Line 2-3 represents condensation process of steam at constant pressure and temperature.

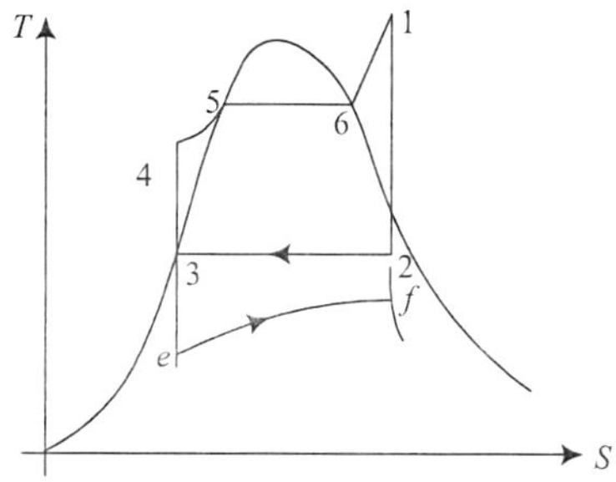

Fig. 11.5 Condenser irreversibility

Line *e*-*f* represents the cooling water temperature rise in the condenser. The pinch point *f*-2 controls the condenser pressure which should be as low as possible and also the size and cost of condenser and pumping cost will increase with the decrease of *f*-2.

11.6.3 Friction Losses in Steam Turbine and Feed Pump

For ideal Rankine cycle, 1-2′ is the reversible adiabatic and isentropic expansion process in steam turbine and 3-4′ is the isentropic compression in feed pump. The flow rates in turbine and pump are large and the process may be considered adiabatic as heat losses can be neglected. But due to fluid friction, the entropy of fluid will increase in both cases. This is called internal irreversibility and can be expressed by isentropic efficiency.

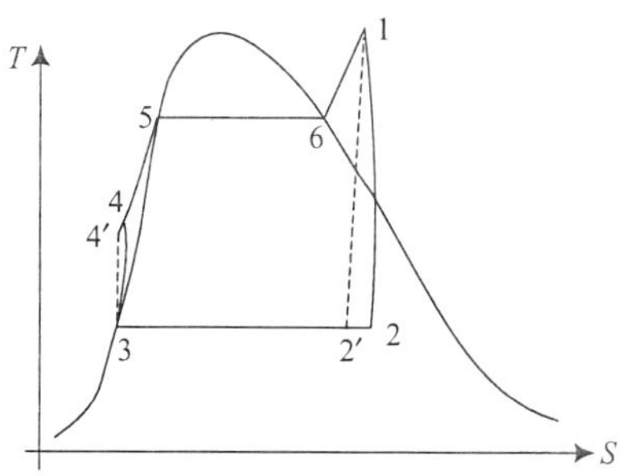

Fig. 11.6 Turbine and pump friction

The isentropic efficiency of steam turbine,

$$\eta_T = \frac{h_1 \quad h_2}{h_1 - h_2'}$$

The isentropic efficiency of boiler feed pump,

$$\eta_p = \frac{h_4' - h_3}{h_4 - h_3}$$

The actual pump work,

$$W_p = \frac{h'_4 - h_3}{\eta_p} = \frac{v_3(p_4 - p_3) \times 10^2}{\eta_p}$$

The turbine produces less work and pump consumes more work due to irreversibility.

11.6.4 Pressure Losses

In an ideal Rankine cycle, p_1 is the pressure at turbine inlet, p_5, the pressure at boiler outlet and p_4, the pressure at pump exit.

$$\therefore \quad p_1 = p_5 = p_4$$

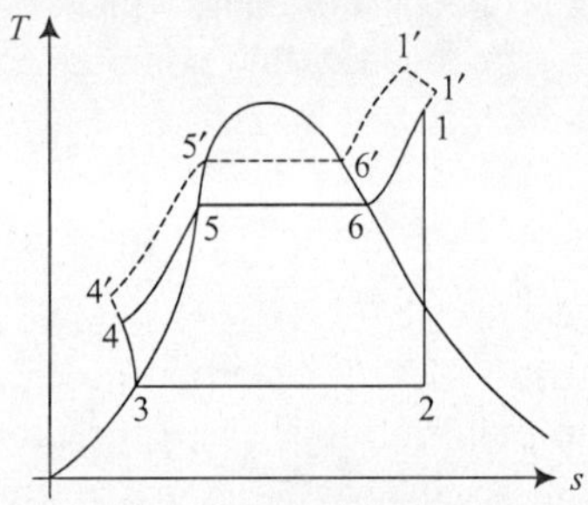

Fig. 11.7 Pressure losses

The liquid leaving the pump must be at a higher pressure (p'_4) than pressure (p_1) at turbine inlet because of pressure drops due to friction in the boiler heat exchangers, feedwater heaters, pipes, bends, valves, etc. In actual cycle $p'_4 > p_5 > p_1$ due to friction in the pipeline and entropy decreases from 1" to 1 due to heat loss.

11.7 IMPROVEMENT OF CYCLE EFFICIENCY

The Rankine cycle efficiency increases with the increase of mean temperature of heat supply in the cycle. This is achieved by the following methods:

1. Use of high steam parameters, i.e., high pressure and temperature of superheated steam.
2. Reheating of partially expanded steam.
3. Regenerative heating of boiler feed water.

11.7.1 High Steam Parameters

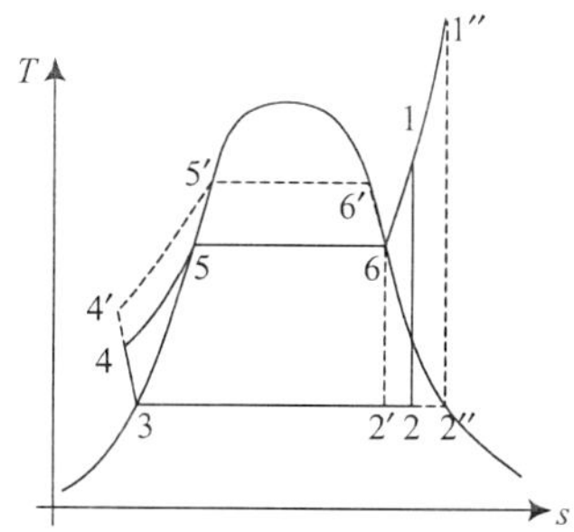

Fig. 11.8 High steam parameters

A higher pressure in the boiler ($p'_5 > p_5$) raises the temperature at which latent heat (h_{fg}) is added which increases mean temperature of heat supply and hence cycle efficiency. However, the moisture content at turbine exhaust ($X'_2 > X''_2$) increases which is limited to 12% to avoid blade erosion and other losses.

If the temperature of superheated steam is increased keeping the pressure constant, it has double advantage.

1. Improvement of cycle efficiency due to increase of mean temperature of heat supply.

$$\left[\frac{T_1'' + T_4}{2}\right] > \left[\frac{T_1 + T_4}{2}\right]$$

2. Final witness in steam is less ($X''_2 > X_2$) which avoids blade erosion.

   The turbine inlet temperature is limited due to metallurgical considerations. The superheater, piping, valves, steam chest and inlet turbine blades are made from ferrite steels for which temperature limit is 550°C. Austenitic steels have to be used above this temperature which are high alloy steels and four times costlier. For a temperature of 550°C and 12% exhaust wetness, the initial steam pressure would be about 150 bar.

The high steam parameters of the order of 150 bar and 550°C result in the following advantages if used in *large capacity* steam turbines.

1. Extra cost of high temperature components is offset by a general saving in the number of components per MW.
2. Losses are proportionally smaller.
3. High flow rates ensure large and efficient high pressure blading.

11.7.2 Reheating

On large turbines, it becomes economical to increase the cycle efficiency by using reheat, which is a way of partially overcoming the temperature limitations. By returning partially expanded steam from high pressure cylinder to the reheater in the boiler, the mean temperature of heat supply is increased and exhaust wetness is considerably reduced. However, reheating leads to complexity of the plant and higher initial cost. Therefore, reheating is adopted for large turbines above 100 MW.

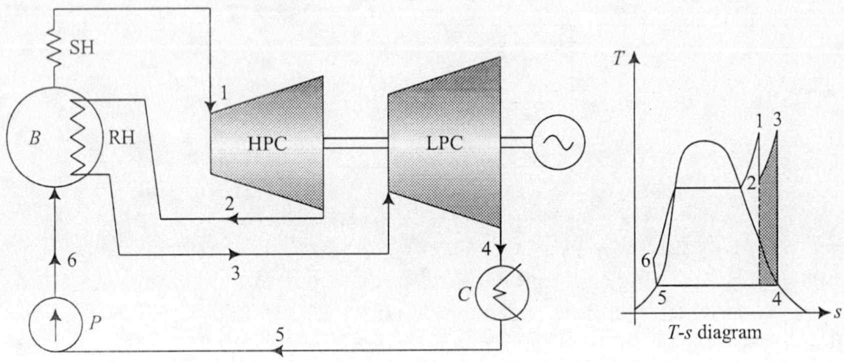

Fig. 11.9 Reheat cycle

11.7.3 Regenerative Heating

Regenerative heating of boiler feedwater is widely used in modern steam power plants which enhances the mean temperature of heat supply in the boiler. The cycle efficiency increases. The amount of heat supplied to the cycle in low temperature region (economizer) by flue gases is reduced. This heating is achieved by steam extracted from turbine.

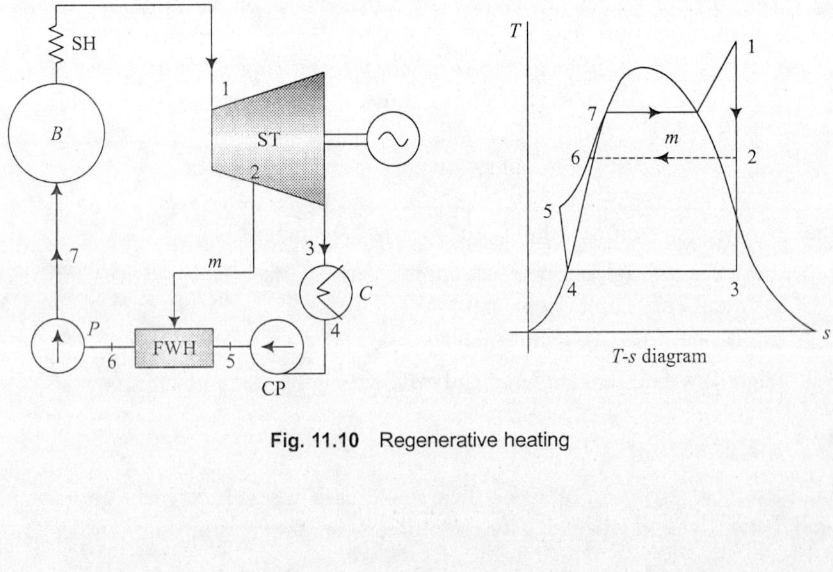

Fig. 11.10 Regenerative heating

11.7.4 Exhaust Pressure

A lower exhaust pressure lowers the temperature at which heat is rejected, which increases cycle efficiency. For condensing turbines, the vacuum obtainable is a function of cooling water temperature at plant site. Any improvement in vacuum is very effective in increasing the work done since a narrow but large addition is made to *T-s* diagram area.

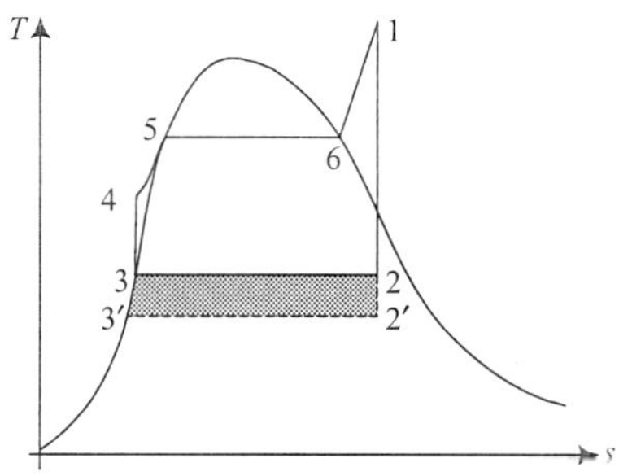

Fig. 11.11 Effect of reduced exhaust pressure

The additional work done = Area 2′-3′-3-2-2′.

11.8 REHEAT CYCLE

A reheat cycle along with *T-s* diagram is shown in Fig. 11.12.

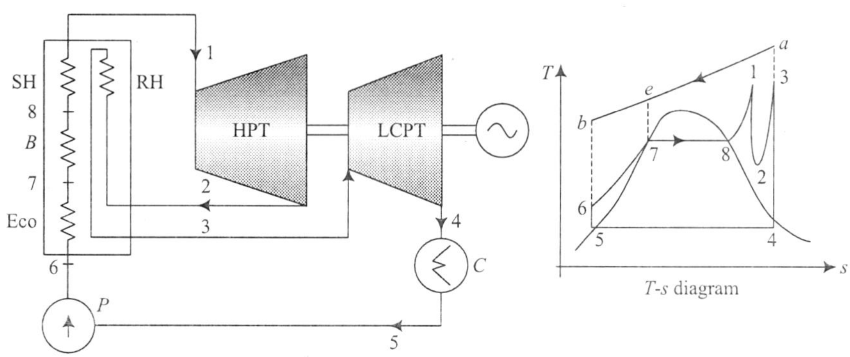

Fig. 11.12 Rankine cycle with superheat and reheat

The pressure p_2 at which steam is reheated affects the cycle efficiency. The optimum reheat pressure for most of the modern power plant is 0.2 to 0.25 of initial steam pressure.

The analysis of a reheat cycle involves two turbine work terms as well as two heat addition terms. For unit flow rate

$$W_T = (h_1 - h_2) + (h_3 - h_4)$$
$$W_p = (h_6 - h_5)$$
$$W_N = (h_1 - h_2) + (h_3 - h_4) - (h_6 - h_5)$$
$$Q_S = (h_1 - h_6) + (h_3 - h_2)$$
$$\eta_{Th} = \frac{W_N}{Q_S}$$
$$\text{s.s.c.} = \frac{3600}{W_N} \text{ kg/kWh}$$

Heat rate, $$HR = \frac{3600}{\eta_{Th}} \text{ kJ/kWh}$$

11.9 IDEAL REGENERATION CYCLE

An ideal regeneration cycle is shown in Fig. 11.13. The condensate circulates around the turbine casing and is heated continuously and reversibly by expanding steam. The process 1-2 is the reversible expansion of steam in the turbine. The process 4-5 is reversible heating of water. The slopes of lines 1-2 and 5-4 are identical in contour.

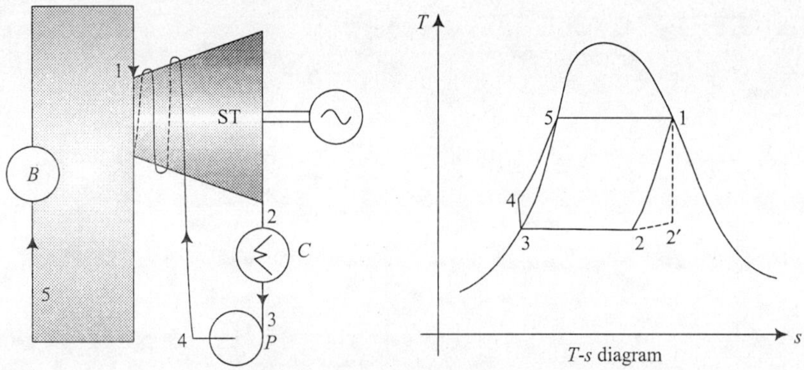

Fig. 11.13 Ideal regeneration cycle

$$Q_1 = (h_1 - h_5) = T_1 (s_1 - s_5)$$
$$Q_2 = (h_2 - h_3) = T_2 (s_2 - s_3)$$

For reversible processes,

$$s_1 - s_5 = s_2 - s_3$$

Also, $$s_5 - s_3 = s_1 - s_2$$

∴ $$\eta = 1 - \frac{Q_2}{Q_1} = 1 - \frac{T_2}{T_1}$$

The efficiency of ideal regeneration cycle is equal to that of Carnot cycle.

$$W_T = (h_1 - h_2) - (h_5 - h_4)$$

$$W_p = (h_4 - h_3)$$

$$W_N = W_T - W_p$$

The network output of ideal regeneration cycle is less than that for Rankine cycle.

The ideal regeneration cycle is not practical because:

1. reversible heat transfer cannot be realized
2. heat exchanger in the turbine is mechanically impractical
3. the moisture content (X_2) in the turbine is excessive which leads to high erosion of turbine blades

11.10 REGENERATIVE FEEDWATER HEATING CYCLES

The efficiency of a steam power cycle can be improved by raising the mean temperature at which heat is supplied in the boiler. This can be done in two ways:

1. Increasing the temperature of steam at exit from the boiler. This is done by increasing the pressure of steam, use of superheating and reheating.
2. Increasing the temperature of feedwater at inlet to the boiler. This can be done by preheating water in the economizer by use of exhaust flue gases and regenerative feedwater heating by use of steam bled from the steam turbine at proper points. The use of regenerative feedwater heating results in higher thermal efficiency of the order of 4% but there is decrease in power developed and increase of specific steam consumption (kg/kWh).

There are three types of schemes followed for feedwater heating:

1. surface heaters with drains cascaded to condenser
2. surface heaters with drip pumps
3. direct contact heaters

11.10.1 Surface Heaters with Drains Cascaded to Condenser

This scheme is the simplest and most commonly used in power plants. The feedwater heaters are of shell and tube type heat exchangers. The feedwater passes through the tubes and bled steam, on the shell side. The bled steam is completely condensed. The condensate is fed back to the next lower pressure feedwater heater. The condensate of the lowest pressure feedwater heater is led back to main condenser. The amount of bled steam required for each feedwater heater can be calculated from energy balance that heat gained by feedwater is equal to heat lost by bled steam in condensing.

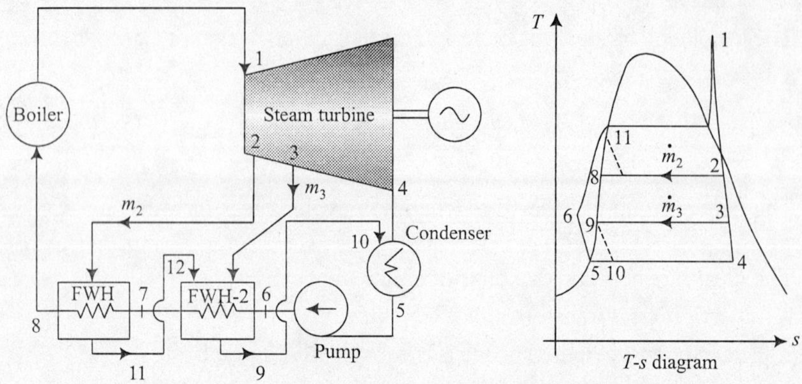

Fig. 11.14 Surface heaters with drains cascaded back to condenser

The system requires throttling of drains before feeding to lower pressure heater and this results in energy loss.

A mass balance based on a unit flow rate at turbine inlet (point 1) is given below:

Mass flow between 1 and 2 = 1

Mass flow between 2 and 3 = $1 - \dot{m}_2$

Mass flow between 3 and 10 = $1 - \dot{m}_2 - \dot{m}_3$

Mass flow between 10 and 1 = 1

Mass flow between 2 and 12 = $\dot{m}_2$

Mass flow between 3 and 12 = $\dot{m}_3$

Mass flow between 12 and 10 = $\dot{m}_2 + \dot{m}_3$

The energy balance on feedwater heater no. 1,

$$\dot{m}_2 (h_2 - h_{11}) = h_8 - h_7$$

The energy balance on feedwater heater no. 2,

$$\dot{m}_3 (h_3 - h_9) + \dot{m}_2 (h_{12} - h_9) = h_7 - h_6$$

Throttling process is a constant enthalpy process.

$$\therefore \qquad h_{12} = h_{11} \text{ and } h_{10} = h_9$$

$\dot{m}_2$ and $\dot{m}_3$ can be calculated as various enthalpies can be calculated on the basis of known steam bleeding pressures p_2 and p_3.

Heat added,

$$q_{8-1} = h_1 - h_8$$

Turbine work,

$$W_T = (h_1 - h_2) + (1 - m_2)(h_2 - h_3) + (1 - \dot{m}_2 - \dot{m}_3)(h_3 - h_4)$$

Pump work,

$$W_p = h_6 - h_5 = \frac{v_5(p_6 - p_5)}{\eta_p}$$

Heat rejected,

$$q_{4-5} = (1 - \dot{m}_2 - \dot{m}_3)(h_4 - h_5) + (\dot{m}_2 + \dot{m}_3)(h_{10} - h_5)$$

Net cycle work,

$$W_N = W_T - W_p$$

Cycle thermal efficiency,

$$\eta_{Th} = \frac{W_N}{q_{8-1}}$$

Work ratio, $WR = \dfrac{W_N}{W_T}$

11.10.2 Feedwater Heaters with Drains Pumped Forward

In order to avoid losses due to throttling, small drip pumps are used for each feedwater heater. The drain from the heater is pumped forward into main feedwater line. The plant is more complex due to addition of drain pump after every heater.

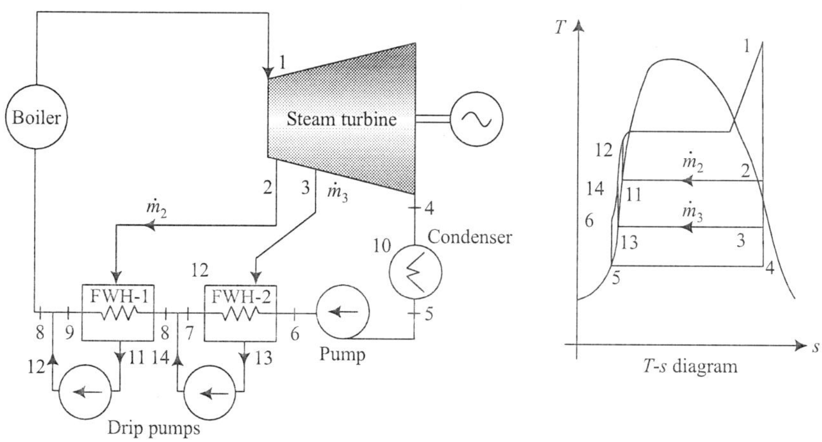

Fig. 11.15 Surface feedwater heaters with drip pumps

A mass balance based on unit mass flow rate at turbine inlet (point 1) is given below:

Mass flow between 1 and 2 $= 1$

Mass flow between 2 and 12 $= \dot{m}_2$

Mass flow between 2 and 3 $= 1 - \dot{m}_2$

Mass flow between 3 and 14 $= \dot{m}_3$

Mass flow between 3 and 7 $= 1 - \dot{m}_2 - \dot{m}_3$

Mass flow at 14 $= \dot{m}_3$

Mass flow between 8 and 9 $= 1 - \dot{m}_2$

Mass flow at 12 $= \dot{m}_2$

Mass flow between 10 and 1 $= 1$

The energy balance on high pressure heater, FWH – 1,

$$\dot{m}_2 (h_2 - h_{11}) = (1 - \dot{m}_2)(h_9 - h_8)$$

The energy balance on low pressure feedwater heater, FWH – 2,

$$\dot{m}_3 (h_3 - h_{13}) = (1 - \dot{m}_2 - \dot{m}_3)(h_7 - h_6)$$

The values of various enthalpies are found out at various temperatures.

$$t_9 = t_{11} - TTD$$

$$t_7 = t_{13} - TTD$$

$$h_{12} = h_{11} + v_{11} \frac{(p_{12} - p_{11})}{\eta_p}$$

$$h_{14} = h_{13} + v_{13} \frac{(p_{14} - p_{13})}{\eta_p}$$

$$h_{10} = \dot{m}_2 h_{12} + (1 - \dot{m}_2) h_9$$

$$(1 - \dot{m}_2) h_8 = \dot{m}_3 h_{14} + (1 - \dot{m}_2 - \dot{m}_3) h_7$$

The turbine work,

$$W_T = (h_1 - h_2) + (1 - \dot{m}_2)(h_2 - h_3) + (1 - \dot{m}_2 - \dot{m}_3)(h_3 - h_4)$$

The total pump work,

$$\Sigma W_p = (1 - \dot{m}_2 - \dot{m}_3)(h_6 - h_5) + \dot{m}_3 (h_{14} - h_{13}) + \dot{m}_2 (h_{12} - h_{11})$$

Heat supplied in the boiler,

$$q_{10-1} = h_1 - h_{10}$$

Thermal efficiency,

$$\eta_{Th} = \frac{W_T - \Sigma W_p}{q_{10-1}}$$

11.10.3 Direct Contact Heaters

The bled steam is mixed directly with the incoming subcooled feedwater to produce saturated water at the extraction pressure. The schematic diagram along with its *T-s* diagram are shown in Fig. 11.16. In addition to condensate pump, as many additional pumps are required as feedwater heaters. Each pump carries full flow of feedwater. Such large pumps increase plant complexity and cost and result in operational, service and noise problems.

In modern power plants only surface type heaters are used. Only one deaerator is used as direct contact heater. The dissolved air in feedwater is removed as water has minimum solubility of air when heated to saturation temperature.

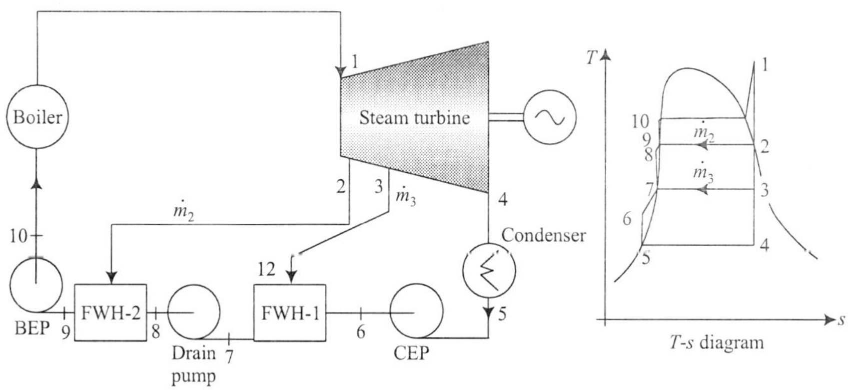

Fig. 11.16 Direct contact heaters

Mass balance as well as energy balance are carried out for the analysis of the system.

The mass balance based on unit flow rate at turbine inlet (point 1) is given below:

Mass flow between 1 and 2 = 1

Mass flow between 2 and 9 = $\dot{m}_2$

Mass flow between 2 and 3 =1 – $\dot{m}_2$

Mass flow between 3 and 7 = $\dot{m}_3$

Mass flow between 4 and 7 = 1 – $\dot{m}_2$ – $\dot{m}_3$

Mass flow between 7 and 9 = 1 – $\dot{m}_3$

Mass flow between 9 and 1 = 1

The energy balance on feedwater heater, FWH – 2 gives:

$$\dot{m}_2 (h_2 - h_9) = (1 - \dot{m}_2)(h_9 - h_8)$$

The energy balance on feedwater heater, FWH – 1 gives:

$$\dot{m}_3 (h_3 - h_7) = (1 - \dot{m}_2 - \dot{m}_3)(h_3 - h_7)$$

The various enthalpies can be found out corresponding to bled steam pressures (which are known) and $\dot{m}_2$ and $\dot{m}_3$ can be estimated from above equations.

Heat added in the boiler,

$$q_{10-1} = (h_1 - h_{10})$$

Turbine work output,

$$W_T = (h_1 - h_2) + (1 - \dot{m}_2)(h_2 - h_3) + (1 - \dot{m}_2 - \dot{m}_3)(h_3 - h_4)$$

Pump work,

$$\Sigma W_p = (1 - \dot{m}_2 - \dot{m}_3)(h_6 - h_5) + (1 - \dot{m}_2)(h_8 - h_7) + (h_{10} - h_9)$$

$$\approx (1 - \dot{m}_2 - \dot{m}_3)\frac{v_5(p_6 - p_5)}{\eta_p} + (1 - \dot{m}_2)\frac{v_7(p_8 - p_7)}{\eta_p} + \frac{v_4(p_{10} - p_9)}{\eta_p}$$

Heat rejected in the condenser,

$$q_{4-5} = (1 - \dot{m}_2 - \dot{m}_3)(h_4 - h_5)$$

Net cycle work,

$$W_N = W_T - \Sigma W_p$$

Cycle thermal efficiency,

$$\eta_{Th} = \frac{W_N}{q_{10-1}}$$

Work ratio,

$$WR = \frac{W_N}{W_T}$$

where, η_p= Pump efficiency.

11.11 COGENERATION CYCLES

Cogeneration is the simultaneous generation of electricity and steam (or heat) in a single power plant. Chemical industries, paper mills, municipalities that use distinct heating need process steam (or heat) as well as electricity. These industries and municipalities can produce electricity more conveniently and cheaply by cogeneration.

The cogeneration plant efficiency,

$$\eta_{CO} = \frac{W_T + Q_H}{Q_S}$$

where, W_T = electric energy generated

Q_H = heat energy in process steam

Q_S = heat added to plant

11.11.1 Cogeneration Cycle with Back Pressure Turbine

A cogeneration plant with back pressure turbine along with T-s diagram are shown in Fig. 11.17.

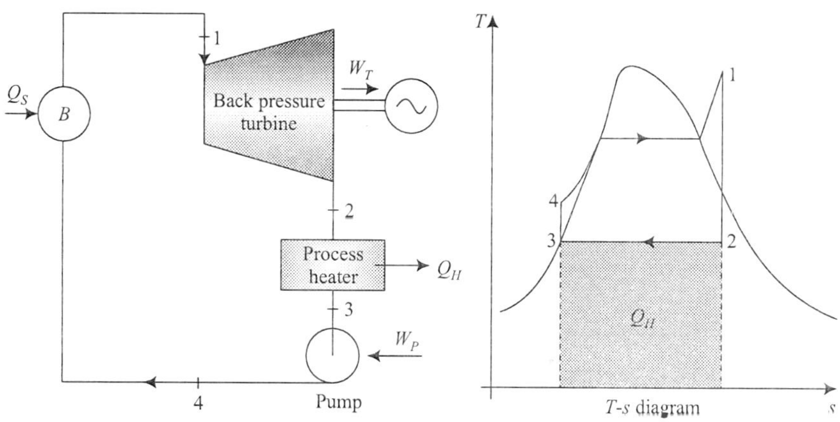

Fig. 11.17 Cogeneration plant with back pressure turbine

The power generated by turbine,

$$W_T = \dot{m}(h_1 - h_2) \quad \text{[kW]}$$

The process heat required,

$$Q_H = \dot{m}(h_2 - h_3) \quad \text{[kJ/s]}$$

where, $\dot{m}$ = mass flow rate of steam [kg/s]

The cogeneration plant efficiency,

$$\eta_{CO} = \frac{W_T + Q_H}{Q_S}$$

where, Q_S = Heat supplied in the boiler.

11.11.2 Cogeneration Cycle with Extraction Turbine

A certain quantity of steam is extracted continuously from an intermediate stage of turbine for heating purposes at a required temperature and pressure.

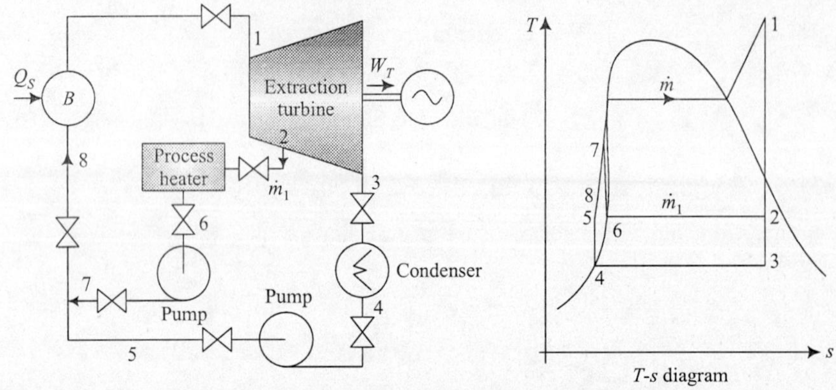

Fig. 11.18 Cogeneration plant with extraction turbine

Heat supplied in the boiler,

$$Q_S = \dot{m}(h_1 - h_8)$$

Heat rejected in the condenser,

$$Q_{3-4} = (\dot{m} - \dot{m}_1)(h_3 - h_4)$$

Process heat in the extracted steam,

$$Q_H = \dot{m}_1(h_2 - h_6)$$

The turbine work,

$$W_T = \dot{m}(h_1 - h_2) + (\dot{m} - \dot{m}_1)(h_2 - h_3)$$

The pump work,

$$\Sigma W_p = (\dot{m} - \dot{m}_1)(h_5 - h_4) + \dot{m}_1(h_7 - h_6)$$

The energy balance for process heater,

$$\dot{m}_1 h_7 + (\dot{m} - \dot{m}_1)h_5 = \dot{m}_1 h_8$$

11.12 CYCLES FOR NUCLEAR POWER PLANTS

The nuclear energy liberated by the chain reaction of nuclear fission is transformed into heat which is removed from the nuclear reactor by a coolant. Fuels for nuclear reactors include natural isotopes such as U^{235} and synthetic isotopes as Pu^{239}, U^{233}, etc. created by breeding. The coolant can be light water, heavy water or liquid metal.

11.12.1 Single Loop Cycle

In a single loop cycle using boiling water reactor, steam is directly generated in the nuclear reactor which also acts as steam generator. Although such nuclear power plants are simple in construction and low cost but the working fluid, i.e. water and steam are irradiated in the reactor and become radioactive requiring biological shielding of reactor and steam turbine plant. The repair and maintenance of such equipment becomes difficult. The single loop cycle is shown in Fig. 11.19.

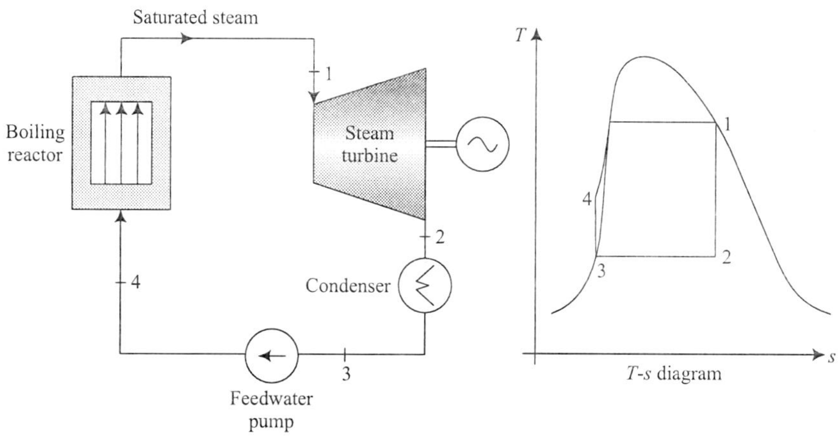

Fig. 11.19 Direct cycle BWR plant

11.12.2 Double Loop Cycle

A double loop cycle consists of pressurized-water reactor and the power plant is composed of two loops in series. The coolant loop is called primary loop and the water-steam is called working fluid loop. The coolant (heavy water) picks up reactor heat and transfers it to the working fluid in the steam generator. The steam is then used in a Rankine-type cycle to generate electricity.

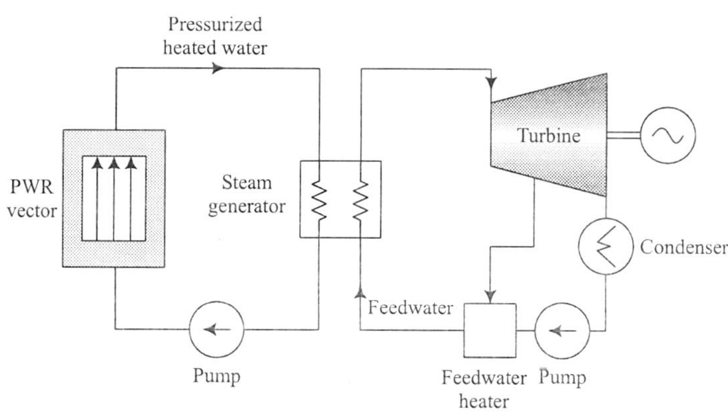

Fig. 11.20 PWR power plant

11.12.3 Triple Loop Cycle

Sodium and other liquid metals are used as coolant in fast breeder reactors. These coolants are generally chemically active and suffer from high induced radioactivities. An intermediate coolant loop is used between the primary radioactive coolant and the steam cycle as shown in Fig. 11.21. The intermediate coolant is usually also a liquid metal, often Na or NaK. The intermediate loop guards against reaction between the radioactive primary coolant and water.

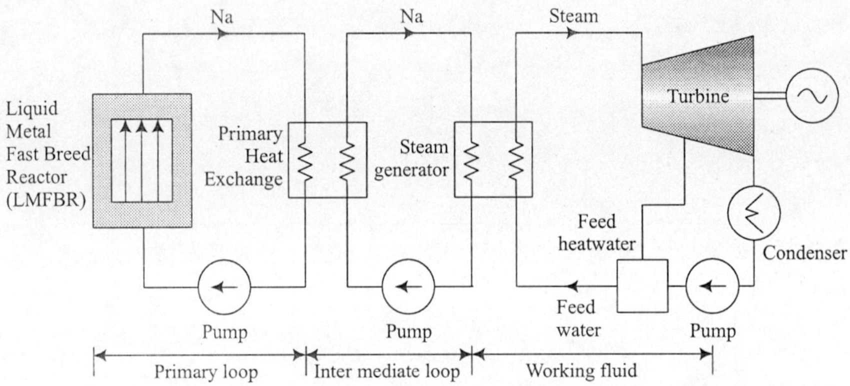

Fig. 11.21 LMFBR power plant

11.13 CYCLES FOR GEOTHERMAL POWER PLANTS

Geothermal energy is primarily energy from the earth's own interior filled with hot molten mass called magma. The average energy flow density towards earth's surface is 0.063 W/m^2 and average temperature gradient is 30 K per kilometre of depth. Locations with very high temperature gradients are suitable for geothermal energy utilization. It can be used for heat and power generation. Geothermal sources are classified as:

1. *Hydrothermal systems*: Water is heated by interior hot rocks and steam reaches the earth's surface at 8 bar and 205°C in dry condition. This steam can be used to produce power by Rankine cycle with minimum costs. At certain locations, hot water is trapped underground and is brought up by pumping through drilled wells. The hot water is flashed into steam to produce power by Rankine cycle.
2. *Geopressured systems*: Hot brine at about 160°C is trapped underground at about 2400 to 9100 m depth at 1000 bar. This brine has recoverable methane. It is economical and technically feasible to generate electricity involving combustion of methane as well as heat from the thermal content of water.
3. *Petrothermal systems*: There is no underground water but only hot dry rocks at 150 to 298°C near the earth's surface. Water is pumped into the earth and brought out to the surface to generate power.

A number of cycles have been developed based on the above systems.

11.13.1 Vapour Dominated Power Plant

The basic layout and *T-s* diagram are shown in Fig. 11.22. Dry steam at 35 bar and 200°C is available at the bottom of the well which is superheated by throttling to 7 bar. After passing through steam turbine and condenser, it is reinjected back to earth.

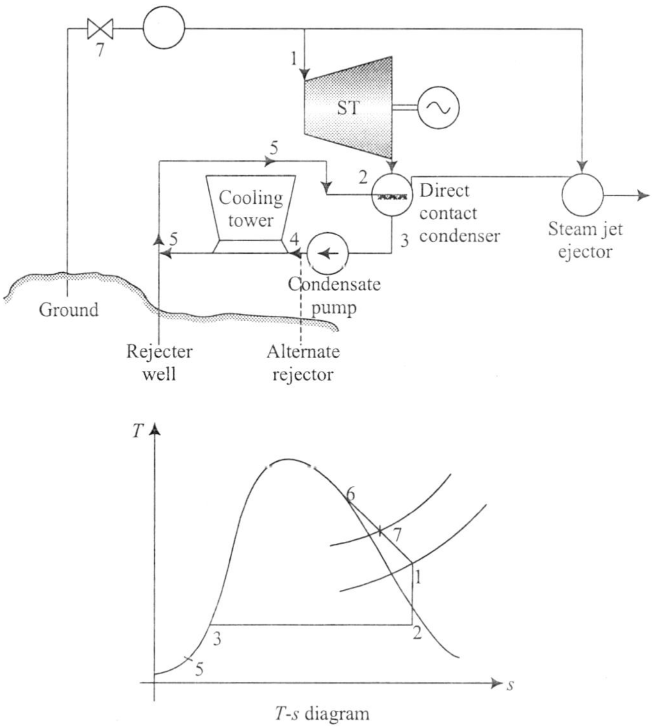

Fig. 11.22 Vapour-dominated power plant

11.13.2 The Flashed Steam System

Hot water available above 150°C to 315°C underground is trapped and flows under its own pressure to the surface of the earth. The pressure drop causes partial flashing into steam. Hot water from reservoir (1) reaches the well head (2) Pressure p_2 is lower than p_1 and process 1-2 is a constant enthalpy throttling process. The two-phase mixture of low quality (2) is passed through a flash separator (3). The quality of steam is higher at point 3. The dry saturated steam (4) at pressure of about 8 bar is expanded in the steam turbine. The separated brine (5) is reinjected into the ground.

The exhaust steam from turbine is mixed with cooling water in a direct contact condenser. The mixture is cooled in a cooling tower.

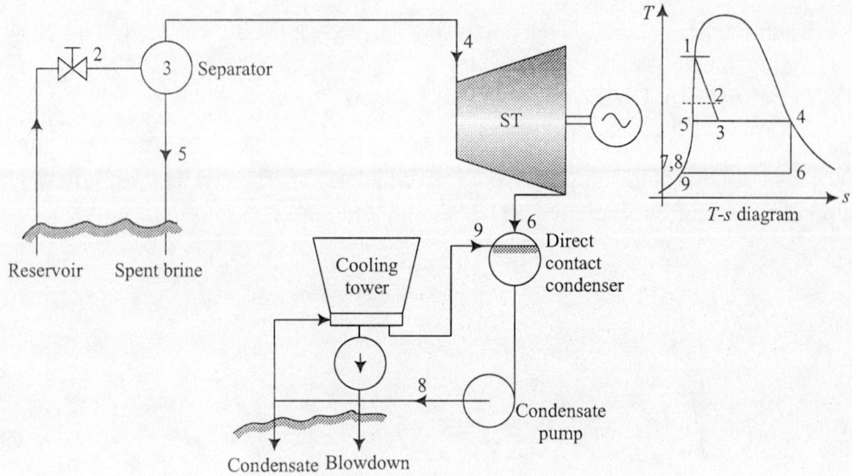

Fig. 11.23 Single flash steam power plant

In order to recover maximum quantity of heat energy from brine, some improvements in the cycle are carried out. A 50 MW double-flash system is working as Hatchobane plant, Kyushu in Japan.

11.13.3 Total Flow Power Plant

Hot brine from the well is throttled and full flow is expanded in a two-phase expander. A barometric condenser is used. As seen from *T-s* diagram, thermodynamically maximum available energy can be converted into mechanical work.

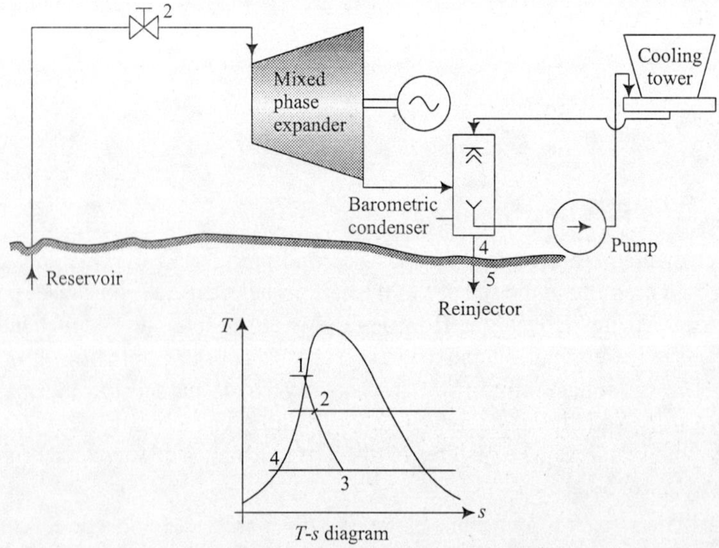

Fig. 11.24 Total flow geothermal power plant

11.13.4 Hybrid System

The geothermal fluid (brine) can be separated into steam and superheated in a fossil-fuel superheater. A flow diagram and *T-s* diagram of such a plant are shown in Fig. 11.25.

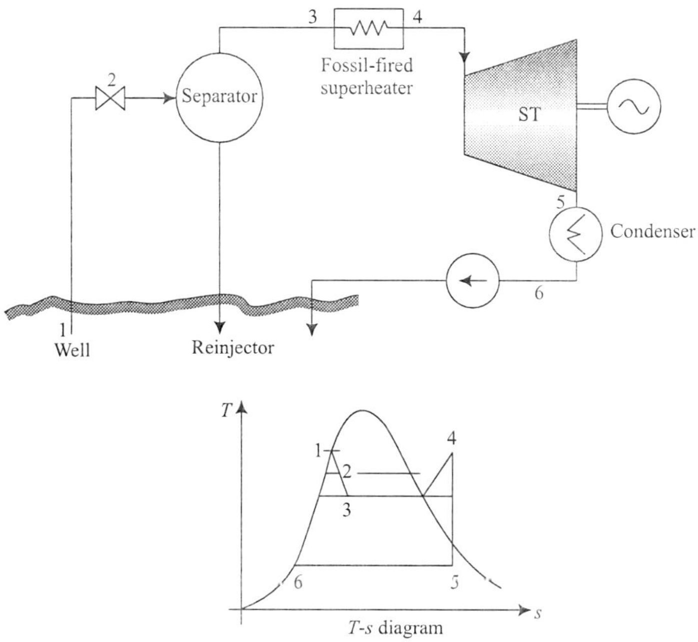

Fig. 11.25 Fossil-superheated hybrid system

11.14 SOLAR THERMAL POWER PLANTS

Solar radiations are collected and concentrated to heat a working fluid (water) for power cycles. The solar heat can be collected on a concentrating solar collector or solar pond or sea water and then used to heat the working fluid for operating a vapour Rankine cycle for the production of electricity. The flow diagram of the most popular type central receiver solar power plant is shown in Fig. 11.26.

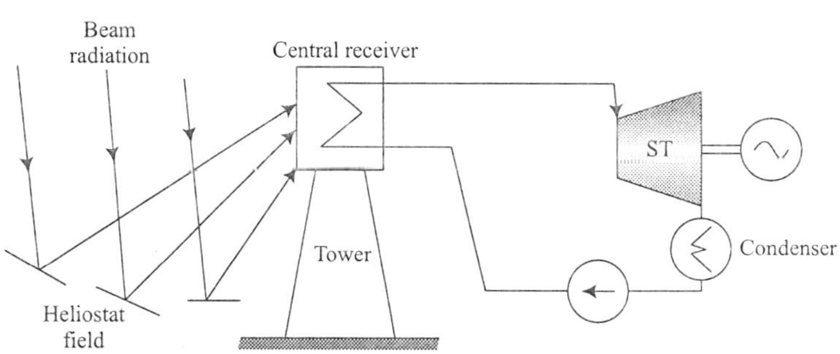

Fig. 11.26 Central receiver solar power plant

A heliostat field consists of a large number of flat mirrors 25 to 150 m^2 area which reflects the beam radiations onto a central receiver mounted on a tower. Each mirror is tracked on two axis. The absorber surface temperature may be 400 to 1000°C. Steam is raised for the conventional steam power plant.

QUESTION BANK NO. 11

1. Draw a basic layout of a steam power plant and explain the four basic components.
2. Explain the working of a simple Rankine cycle with the help of *T-s*, *p-v* and *h-s* diagrams.
3. Draw the Carnot cycle on *T-s* diagram and mark the different thermodynamic processes.
4. Give reasons why the Carnot cycle cannot be used as practical cycle for steam power plant though its efficiency is maximum.
5. How does Rankine cycle overcome the limitations of Carnot cycle in reference to steam power plants?
6. Compare Rankine cycle and Carnot cycle for a steam power plant.
7. What are the main deviations made in the analysis of ideal Rankine cycle?
8. What is mean temperature of heat supply? How can this temperature be increased?
9. Discuss various methods of improvement of Rankine cycle efficiency.
10. Explain the following:
    (a) Reheating of steam
    (b) Ideal regenartion
11. Discuss the merits and demerits of various Regenerative Feedwater heating cycles.
12. What is cogeneration? Discuss various cogeneration cycles.
13. Explain various cycles used for nuclear power plants giving suitable examples of nuclear reactors used in various cycles.
14. What are various cycles used for geothermal power plants?
15. Discuss central receiver solar power plant.

CHAPTER 12

Combined and other Vapour Power Cycles

12.1 INTRODUCTION

A combined-cycle system consists of two or more cycles operating at different temperatures. Heat rejected by the higher temperature cycle is used to drive the lower temperature cycle to produce additional power and operate at a higher overall efficiency than either cycle by itself could achieve. The higher temperature cycle is commonly called the topping cycle. The lower temperature cycle is called the bottoming cycle.

Topping cycles have operated as Otto (diesel engine), Brayton (gas turbine, MHD), and Rankine cycle (liquid metal); gas turbines being the most popular ones. Bottoming cycles have been Rankine cycles (steam, organic fluids) and vapour absorption refrigeration cycles. Triple Rankine cycles have also been proposed with Na-cycle as topping plant, Hg-cycle as intermediate plant and steam-cycle as bottoming plant.

It is quite possible that steam will remain as the dominant working fluid for foreseeable future for the bottoming Rankine cycle, except in cases where the use of water is either not feasible or inefficient as in low temperature processes as hot water geothermal, solar and other applications.

Combined cycle plant is very promising for the following advantages:

1. High thermal efficiency
2. Low specific capital cost
3. Low cooling water requirements
4. Waste heat recovery and energy conservation potentials

Gas turbine-steam turbine combined-cycle plants are becoming increasingly popular especially in locations where inexpensive low-ash fuel like natural gas is available for driving gas turbine and also for supplementary heating in the waste-heat boiler.

12.2 COMBINED CYCLE EFFICIENCY

A Carnot engine E_1 operates between temperatures T_1 and T_2 and has an efficiency of η_1. The engine E_2 operates between temperatures T_2 and T_3 receiving heat rejected from engine E_1. It has an efficiency of η_2. The combined cycle engine effectively works between temperatures T_1 and T_3.

$$\eta_1 = \frac{T_1 - T_2}{T_1} = 1 - \frac{T_2}{T_1}$$

$$\therefore \quad 1 - \eta_1 = \frac{T_2}{T_1} \quad \ldots(1)$$

Similarly,
$$1 - \eta_2 = \frac{T_3}{T_2} \quad \ldots(2)$$

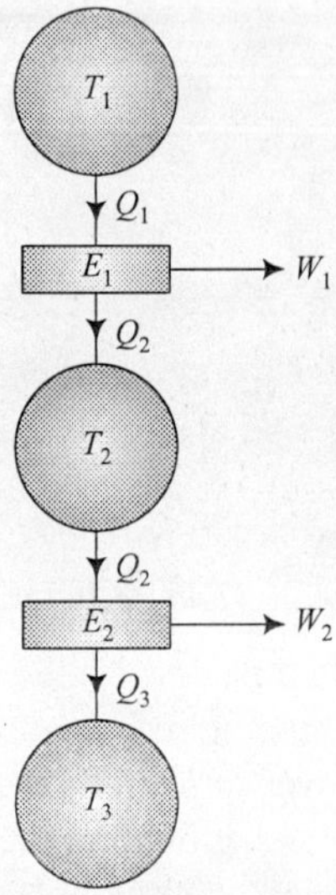

Fig. 12.1 Combined-cycle engine

Multiply equations (1) and (2),

$$\therefore \quad (1 - \eta_1)(1 - \eta_2) = \frac{T_2}{T_1} \cdot \frac{T_3}{T_2} = \frac{T_3}{T_1} \quad \ldots(3)$$

The efficiency of combined cycle,

$$\eta_3 = \frac{T_1 - T_3}{T_1} = 1 - \frac{T_3}{T_1}$$

$$\therefore \quad (1 - \eta_3) = \frac{T_3}{T_1} \quad \ldots(4)$$

From equations (3) and (4)

$$(1 - \eta_3) = (1 - \eta_1)(1 - \eta_2) = \frac{T_3}{T_1}$$

$$\therefore \quad 1 - \eta_3 = 1 - \eta_2 - \eta_1 + \eta_1\eta_2$$

$$\therefore \quad \eta_3 = \eta_1 + \eta_2 - \eta_1\eta_2$$

For triple Rankine cycle,

$$\eta = 1 - (1 - \eta_1)(1 - \eta_2)(1 - \eta_3)$$

Examples If a simple gas turbine with cycle efficiency of 30% is superimposed on a steam turbine plant with overall efficiency of 40%, a combined plant efficiency will be

$$\eta = \eta_1 + \eta_2 - \eta_1\eta_2$$
$$= 0.3 + 0.4 - 0.3 \times 0.4 = 0.58 = 58\%$$

For sodium-mercury-steam triple cycle, if $\eta_1 = 0.5$, $\eta_2 = 0.4$ and $\eta_3 = 0.4$, the combined cycle efficiency will be

$$\eta = 1 - (1 - \eta_1)(1 - \eta_2)(1 - \eta_3)$$
$$= 1 - (1 - 0.5)(1 - 0.4)(1 - 0.4)$$
$$= 1 - 0.5 \times 0.6 \times 0.6 = 0.82 \text{ or } 82\%$$

There is a great scope for multi-fluid combined cycle plants for achieving high thermal efficiency and fuel economy and utilizing favourable characteristics of working fluid in different temperature ranges for bulk power generation.

12.3 COMBINED CYCLE POWER PLANT WITHOUT SUPPLEMENTARY FIRING

A combined cycle power plant without supplementary firing consists of a topping gas turbine plant (Brayton cycle) and a bottoming steam turbine plant (Rankine cycle). The fuel is burnt only in the gas turbine combustor. Superheated steam is produced in the heat recovery steam generator (HRSG) using the heat of gas turbine exhaust gas. Power is produced both in gas turbine as well as in steam turbine. The main scheme of the plant is shown in Fig. 12.2.

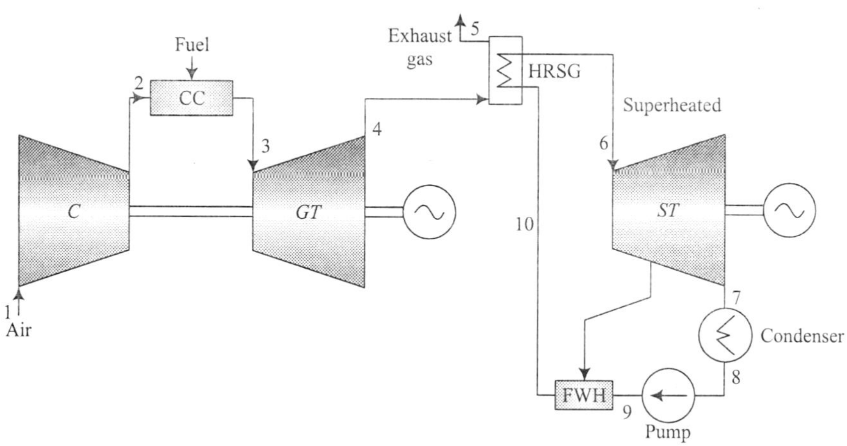

Fig. 12.2 Combined cycle power plant without supplementary firing

The thermodynamics cycle of the power plant is shown on *T-s* diagram in Fig. 12.3.

The representative energy flow diagram of the cycle is shown in Fig. 12.4.

Q_{in} = Heat supplied in the combustion chamber

$Q_1 = Q_{out}$ = Heat rejected in the condenser

Q_2 = Heat loss with stack gases

Q_3 = Heat loss in HRSG

Q_4 = Flue gas by-pass gas

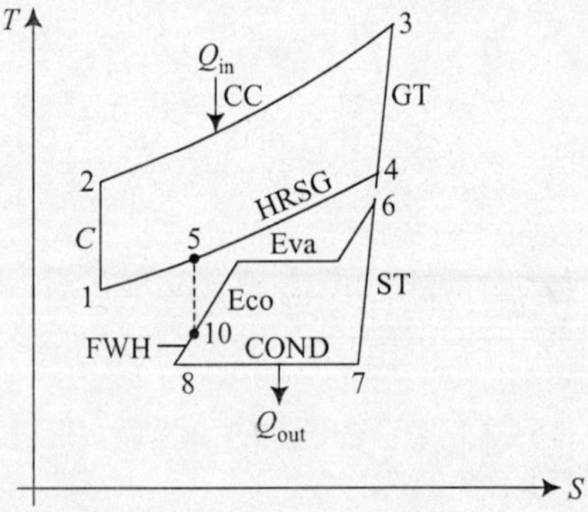

Fig. 12.3 Combined cycle on *T-s* diagram

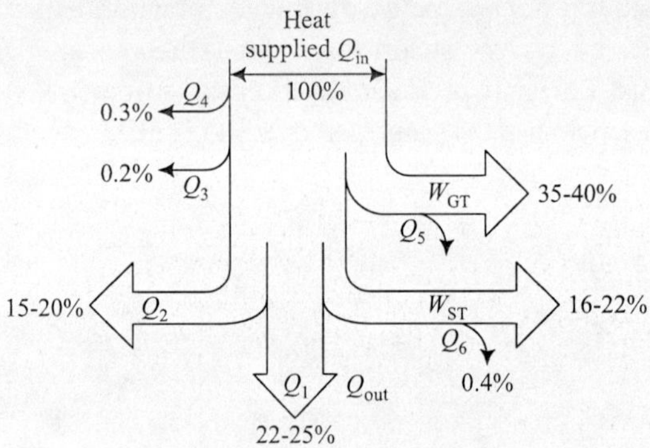

Fig. 12.4 Energy flow diagram

Q_5 = Gas turbine energy loss

Q_6 = Steam turbine energy loss

W_{GT} = Power generated in gas turbine plant

W_{ST} = Power generated in steam turbine plant

The combined cycle efficiency is the ratio of total cycle power output and rate of heat addition to the cycle.

$$\eta_{CC} = \frac{(W_{GT} + W_{ST})}{\dot{Q}_{in}}$$

The thermal efficiency of gas turbine plant,

$$\eta_{GT} = \frac{W_{GT}}{\dot{Q}_{in}}$$

The rate of heat removal from topping plant,

$$\dot{Q}_{out,\,GT} = \dot{Q}_{in} - W_{GT} = Q_{in}\,(1 - \eta_{GT})$$

This is also the rate of heat input to HRSG of the bottoming steam turbine plant.

$$\dot{Q}_{in,\,ST} = \dot{Q}_{in}(1 - \eta_{GT})$$

The thermal efficiency of steam turbine plant,

$$\eta_{ST} = \frac{W_{ST}}{\dot{Q}_{in},\,ST}$$

The power output of steam turbine plant,

$$W_{ST} = \eta_{ST}\,\dot{Q}_{in,\,ST} = \eta_{ST}\,\dot{Q}_{in}\,(1 - \eta_{GT})$$

$$\therefore \qquad \eta_{CC} = \frac{W_{GT} + \eta_{ST}\dot{Q}_{in}\,(1 - \eta_{GT})}{\dot{Q}_{in}}$$

$$= \frac{W_{GT}}{\dot{Q}_{in}} + \eta_{ST}\,(1 - \eta_{GT})$$

$$= \eta_{GT} + \eta_{ST}\,(1 - \eta_{GT})$$

$$= \eta_{GT} + \eta_{ST} - \eta_{ST}\eta_{GT}$$

Example 12.1 A combined cycle power plant has total power output of 350 MW. The gas turbine and steam turbine efficiencies are respectively 38% and 25%. Calculate the combined cycle efficiency. Also find out the fuel consumption rate with lower heating value of 42.7 MJ/kg.

Solution

$$\eta_{GT} = 0.38;\ \eta_{ST} = 0.25$$

$$\eta_{CC} = \eta_{GT} + \eta_{ST}\,(1 - \eta_{GT})$$

$$\eta_{CC} = 0.38 + 0.25\,(1 - 0.38) = 0.535 \text{ or } 53.5\%$$

$$\eta_{CC} = \frac{W_{CC}}{\dot{m}_f CV}$$

$$\therefore \qquad \dot{m}_f = \frac{W_{CC}}{\eta_{CC} CV} = \frac{350}{0.535 \times 42.7} = 15.3 \text{ kg/s}$$

12.3.1 Single-pressure Combined Cycle Power Plant

The combined cycle power plant with single pressure loop is shown in Fig. 12.5. The plant consists of:

1. a gas turbine plant consisting of a compressor, combustion chamber and a gas turbine with electric generator.

2. a natural circulation HRSG consisting of economizer, evaporator, steam drum and superheater.
3. a steam turbine plant consisting of steam turbine with electric generator, condenser, condensate pump, deaerator and feedpump.

The temperature distribution curve of HRSG is shown in Fig. 12.6. The gas turbine exhaust gas is not sufficiently cooled in the HRSG. The efficiency of steam turbine plant is also low as compared to plants with multi-stage feedwater heating and reheating of steam.

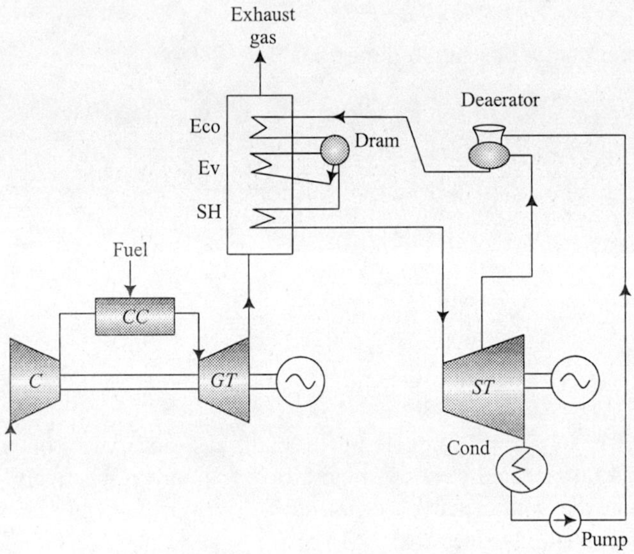

Fig. 12.5 Single pressure plant

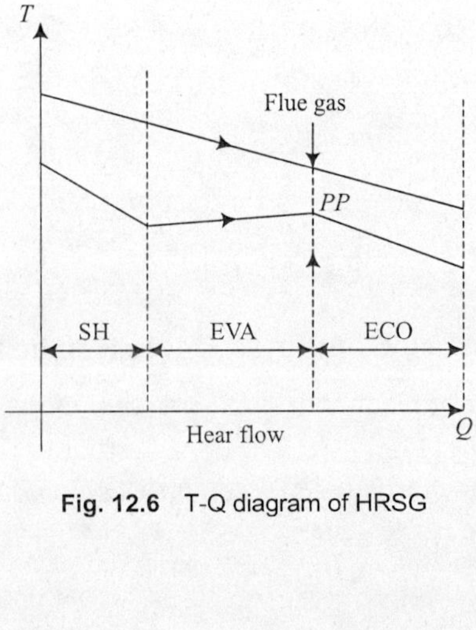

Fig. 12.6 T-Q diagram of HRSG

12.3.2 Dual Pressure Combined Cycle Plant

The combined cycle efficiency is improved in the multi-pressure schemes by generating steam at various pressures thereby reducing the flue gas exit temperature in the HRSG. The low pressure steam in dual-pressure cycle is used for feedwater heating and deaeration. The high pressure steam generator is used in steam turbine.

The dual-pressure combined cycle plant is shown in Fig. 12.7 and its *T-Q* diagram is shown in Fig. 12.8.

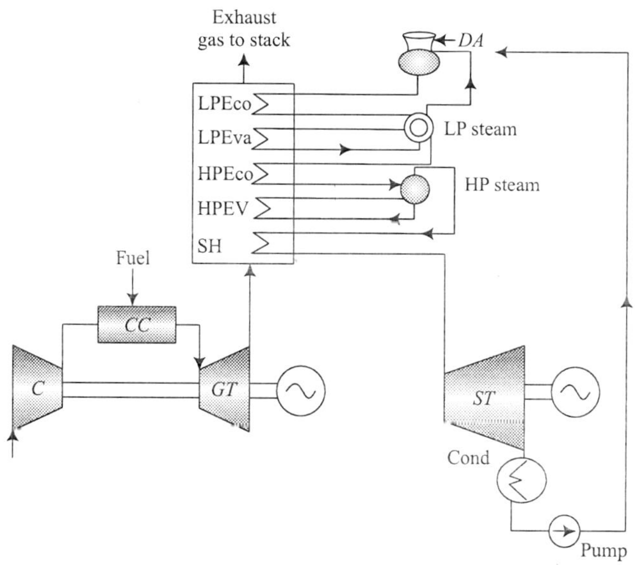

Fig. 12.7 Dual-pressure combined cycle plant

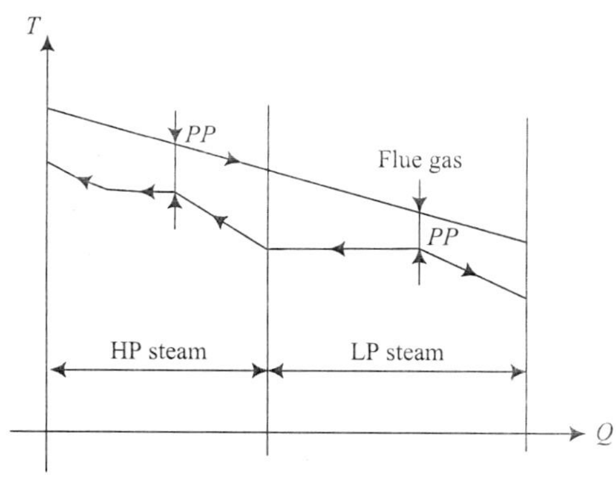

Fig. 12.8 T-Q diagram for dual-pressure HRSG

12.4 COMBINED CYCLE POWER PLANT WITH SUPPLEMENTARY FIRING

The gas turbine plants work with high quality fuels or natural gas. In combined cycle plants without supplementary firing, whole of high quality fuel is burnt in the combustion chamber and the heat released is used to produce the total cycle power. In order to save fuel cost, high-quality fuel may be burnt in gas turbine combustion chamber and the heat released is used to produce power mainly in gas turbine plant and the exhaust gases released to HRSG is further heated in the HRSG by burning coal or low quality fuel. Such a scheme is called combined cycle power plant with supplementary firing. The main scheme of the plant is shown in Fig. 12.9.

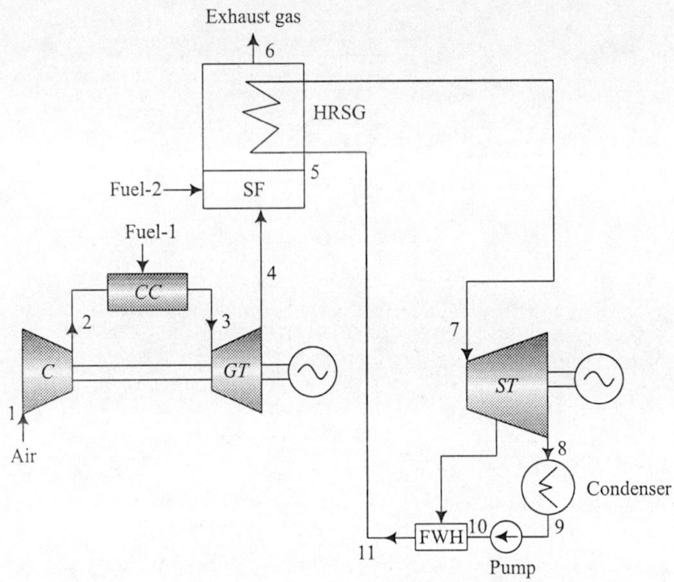

Fig. 12.9 Combined plant with supplementary firing

The *T-s* diagram of the cycle is shown in Fig. 12.10.

The total heat supplied to the combined cycle,

$$Q_{in} = Q_{CC} + Q_{SF}$$

where, Q_{CC} = Heat input through combustion chamber of gas turbine.

Q_{SF} = Heat input through supplementary furnace.

$\frac{Q_{CC}}{Q_{ST}} = f_{SF} =$ Ratio of heat supplied in supplementary furnace and gas turbine combustion chamber

$$\therefore \quad Q_{in} = Q_{CC}\,(1 + f_{SF})$$

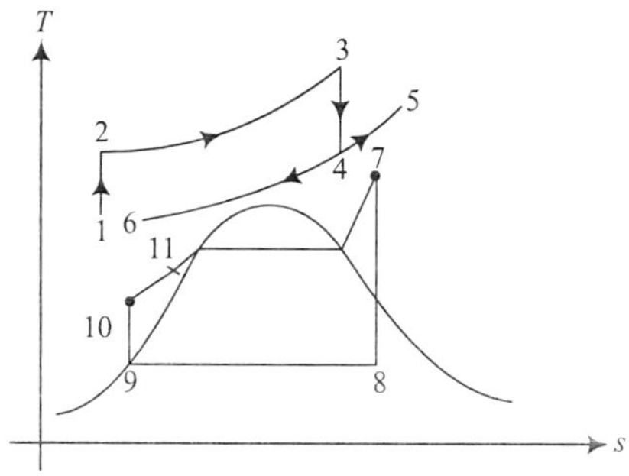

Fig. 12.10 *T-s* diagram

$$\eta_{CC} = \frac{W_{GT} + W_{ST}}{Q_{in}} = \frac{W_{GT} + W_{ST}}{Q_{CC}(1 + f_{SF})}$$

$$\eta_{GT} = \frac{W_{GT}}{Q_{CC}} \text{ and } \eta_{ST} = \frac{W_{ST}}{Q_{in,ST}}$$

where heat input to steam turbine cycle,

$$Q_{in,\,ST} = Q_{out,\,ST} + Q_{SF}$$

The exhaust heat from gas turbine,

$$Q_{out,\,GT} = Q_{CC}(1 - \eta_{GT})$$

$$\therefore \quad \eta_{ST} = \frac{W_{ST}}{Q_{CC}(1 - \eta_{GT} + f_{SF})}$$

$$\eta_{CC} = \frac{W_{GT} + \eta_{ST} Q_{CC}(1 - \eta_{GT} + f_{SF})}{Q_{CC}(1 + f_{SF})}$$

$$= \frac{\eta_{GT} + \eta_{ST}(1 - \eta_{GT} + f_{SF})}{(1 + f_{SF})}$$

$$= \frac{[(\eta_{GT} + \eta_{ST} - \eta_{GT}\eta_{ST}) + f_{SF}\eta_{ST}]}{(1 + f_{SF})}$$

Remarks:

1. The efficiency of combined cycle depends upon mainly efficiency of topping cycle.
2. The efficiency of combined cycle without supplementary firing is more than that with supplementary firing.
3. The construction of plant is simpler and cheaper in case of combined cycle without supplementary firing as compared to that with supplementary firing.
4. There is more flexibility in the choice of fuels for plants with supplementary firing as solid fuels (mainly coal) can be used for supplementary firing and result in partial saving of expensive natural gas or fuel oil used for gas turbine combustion chamber.

Example 12.2 In example 12.1, 40% of fuel requirement is burned in supplementary firing furnace of HRSG, what are the changes in the results?

Solution

$$\eta_{CC} = \frac{(\eta_{GT} + \eta_{ST} - \eta_{GT}\eta_{ST} + f_{SF}\eta_{SF})}{(1 + f_{SF})}$$

$$= \frac{(0.38 + 0.25 - 0.38 \times 0.25 + 0.4 \times 0.25)}{(1 + 0.4)}$$

$$= 0.454 \text{ or } 45.4\%$$

$$\dot{m}_f = \frac{W_{CC}}{\eta_{CC} LHV} = \frac{350}{0.454 \times 42.7} = 18.05 \text{ kg/s}$$

There is decrease in cycle efficiency from 53.5% to 45.4% and increase in fuel consumption from 15.3 kg/s to 18.05 kg/s. Combined cycle power plants without supplementary firing more efficient and economical in fuel consumption as compared to those with supplementary firing.

12.5 COMBINED CYCLE FOR NUCLEAR POWER PLANTS

A high-temperature gas-cooled reactor (HTGR) is the heat source for the gas turbine cycle. Reactor-turbine combination uses helium gas as the reactor coolant and gas-turbine cycle working fluid in a closed cycle. The steam cycle is fairly standard. The gas turbine exhaust heat is used for a regeneration and feedwater heating. Such a cycle is shown in Fig. 12.11 and *T-s* diagram in Fig. 12.12.

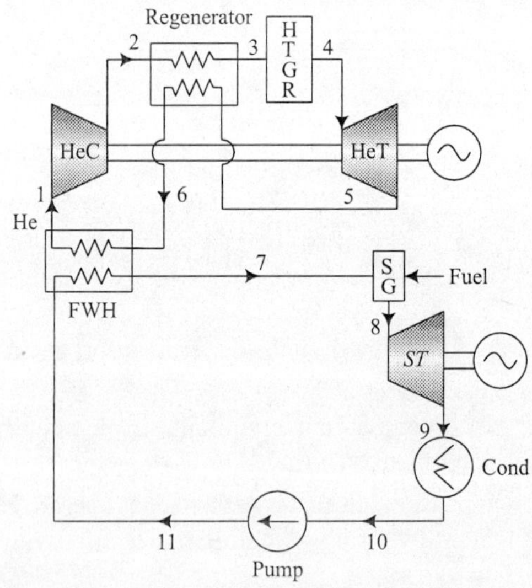

Fig. 12.11 Combined cycle with a nuclear gas turbine and fossil-fuel fired steam turbine

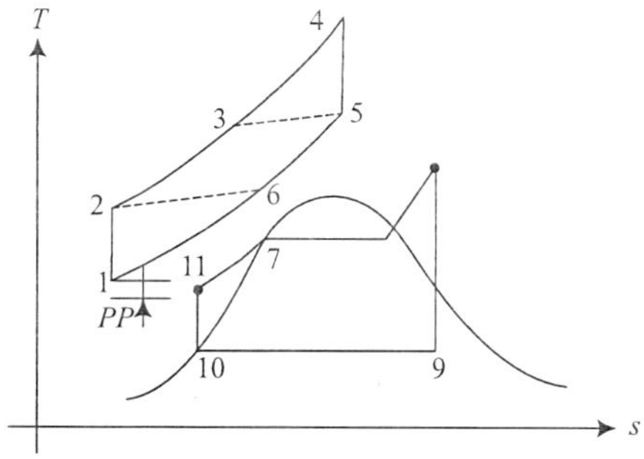

Fig. 12.12 T-s diagram

The thermodynamic analysis is similar to conventional combined cycle with supplementary firing.

12.6 MHD POWER CYCLES

Magnetohydrodynamic (MHD) is a direct heat-to-electricity conversion technique based on Faraday's Law that when an electric conductor moves across a magnetic field, a voltage is induced in it which produces an electric current. Atmospheric air is compressed in a compressor, preheated in a boiler to 1000–1200°C and fed into a combustion chamber where the combustion products form at a temperature of 2500°C and are ionized. Compounds of potassium, calcium and other alkali metals are added in the combustion chamber to intensify ionization.

Hot ionized gases (plasma) possess properties of an electric conductor and are accelerated in a nozzle to a velocity of about 700 m/s and passed through a duct surrounded by a powerful magnetic field. The ionized gas particles induce a direct current which is converted into alternating current. The exhaust gases at duct exit have a temperature of 1500–2000°C and is utilized for generation of steam for a bottoming steam turbine plant.

The electric power produced in the MHD plant is proportional to magnetic flux density (B), gas velocity (U) and gas conductivity (σ).

A schematic diagram of MHD power cycle along with its T-s diagram are shown in Fig. 12.13 and Fig. 12.14 respectively.

The total current,

$$I = \sigma(E + U \times B)$$

where,

σ = Gas conductivity [mhos/m]

E = Applied electrical field [V]

U = Gas velocity [m/s]

B = Magnetic field density [Wb/m^2]

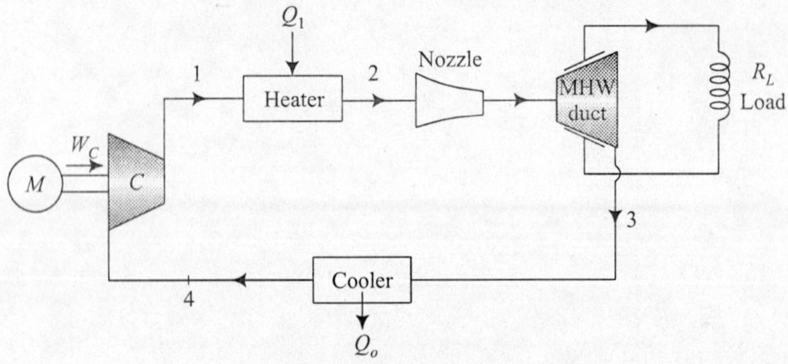

Fig. 12.13 MHD power cycle

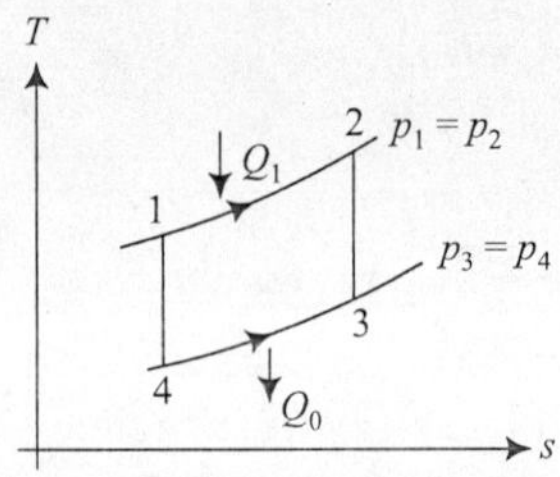

Fig. 12.14 *T-s* diagram of MHD cycle

$$\text{The total output} = W_{MHD} - W_{Comp}$$

$$\text{Heat input} = Q_1$$

$$W_{MHD} = h^*_2 - h^*_3$$

$$W_{Comp} = h^*_1 - h^*_4$$

$$Q_1 = h^*_2 - h^*_1$$

where * denotes enthalpy at stagnation values which take into account kinetic energy of flow,

$$h^* = h + \frac{U^2}{2}$$

The thermal efficiency,

$$\eta_{Th} = \frac{(h_2^* - h_3^*) - (h_1^* - h_4^*)}{(h_2^* - h_1^*)}$$

Remarks:

The following factors can adversely affect the thermal efficiency:

1. Dissipation of energy in the internal resistance of ionized gas.
2. A space charge barrier at the electrode surface.
3. Heat loss through the electrode and insulator walls.

4. Losses associated with fluid friction.
5. Hall effect losses due to current induction in the direction of flow gases.

12.7 COMBINED MHD-STEAM POWER PLANT

MHD generator is the topping plant and steam turbine plant as the bottoming plant. The overall efficiency of about 60% can be achieved in the combined cycle. The schematic diagram is shown in Fig. 12.15.

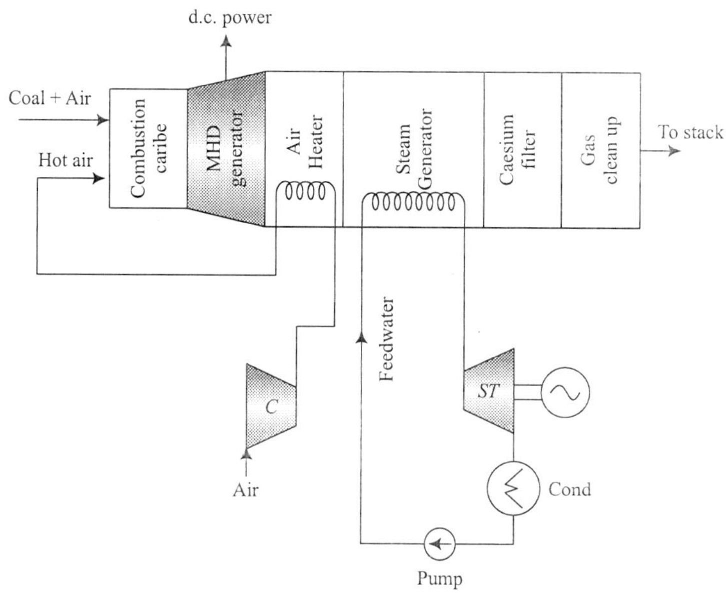

Fig. 12.15 MHD generator-steam turbine combined cycle plant

The exhaust gas from MHD generator at about 2000°C is used to raise steam for a conventional steam power plant. If the fraction Z of the fuel energy is directly converted to electricity in the MHD generator, the remainder $(1 - Z)$ is converted with an efficiency η' in the bottoming steam plant so that overall efficiency is

$$\eta = Z + \eta'(1 - Z)$$

If $Z = 0.3$ and $\eta' = 0.4$, then $\eta = 0.58$ which is a good power plant efficiency.

12.8 BINARY VAPOUR CYCLE

In binary vapour cycle, two Rankine cycles working on different working fluids are combined to increase the overall efficiency of the plant. Water has critical temperature of 374°C at 221 bar and is not a good working fluid for high temperatures. Mercury vapour is very useful working fluid for high temperatures which has a critical temperature of 588.4°C at a pressure of 21 bar only.

By using mercury cycle as topping Rankine cycle and steam Rankine cycle as bottoming plant, the cycle temperature can be increased without corresponding increase of pressure. A binary vapour cycle along with *T-s* diagram is shown in Fig. 12.16.

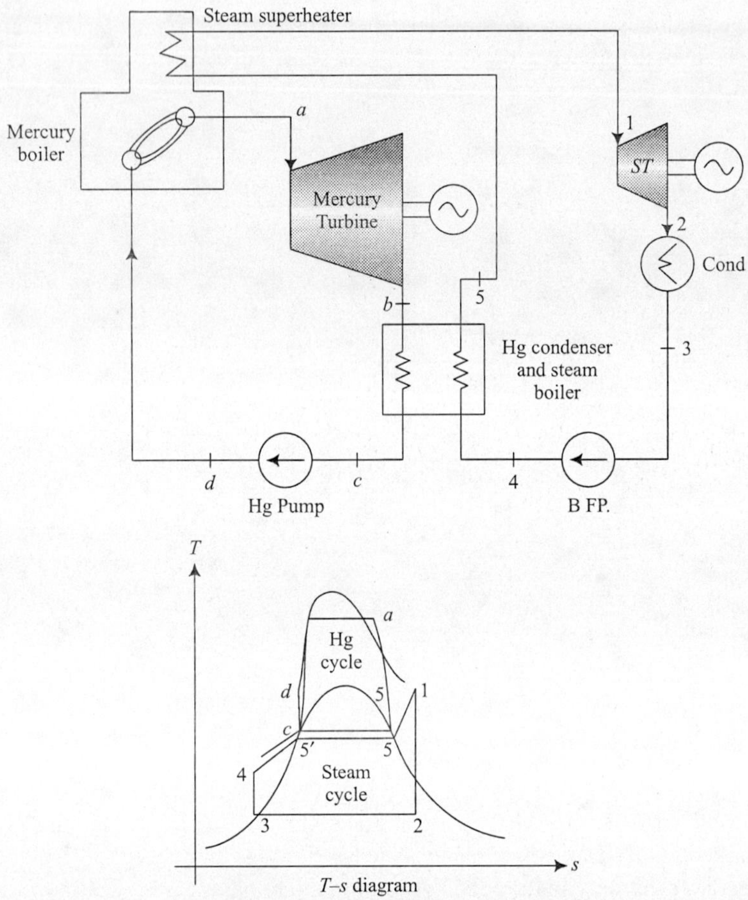

Fig. 12.16 Binary vapour cycle

Let *m* kg be the flow rate of mercury for 1 kg of steam.

$$Q_1 = m\,(h_a - h_d) + (h_1 - h_5) + (h'_5 - h_4)$$

$$Q_2 = (h_2 - h_3)$$

$$W_T = m(h_a - h_b) + (h_1 - h_2)$$

$$\sum W_p = m(h_d - h_c) + (h_4 - h_3)$$

$$\eta_{cycle} = \frac{Q_1 - Q_2}{Q_1} = \frac{W_T - W_p}{Q_1}$$

$$\text{Steam rate } (ssc) = \frac{3600}{W_T - W_p} \text{ kg/kWh}$$

The energy balance of Hg-condenser steam boiler gives,

$$m(h_b - h_c) = (h_5 - h_4)$$

$$\therefore \qquad m = \frac{(h_5 - h_4)}{(h_b - h_c)} \text{ kg Hg/kg } H_2O$$

m may be 7-8.

Remarks:

1. Efficiency of binary vapour cycle is very high.
2. Special plant design is needed to prevent leakage of poisonous fumes of mercury.
3. High mercury inventory is very costly.
4. Pumping problem with mercury.
5. Mercury does not wet heat transfer surface and hence there is difficulty in heat transfer.

12.9 TRIPLE RANKINE CYCLE

The idea of binary vapour cycle has been extended to triple Rankine cycle where Na-cycle, Hg-cycle and steam cycle are coupled in series. The *T-s* diagram of the cycle is shown in Fig. 12.17.

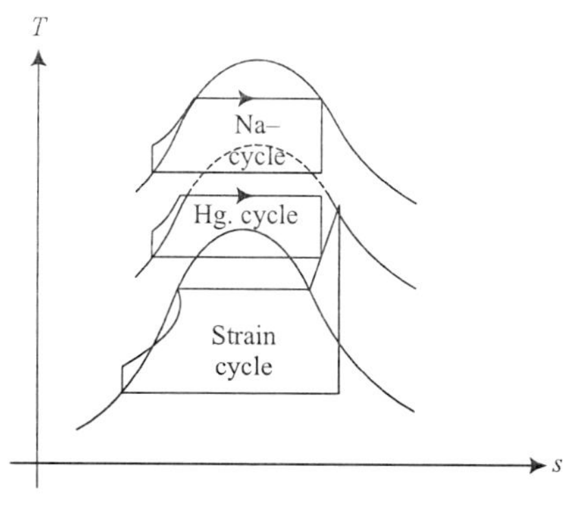

Fig. 12.17 Triple Rankine cycle

The efficiency of the cycle will be very high but with added plant complexity and cost.

QUESTION BANK NO. 12

1. Derive combined cycle efficiency.
2. Draw a schematic diagram of a gas turbine steam turbine plant without supplementary firing. Explain its working with the help of a *T-s* diagram.
3. Show the energy flow diagram of a combined cycle plant with the help of Sankey diagram.
4. Compare the performance of multi-pressure combined cycle plant with a single-power plant.
5. What is supplementary firing? What is the efficiency of a combined cycle with supplementary firing?
6. With the help of a line diagram and *T-s* diagram, explain the working of a combined cycle with nuclear reactor.
7. Describe MHD power cycle. Draw its line diagram and *T-s* diagram.
8. Draw the line diagram of a combined MHD-steam power cycle.
9. What is binary vapour cycle? What are its advantages?
10. Draw Hg-steam binary vapour cycle. How this idea has been extended to Triple Rankine cycle?

CHAPTER 13

Refrigeration Cycles

13.1 INTRODUCTION

There are many types of refrigeration systems:

1. Air/gas refrigeration system
2. Vapour compression refrigeration system
3. Vapour absorption refrigeration system
4. Steam jet refrigeration system

These systems are different in principle of operation, plant and equipment, temperature range and applications. These systems work on different cycles of operation.

13.2 AIR REFRIGERATION CYCLE

In an air refrigeration cycle, the air is the refrigerant. The heat carrying capacity per kg of air is small as there is no change of phase. However, this system is favoured for aircraft cooling because of low weight and volume of equipment.

The ideal air-refrigeration cycle is the reversed Brayton cycle called Bell-Coleman cycle.

The line diagram and *T-s* diagram are shown in Fig. 13.1.

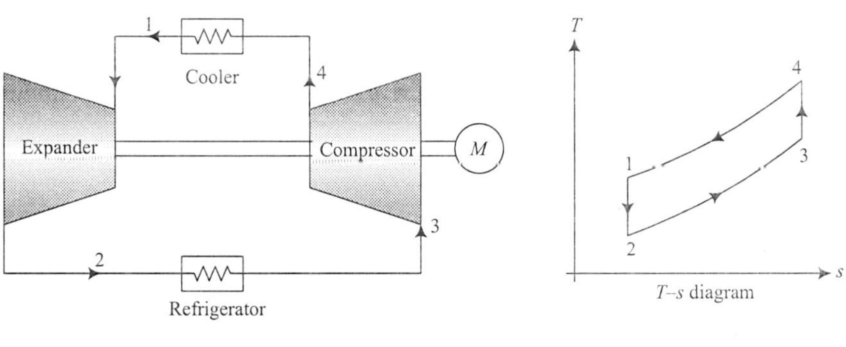

Fig. 13.1 Bell-Coleman air refrigeration cycle

The *COP* of air refrigeration cycle,

$$COP = \frac{Q_2}{W_{net}} \frac{(h_3 - h_2)}{(h_4 - h_3) - (h_1 - h_2)} = \frac{(T_3 - T_2)}{(T_4 - T_3) - (T_1 - T_2)}$$

$$= \frac{(T_3 - T_2)}{T_3\left(\frac{T_4}{T_3} - 1\right) - T_2\left(\frac{T_1}{T_2} - 1\right)}$$

For isentropic compression and expansion,

$$\frac{T_4}{T_3} = \left(\frac{p_1}{p_2}\right)^{\frac{\gamma-1}{\gamma}} = \frac{T_1}{T_2}$$

where, p_1 = pressure after compression

p_2 = pressure before compression

$$COP = \frac{(T_3 - T_2)}{(T_3 - T_2)\left(\frac{T_1}{T_2} - 1\right)} = \frac{T_2}{(T_1 - T_2)}$$

Also,
$$COP = \frac{1}{\left(\frac{p_1}{p_2}\right)^{\frac{\gamma-1}{\gamma}} - 1}$$

Example 13.1 In an aircraft cooling system, air enters the compressor at 1 bar and 4°C and is compressed to 3 bar with an isentropic efficiency of 72%. The compressed air is cooled in a heat exchanger at constant pressure to 55°C and expands in an expander to 1 bar with an isentropic efficiency of 78%. The low temperature air absorbs heat load of 3 tonnes of refrigeration at constant pressure before re-entering the compressor. The compressor is driven by the air turbine. Determine the *COP* of refrigerator, the power required and air mass flow rate.

Solution Temperature $T_1 = 4°C + 273 = 277$ K

$$T_3 = 55°C + 273 = 328 \text{ K}$$

$$\frac{T'_2}{T_1} = \left(\frac{p_2}{p_1}\right)^{\frac{\gamma-1}{\gamma}}$$

$$T'_2 = 277\ (3)^{\frac{1.4-1}{1.4}} = 379 \text{ K}$$

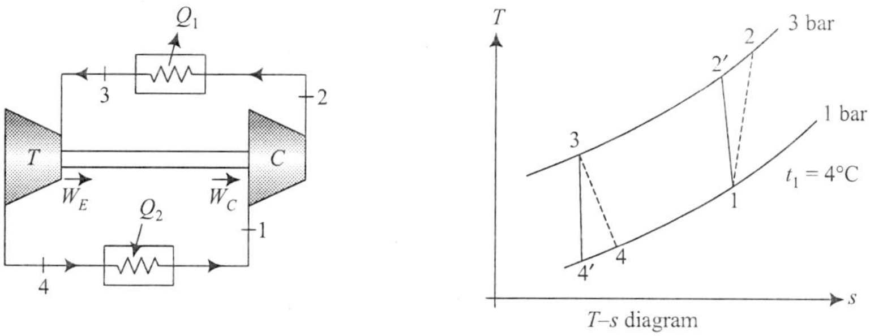

Fig. 13.2 Aircraft cooling system

$$\eta_C = \frac{T'_2 - T_1}{T_2 - T_1} = 0.72$$

$$\therefore \qquad T_2 - T_1 = \frac{379 - 277}{0.72} = 141.8 \text{ K}$$

$$\frac{T'_4}{T_3} = \left(\frac{p_2}{p_1}\right)^{\frac{\gamma - 1}{\gamma}}$$

$$T'_4 = 328\ (3)^{\frac{1.4 \quad 1}{1.4}} = 240 \text{ K}$$

$$\eta_T = \frac{T_3 - T_4}{T_3 - T'_4}$$

$$\therefore \qquad T_3 - T_4 = \frac{(328 - 240)}{0.78} = 68.6 \text{ K}$$

$$T_4 = 328 - 68.6 = 259.4 \text{ K}$$

Refrigeration effect = $C_p\,(T_1 - T_4) = C_p\,(273 - 259.4) = 17.6\ C_p$ kJ/kg

Net work input = $W_C - W_T = C_p\,[(T_2 - T_1) - (T_3 - T_4)]$

$= C_p\,[141.8 - 68.6] = 73.2\ C_p$ kW/kg

$$\therefore \qquad COP = \frac{17.6\,C_p}{73.2\,C_p} = 0.24$$

1 tonne of refrigeration = 3.5 kW

$$COP = \frac{\text{Refrigeration effect}}{\text{Power required}}$$

$$\therefore \qquad \text{Power required} = \frac{3 \times 3.5}{0.24} = \mathbf{43.75\ kW} \quad \textbf{Ans.}$$

Again refrigeration effect,

$$3.5 \times 3 = mC_p(T_1 - T_4)$$

$$\therefore \qquad \dot{m} = \frac{3.5 \times 3}{1.005(277 - 259.4)} = 0.5936 \text{ kg/s}$$

$$= 2137 \text{ kg/hr} \quad \textbf{Ans.}$$

13.3 VAPOUR COMPRESSION REFRIGERATION CYCLE

The working fluid in a refrigeration cycle is called a refrigerant. The basic refrigeration cycle along with *T-s* diagram and *p-h* diagram are shown in Fig. 13.3.

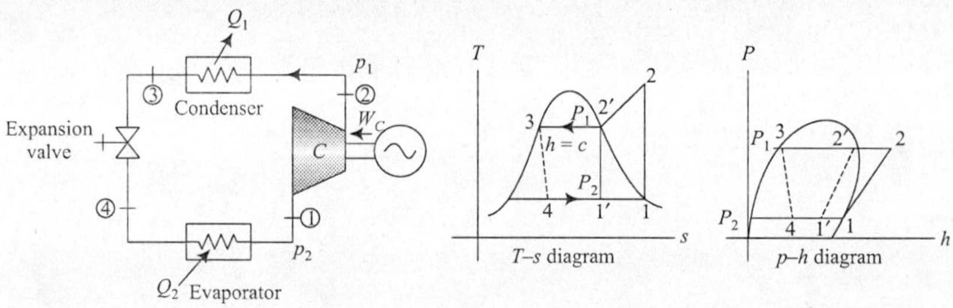

Fig. 13.3 Vapour compression refrigeration cycle

The basic operations for an ideal plants are:

1. *Compression: Process 1-2*: The saturated vapour of refrigerant is compressed isentropically. Process 1-2 is called dry compression but process 1′-2′ is called wet compression. The compressor can be reciprocating, rotary or centrifugal type.

$$W_C = (h_2 - h_1) \quad \text{[kJ/kg]}$$

2. *Condensing: Process 2-3*: The high pressure refrigerant gas is first desuperheated and condensed. The heat rejected,

$$Q_1 = (h_2 - h_3) \quad \text{[kJ/kg]}$$

3. *Expansion: Process 3-4*: The high pressure liquid refrigerant is expanded in an expansion valve or capillary tube. The process of throttling is adiabatic

$$h_3 = h_4$$

The quality of refrigerant at inlet to evaporator.

$$X_4 = \frac{h_{f_3} - h_{f_4}}{h_{fg_4}}$$

4. *Evaporation: Process 4-1*: The evaporation process is a constant pressure process. The heat is picked up from surrounding at high temperature by refrigerant at low temperature after throttling in expansion valve. The heat picked up is called refrigeration effect Q_2. The liquid refrigerant is evaporated into low pressure gas and enters the compressor to start a new cycle.

$$Q_2 = (h_1 - h_4) \quad [\text{kJ/kg}]$$

The values of h_1, h_2, h_3 and h_4 can be obtained from the property tables and charts of the refrigerant.

$$COP = \frac{Q_2}{Wc} = \frac{(h_1 - h_4)}{(h_2 - h_1)}$$

13.4 ACTUAL VAPOUR COMPRESSION CYCLE

A small degree of subcooling of liquid refrigerant after the condenser is used. This helps to reduce the mass of vapour formed during expansion. There would be less formation of vapour bubbles which may obstruct the flow of liquid refrigerant through the expansion valve by vapour locking.

In order to avoid the carry over of liquid droplets into the compressor, some degree of superheating may be used after the evaporator.

Both these effects of subcooling of liquid refrigerant and superheating of vapour are shown on the *p-h* diagram in Fig. 13.4.

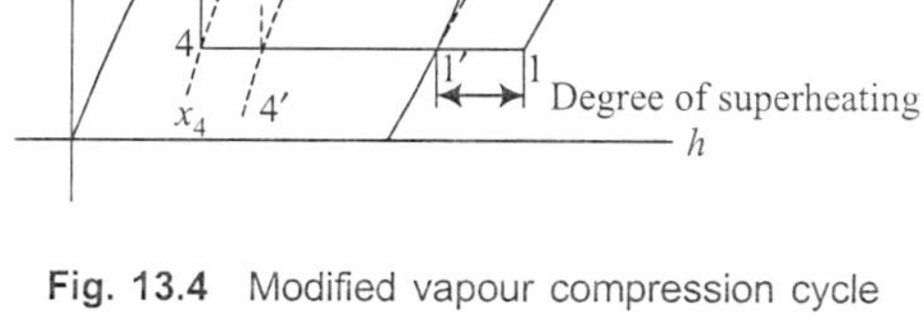

Fig. 13.4 Modified vapour compression cycle

The cycle 1′-2′-3′-4′ is the ideal refrigeration cycle. The cycle 1-2-3-4 is the actual refrigeration cycle including subcooling of refrigerant liquid and superheating of refrigerant vapour.

The degree of subcooling = $h'_3 - h_3$

The degree of superheat = $t_1 - t'_1$

The following effects are noticed:

1. Refrigeration effect $(h_1 - h_4)$ increases.
2. The compression discharge temperature $(t_2 > t'_2)$ increases which adversely affect the load on the condenser.
3. The quality of refrigerant at inlet to evaporator decreases $(X_4 < X'_4)$

Example 13.2 A 5-tonne refrigeration plant is working on R-12 with condenser and evaporator temperatures as 40°C and –10°C respectively. Determine

(i) The refrigerant flow rate in kg/s.
(ii) The volumetric flow rate of the compressor in m^3/s.
(iii) The compressor outlet temperature.
(iv) The pressure ratio.
(v) The heat rejected in the condenser in kW.
(vi) The quality of refrigerant after throttling.
(vii) The *COP* of the plant.
(viii) The power required to drive the compressor.
(ix) Compare the *COP* with that of a Carnot refrigerator working between 40°C and –10°C.

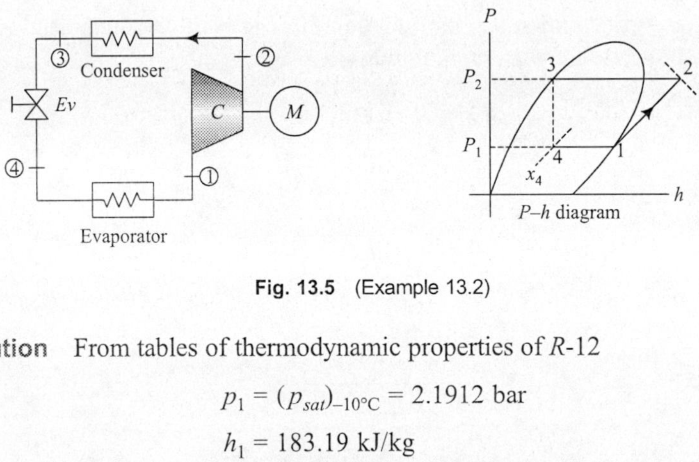

Fig. 13.5 (Example 13.2)

Solution From tables of thermodynamic properties of *R*-12

$$p_1 = (p_{sat})_{-10°C} = 2.1912 \text{ bar}$$

$$h_1 = 183.19 \text{ kJ/kg}$$

$$s_1 = 0.7019 \text{ kJ/kg}$$

$$v_1 = 0.077 \text{ m}^3/\text{kg}$$

$$p_2 = (p_{sat})_{40°C} = 9.6066 \text{ bar}$$

$$h_3 = 74.59 \text{ kJ/kg}$$

$$h_4 = h_3$$

From tables of superheated *R*-12, for p_2 = 9.6066 bar and $s_2 = s_1$ = 0.7019 kJ/kgK, t_2 = 48°C and h_2 = 209.41 kJ/kg.

(i) The plant capacity = 5 × 14000 = 70,000 kJ/hr.

$$\dot{m}(h_1 - h_4) = \frac{70,000}{3600} = 19.44 \text{ kW}$$

$$\therefore \quad \dot{m} = \frac{19.44}{183.19 - 74.59} = \mathbf{0.18 \text{ kg/s}} \quad \textbf{Ans.}$$

(ii) Volumetric flow rate = $\dot{m}\, v_1 = 0.18 \times 0.077 =$ **0.0139 m³/s** **Ans.**

(iii) Compressor discharge temperature = **48°C Ans**

(iv) Pressure ratio $= \dfrac{p_2}{p_1} = \dfrac{9.6066}{2.1912} = \mathbf{4.39}$ **Ans**

(v) Heat rejected in the condenser,

$$Q_1 = \dot{m}(h_2 - h_3) = 0.18\,(209.41 - 74.59)$$

$$= \mathbf{24.27 \text{ kW}} \quad \textbf{Ans.}$$

(vi) $h_4 = h_f + x_4\, h_{fg}$

$$= 26.87 + x_4\,(156.31) = 74.59$$

$$\therefore \quad x_4 = 0.305$$

Quality = **30.5%** **Ans.**

(vii) Plant COP $= \dfrac{h_1 - h_4}{h_2 - h_1} = \dfrac{183.19 - 74.59}{209.41 - 183.19} = \mathbf{4.14}$ **Ans**

(viii) Power required to drive the compressor,

$$W_C = \dot{m}(h_2 - h_1) = 0.18 \times 26.22 = \mathbf{4.72 \text{ kW}} \quad \textbf{Ans.}$$

(ix) Carnot COP $= \dfrac{T_2}{T_1 - T_2} = \dfrac{263}{50} = 5.26$

$$\therefore \quad \frac{\text{COP plant}}{\text{Carnot COP}} = \frac{4.14}{5.26} = \mathbf{0.787} \quad \textbf{Ans.}$$

13.5 ABSORPTION REFRIGERATION CYCLE

The absorption refrigeration avoids the use of a compressor. The refrigerant gas from evaporator is absorbed in a chamber containing a homogeneous mixture of ammonia and water called aqua-ammonia. The chamber is maintained at a low pressure and a temperature slightly above ambient temperature, so that some heat can be rejected to the surrounding. The strong ammonia solution so obtained is pumped to high pressure and high temperature generator. The ammonia vapour is liberated from the solution by the process of heating from a high temperature source. The ammonia vapours are further condensed in the condenser. The liquid ammonia is expanded in an expansion valve and passed to evaporator. The low pressure ammonia vapour is sent to the absorber to complete the cycle.

The line diagram of an absorption refrigeration cycle is shown in Fig. 13.6.

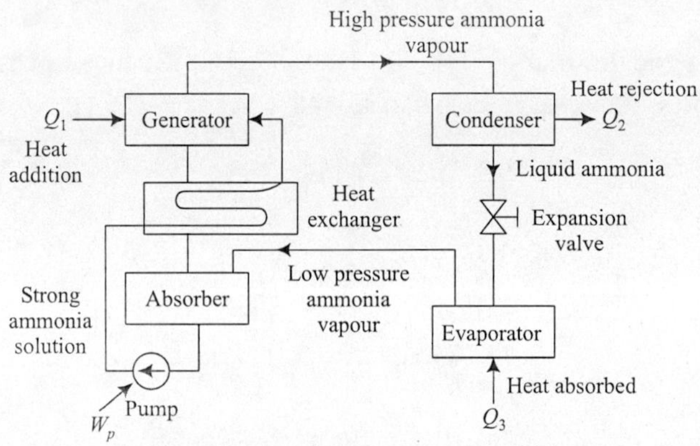

Fig. 13.6 Vapour absorption refrigeration cycle

This system is favoured where waste heat is available.

The coefficient of performance of refrigerator,

$$COP = \frac{\text{Heat absorbed during evaporation}\,(Q_3)}{\text{Work done by pump}\ (W_P)\ +\ \text{Heat supplied in HX}}$$

$$= \frac{T_3(T_1 - T_2)}{T_1(T_2 - T_3)}$$

where, T_1 = Temperature of heat reception by refrigerant

T_2 = Temperature of cooling water

T_3 = Evaporator temperature

13.5.1 Comparison of Vapour Absorption and Compression System

Some important differences between the two systems are compared in Table 13.1.

Table 13.1 Comparison of vapour absorption and compression system

S.No. Absorption system	Compression system
1. Uses low grade heat, i.e., waste heat in the generator.	1. Uses high grade mechanical energy to drive compressor.
2. Operation is quiet as a small pump is the only moving equipment.	2. Operation is noisy due to big compressor.
3. Low maintenance cost.	3. High maintenance cost due to wear and tear of compressor.
4. *COP* can be maintained high even operating the evaporator at low pressure.	4. The *COP* decreases with the decrease of evaporator pressure.
5. Performance is not affected by reducing refrigeration effect.	5. Performance is adversely affected at part loads.
6. There is no damage to equipment due to the presence of liquid drops in the system.	6. Compressor is damaged due to presence of liquid drops in the suction line.
7. Capacity can be controlled easily by automatic operation.	7. Automatic operation is more complicated.

Example 13.3 A vapour absorption refrigerator working with aqua-ammonia refrigerant is receiving heat by condensation of waste steam at 2 bar and 90% quality. A temperature of –10°C is to be maintained while ambient temperature is 30°C. Calculate the *COP* of the refrigerator.

What will be the required steam flow rate for refrigerator load of 20 tonnes and actual *COP* being 40% of maximum *COP*?

Solution

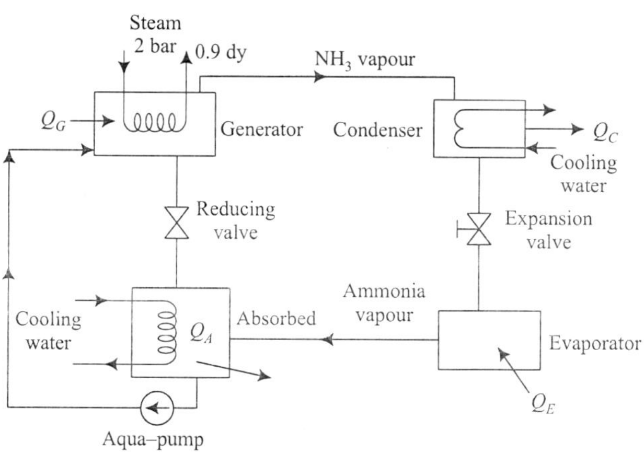

Fig. 13.7 (Example 13.3)

From steam tables at 2 bar,

$$t_{sat} = 120.2°\text{C}$$
$$h_{fg} = 2201.9 \text{ kJ/kg}$$

The maximum $$COP = \frac{(T_1 - T_2)T_R}{(T_2 - T_R)T_1}$$

where, T_1 = Generator temperature
$= 120.2°C + 273 = 393.2$ K
T_2 = Condenser temperature
$= 30°C + 273 = 303$ K
T_R = Evaporator temperature
$= -10°C + 273 = 363$ K

$$\therefore \quad (COP)_{max} = \frac{(393.2 - 303) \times 263}{(303 - 263) \times 393.2} = \mathbf{1.5} \quad \textbf{Ans.}$$

Actual $COP = 1.5 \times 0.4 = 0.60$

Now, $$COP = \frac{Q_E}{Q_G}$$

$$\therefore \quad Q_G = \frac{Q_E}{COP} = \frac{20 \times 14000}{0.60 \times 3600} = 129.6 \text{ kW}$$

Heat transferred by 1 kg of steam on condensation

$$= (h_f + xh_{fg}) - h_f = 0.9 \times 2201.9 = 1981.71 \text{ kJ/kg}$$

$\therefore$ Steam flow rate required

$$= \frac{129.6}{1981.71} = 0.0654 \text{ kg/s}$$

13.6 PRODUCTION OF DRY ICE

Dry ice in the form of solid carbon dioxide is produced by a special vapour compression cycle shown in Fig. 13.8 along with *p-h* diagram of CO_2. Dry ice is used for low temperature refrigeration for preservation of perishables and ice cream, etc. CO_2 is refrigerant as well as product.

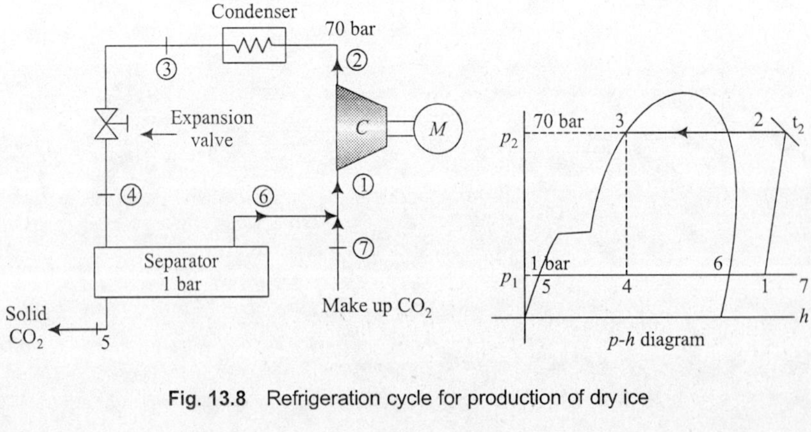

Fig. 13.8 Refrigeration cycle for production of dry ice

QUESTION BANK NO. 13

1. Describe a refrigeration cycle for aircraft cooling.
2. Discuss a vapour compression refrigeration cycle with the help of *T-s* and *p-h* diagram.
3. What is subcooling and superheating of refrigerant in a vapour compression cycle? What are the repercussions of these phenomena?
4. With the help of a line diagram, explain a vapour absorption cycle.
5. Compare the differences between a vapour absorption cycle and vapour compression cycle.
6. How is solid ice produced?

TUTORIAL SHEET NO. 13 NUMERICALS

1. A refrigeration machine is required to produce ice at 0°C from water at 20°C. The machine has a condenser temperature of 298 K while the evaporator temperature is 268 K. The relative efficiency of the machine is 50% and 6 kg of Freon-12 refrigerant is circulated through the system per minute. The refrigerant enters the compressor with a dryness fraction of 0.6. Specific heat of water is 4.187 kJ/kg-K and the latent heat of ice is 335 kJ/kg. Calculate the amount of ice produced in 24 hours. The table of properties of Freon-12 is given below.

Temperature [K]	Liquid heat [kJ/kg]	Latent heat [kJ/kg]	Entropy of liquid [kJ/kg]
298	59.7	138.0	0.2232
268	31.4	154.0	0.1251

[0.661 tonnes]

2. A vapour compression refrigeration system using refrigerant *R*-12 is employed to produce 8640 kg of ice per day. The condensing and evaporating temperatures of the refrigerant are 48°C and 20°C respectively. Saturated liquid leaves the condenser and saturated vapour leaves the evaporator, compression is isentropic. Water at 35°C is used to form ice. The temperature of ice should be –8°C. Heat flows into the brine tank from the surroundings and may be taken to be 10% of total heat removed from the water to form ice at the specified temperature. Determine the power required to drive the compressor in kW. Take specific heat of ice = 2.25 kJ/kg-K, latent heat of ice = 340.72 kJ/kg and specific heat of water = 4.187 kW/kg-K. The following data for *R*-12 may be used:

Temperature	Pressure	Enthalpy liquid	Enthalpy vapour	Entropy liquid	Entropy vapour
[°C]	[bar]	[kJ/kg]	[kJ/kg]	[kJ/kg-K]	[kJ/kg-K]
48	11.64	82.83	205.83	0.2973	0.6802
20	1.51	17.82	178.74	0.0731	0.7087

Specific heat of vapour = 0.82 kJ/kg-K. [21.1 kW]

3. A heat pump working on a reversed Carnot cycle taken in energy from a reservoir maintained at 3°C and delivers it to another reservoir where temperature is 77°C. The heat pump derives power for its operation from a reversible engine operating within the higher and lower temperature limits of 1077°C and 77°C. For 100 kJ/s of energy supplied to the reservoir at 77°C, estimate the energy taken from the reservoir at 1077°C. [25.93 kJ]
4. A cascade refrigeration system of 100 tonnes (350 kW) capacity uses ammonia and carbon dioxide. The evaporating and condensing temperatures of CO_2 are –40°C and 5°C respectively. The evaporating temperature of NH_3 is –7°C. The power supplied to ammonia compressor is 96.5 kW. In the CO_2 circuit, the liquid leaving the evaporator is dry and saturated and compression is isentropic. Calculate the mass flow rate of CO_2 and the *COP* of the system.

   The following data for CO_2 may be used:

Temperature (°C)	Pressure (bar)	Enthalpy (kJ/kg) liquid	Enthalpy (kJ/kg) vapour	Entropy (kJ/kg-K) liquid	Entropy (kJ/kg-K) vapour
–40	10.05	332.7	652.8	3.8531	5.2262
5	39.70	431.0	649.8	4.2231	5.0037

Specific heat of superheated vapour, C_p = 9.85 kJ/kg-K

[*COP* = 1.746]

[$\dot{m}_{CO_2}$ = 94.7 kg/min]

5. A vapour compression heat pump is driven by a power cycle having a thermal efficiency of 25%. For the heat pump, refrigerant 12 is compressed from saturated vaopur at 2.0 bar to the condenser pressure of 12 bar. The isentropic efficiency of the compressor is 80%. Saturated liquid enters the expansion valve at 12 bar. For the power cycle 80% of the heat rejected by it is transferred to the heated space which has a total heating requirement of 500 kJ/min. Determine the power input to the heat pump compressor. The following data for refrigerant 12 may be used.

Pressure (bar)	Temperature (°C)	Enthalpy (kJ/kg) liquid	Enthalpy (kJ/kg) vapour	Entropy (kJ/kg-K) liquid	Entropy (kJ/kg-K) vapour
2.0	–12.53	24.57	182.07	0.0992	0.7035
12.0	49.31	84.21	206.24	0.3015	0.6799

Vapour specific heat at constant pressure = 0.7 kJ/kg [3.2 kW]

6. A Freon-12 refrigerator producing a cooling effect of 20 kJ/s operates on a simple cycle with pressure limits of 1.509 bar and 9.607 bar. The vapour leaves the evaporator dry saturated and there is no undercooling. Determine the power required by the machine.

   If the compressor operates at 300 rpm and has a clearance volume of 3% of stroke volume, determine the piston displacement of the compressor. For compressor, assume that the expansion follows the law $pv^{1.3}$ = constant.

   Given:

Temperature [°C]	p_S [bar]	v_g [m^3/kg]	h_f [kJ/kg]	h_g [kJ/kg]	s_f [kJ/kg-K]	s_g [kJ/kg-K]	C_p [kJ/kg-K]
–20	1.509	1.1088	17.8	178.61	0.073	0.7082	–
40	9.607		74.53	203.05	0.2716	0.682	0.747

[6.29 kW, 477 cm^3]

7. A food storage locker requires a refrigeration capacity of 50 kW. It works between a condenser temperature of 35°C and an evaporator temperature of –10°C, the refrigerant is ammonia. It is subcooled by 5°C before entering the expansion valve by the dry saturated vapour leaving the evaporator. Assuming a single cylinder, single-acting compressor operating at 1000 rpm with stroke equal to 1.2 times the bore, determine:
   (a) the power required, and
   (b) the cylinder dimension

Saturation temp. (°C)	Pressure (bar)	Enthalpy (kJ/kg)		Entropy (kJ/kg)		Specific value [kJ/kg-K]		Sp. heat [kJ/kg-K]	
		Liquid	Vapour	Liquid	Vapour	Liquid	Vapour	Liquid	Vapour
–10	2.9157	154.056	1450.22	0.82965	5.7550	–	0.417477	–	2.492
35	13.522	366.072	148.857	1.56605	5.20086	1.7023	0.095629	4.556	2.903

[10.244 kW, d = 0.106 *m*, L = 0.1278 *m*]

8. A 20 tonne vapour compression refrigeration system using Freon 12 operates between evaporator pressure of 1.004 bar and condenser pressure of 13.663 bar. The system uses a subcooling superheating heat exchanger with 10°C of superheat added. Determine:
   (i) The mass flow rate,
   (ii) The *COP*,
   (iii) The degree of subcooling
   (iv) The power required

The refrigerant leaving the condenser is dry saturated vapour. The compression in compressor is isentropic.

Pressure (bar)	Temperature (°C)	Enthalpy (kJ/kg)		Entropy (kJ/kg-K)		Specific heat [kJ/kg-K]	
		liquid	vapour	liquid	vapour	liquid	vapour
1.004	–30	8.854	174.076	0.0371	0.7165	–	0.5
13.663	55	90.201	207.766	0.3194	0.6777	1.074	–

[0.737 kg/s: 1.932, 5.39°C; 36.41 kW]

9. A 4-cylinder single-acting ammonia compressor with cylinder dimensions as 7.5 × 10 cm operates at 600 rpm. Condenser and evaporator pressures are 12 and 2 bar respectively. The vapour from the evaporator to the suction of compressor is dry and saturated and there is no undercooling in the condenser. Compression takes place according to law $pv^{1.2}$ = constant. If clearance is 2% of the stroke, calculate:
   (i) Refrigeration capacity in tonnes of refrigeration
   (ii) Power required to drive the compressor in kW,
   (iii) Heat rejected to cylinder jacket water in kJ/min.
   (iv) Heat rejected to the condenser in kJ/min.

   Take value of γ for NH_3 as 1.31.

   [7.39 tonnes, 8.98 kW, 31.02 kJ/min, 1958.5 kJ/min]
10. A vapour absorption system has a source temperature of 110°C and refrigeration temperature of –5°C and ambient temperature of 35°C. Calculate its *COP*. [1.315]
11. Dense air is used as refrigerant in Reverse-Brayton Bell-Coleman or Joule cycle. If temperatures at the end of heat absorption and heat rejection are 0°C and 30°C respectively, the pressure ratio is 4 and the pressure in the cooler is 4 bar, determine the temperatures at all state points and volume flow rates at inlet to compressor and at exit of turbine for 1 *TR* cooling capacity. [3.05 kg/min]
12. An ideal vapour compression refrigerator using Freon-12 operates between temperature limits of –10°C and 40°C. The refrigerant leaves the condenser dry saturated. The rate of flow of refrigerant through the unit is 150 kg/hr.

    Calculate the refrigerating affect per kg of refrigerant, the coefficient of performance and capacity of the plant, if,
    (i) The refrigerant leaves the condenser dry saturated as stated.
    (ii) The refrigerant is subcooled to 20°C before throttling. Assume that the enthalpy of refrigerant before throttling is approximately equal to

the enthalpy of liquid refrigerant at undercooled temperature of 20°C and take it as $h_f = 50.59$ kJ/kg.

Give your comments on the result. Solve the problem by using tables.

[(i) 108.53 kJ/kg; 4.02, 1.16

(ii) 132.46 kJ/kg, 4.91, 1.41]

13. One kg of air at a pressure of 1 bar absolute and a temperature of 17°C is compressed to 5 bar absolute. It is then cooled to 26°C in a cooler before entering the expansion cylinder. Determine the *COP* of plant and net refrigerating effect per kg of air. Also show the cycle on *p-v* diagram.

PART IV

HEAT TRANSFER AND HEAT EXCHANGERS

CHAPTER 14

Introduction to Heat Transfer

14.1 INTRODUCTION

Heat is defined as the energy in transit. Heat transfer is the transmission of energy from one region to another as a result of a temperature difference between them. As per the second law of thermodynamics, free flow of heat is possible only from a body of a higher temperature to that at a lower temperature. Since differences in temperatures exist all over the universe, the phenomena of heat flow are as universal as those associated with gravitational attraction.

The science of heat transfer seeks not merely to explain how heat energy may be transferred, but also to predict the rate at which the exchange will take place under certain specified conditions.

14.2 ENGINEERING HEAT TRANSFER

The science of heat transfer is very important in the design and analysis of engineering equipment and plants. The determination of rate of heat transfer at a specified temperature difference is the most important problem.

(i) The dimensions of boilers, heaters, refrigerators and heat exchangers depend not only on the amount of heat to be transmitted, but also on the rate at which the heat is to be transferred under given condition.

(ii) The successful operation of equipment components such as turbine blades or the walls of combustion chambers depends on the possibility of cooling certain metal parts by removing heat continuously at a rapid rate from a surface.

(iii) The design of electrical machines, transformers, bearings requires a detailed heat transfer analysis to avoid conditions that will cause over-heating and damage the equipment.

(iv) The efficient operation of refrigerators and air-conditioning units depends upon effective heat transfer in condensers and evaporators.

(v) The equipment in chemical plants for evaporation, condensation, heating and cooling depend upon heat transfer.

(vi) In civil engineering, cooling of dams, thermal expansion of suspension bridges and railway tracks are based on heat transfer calculation.

(vii) Heat treatment and metallurgy are based on heat transfer rates.

(viii) The problem of large thermal discharges from the power plants and industrial units is associated with heat transfer rates.

Therefore, study of heat transfer is all important for every engineer irrespective of his specialization.

14.3 MODES OF HEAT TRANSFER

Heat transfer takes place by the combination of the following modes.

1. *Conduction*: It is the process of heat transfer by direct contact between particles of a substance. In gases, heat conduction occurs by atomic and molecular interaction and in liquids and solids by elastic molecular collision. In metals, the flow of heat is due mainly to the diffusion of free electrons.
2. *Convection*: It occurs only in liquids and gases by intermixing of particles. Convection is always accompanied by conduction.
3. *Radiation*: The thermal radiation is propagated in the form of electromagnetic waves. There is double transformation of energy. Thermal energy is converted into radiant energy and radiant energy into thermal energy.

14.4 STEADY AND UNSTEADY HEAT TRANSFER

1. *Steady Heat Transfer Process*: The rate of heat flow does not change with time.

   Example: Heating of water in boiler tubes by hot gases.
2. *Unsteady (Transient) Heat Transfer Process:* The temperature of the body and rate of heat transfer change with time.

   Example: Cooling of a casting.
3. *Quasi-steady (Periodic) State Heat Transfer Process*: The variation of temperature in an unsteady state system undergoes a particular cycle.

   Example: Heating and cooling of cylinder wall of an I.C. engine.

14.5 LAWS OF HEAT TRANSFER

The heat transfer processes are governed by certain laws:

- Conduction by Fourier's Law
- Convection by Newton's Law
- Radiation by Stefan-Boltzmann Law

14.5.1 Fourier's Law

The fundamental conditions for the heat transfer by conduction in a solid are:

1. a temperature gradient exists.
2. heat flows in the direction of decreasing temperature.

Fourier's law states that the rate of heat transfer ($\dot{Q}$) is the product of the following factors:

1. Area (A) normal to flow path,
2. Temperature gradient $\left(\dfrac{dt}{dx}\right)$ at the section,
3. Thermal conductivity (k) of the material.

$$\dot{Q} = \frac{dQ}{d\tau} = -kA\frac{dt}{dx} \text{ [kW]}$$

$$k = -\frac{\dot{Q}}{A.\dfrac{dt}{dx}} \text{ [kW/m-°C]}$$

The temperature gradient $\dfrac{dt}{dx}$ is always negative.

Separating the variables,

$$\dot{Q}dx = -kA\,dt$$

Integrating,

$$\dot{Q}\int_{x_1}^{x_2} dx = -kA\int_{t_1}^{t_2} dt$$

$$\dot{Q}(x_2 - x_1) = -kA(t_2 - t_1)$$

$$\dot{Q} = kA\frac{(t_1 - t_2)}{(x_2 - x_1)}$$

$$\dot{Q} = \frac{kA^{\Delta t}}{l}$$

$$\dot{Q} = K(t_1 - t_2)$$

$$\dot{Q} = \frac{t_1 - t_2}{R}$$

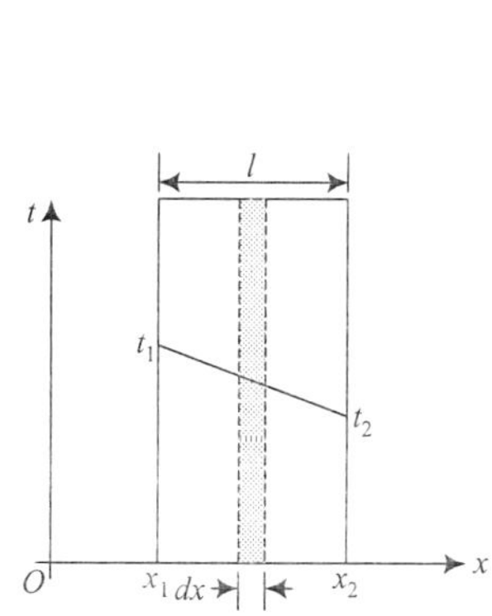

Fig. 14.1 Fourier's law

where $K = \frac{kA}{l}$ and is called thermal conductance [kW/°C] and $R = \frac{l}{kA}$ and is called thermal resistance $\left[\frac{°C}{kW}\right]$.

14.5.2 Newton's Law

The rate of convective heat transfer between a solid boundary and adjacent fluid is given by Newton-Richman Law.

$$\dot{Q} = hA\,\Delta t$$

where, $\dot{Q}$ = Average rate of heat transfer by convection [kW]

A = Heat transfer area [m^2]

Δt = Difference between the temperature of the surface and that of fluid at some specified location. [°C]

h = An average value of proportionality constant called convective heat transfer coefficient or film coefficient of heat transfer [kW/m^2–°C]

$\frac{1}{hA}$ is called the thermal resistance of the film heat transfer process.

$$\dot{Q} = K_C(t_1 - t_2) = \frac{(t_1 - t_2)}{R_C}$$

where, $K_C = hA$ and is called thermal conductance of the film, and

$R_C = \frac{1}{hA}$ and is called thermal resistance of the film.

14.5.3 Stefan Boltzmann Law

The energy radiated by a body is proportional to the fourth power of its absolute temperature. Stefan Boltzmann Law states:

$$\dot{Q} = \sigma A T^4$$

where, $\dot{Q}$ = Maximum rate of heat emission by a body at the given temperature (black body emission) [kW]

A = Area of the body [m^2]

T = Absolute temperature [K]

σ = Stefan-Boltzmann constant

$= 5.67 \times 10^{-8}$ W/m^2 K^4.

A net transfer of heat radiation from a black body of area A_1 at absolute temperature T_1 to another black body at a lower temperature T_2 will be given by:

$$\dot{Q} = \sigma A_1 (T_1^4 - T_2^4)$$

For the case of real bodies,

$$\dot{Q} = \sigma A_1 \varepsilon_1 n_1 F_{12} (T_1^4 - T_2^4)$$

where, $\dot{Q}$ = Net amount of heat exchanged by radiation

ε_1 = Emissivity of surface

F_{12} = A shape factor, depending upon the geometry and relative location of the two surfaces.

or,

$$\dot{Q} = K_r (t_1 - t_2)$$

$$Q = \frac{(t_1 - t_2)}{R_r}$$

where,

$$K_r = \frac{\sigma A_1 \varepsilon_1 F_{12} (T_1^4 - T_2^4)}{(t_1 - t_2)}$$

and

$$R_r = \frac{(t_1 - t_2)}{\sigma A_1 \varepsilon_1 F_{12} (T_1^4 - T_2^4)}$$

14.6 STEADY STATE HEAT CONDUCTION

The steady state heat conduction problems can be solved by:

1. Integration of Fourier's equation. The constants of integration can be evaluated from the boundary conditions which are known temperatures at the surfaces.
2. By the use of analogy between electricity and heat flow. Thermal circuits can be developed and solved.

14.6.1 Conduction of Heat Through a Flat Wall

The heat flow through a small element *dx* of a flat wall under steady condition is given by Fourier's law:

$$d\dot{Q} = -kA \frac{dt}{dx}$$

Separating the variables and integrating,

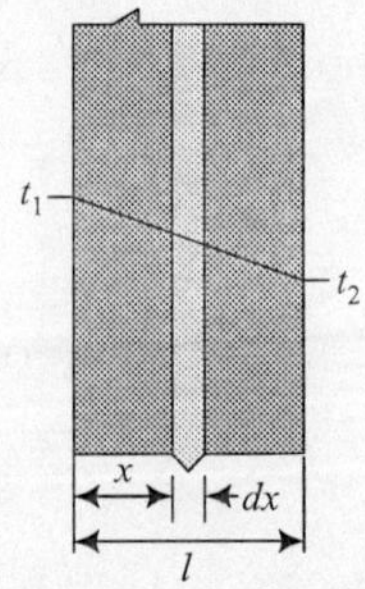

Fig. 14.2 Steady flow conduction through a flat wall

$$\int_0^l d\dot{Q}\,dx = \int_{t_1}^{t_2} -kA\,dt \text{ where } t_1 > t_2.$$

Assuming $\quad k \neq f(t)$

$\therefore \quad \dot{Q}l = -kA(t_2 - t_1) = kA\,(t_1 - t_2)$

$$\dot{Q} = \frac{kA(t_1 - t_2)}{l} = \frac{(t_1 - t_2)}{l/kA} = \frac{(t_1 - t_2)}{R_t}$$

where, $\quad R_t = \dfrac{l}{kA}$ is called thermal resistance of wall.

Using the analogy between electricity and heat flow;
Ohm's law;

$$I = -\frac{\Delta V}{R_e}$$

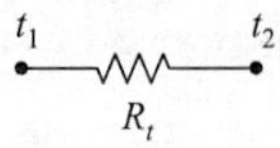

Fourier's law $\quad \dot{Q} = -\dfrac{\Delta t}{R_t} = \dfrac{t_1 - t_2}{R_t}$

14.6.2 Conduction of Heat Through Multilayer Systems

Thermal resistances of flat walls can be found out as:

$$R_1 = \frac{l_A}{k_A A}$$

$$R_2 = \frac{l_B}{k_B A}$$

$$R_3 = \frac{l_C}{k_C A}$$

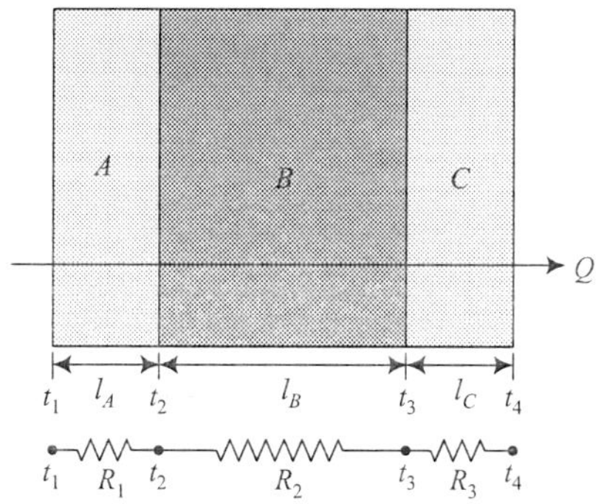

Fig. 14.3 Conduction through a 3-layer system in series

The heat flow rate,

$$\dot{Q} = \frac{(t_1 - t_2)}{R_1} = \frac{(t_2 - t_3)}{R_2} = \frac{(t_3 - t_4)}{R_3}$$

Eliminating t_2 and t_3,

$$\dot{Q} = \frac{(t_1 - t_4)}{R_1 + R_2 + R_3} = \frac{(t_1 - t_4)}{\left(\frac{l}{kA}\right)_A + \left(\frac{l}{kA}\right)_B + \left(\frac{l}{kA}\right)_C} = \frac{\Delta t}{\sum_{i=1}^{3} R_i}$$

For N layers in series,

$$\dot{Q} = \frac{\Delta t}{\sum_{n=1}^{n=N} R_n} = \frac{\Delta t}{\left(\frac{l}{kA}\right)_n}$$

$$= \frac{(t_1) - (t_{N+1})}{\sum_{n=1}^{n=N} (R_t)_n}$$

where Δt = overall temperature difference = thermal potential

14.6.3 Heat Conduction Through a Wall Section with Two Paths in Parallel

$$\dot{Q} = \dot{Q}_1 + \dot{Q}_2$$

$$= \frac{(t_1 - t_2)}{R_1} + \frac{(t_1 - t_2)}{R_2} = \left(\frac{1}{R_1} + \frac{1}{R_2}\right)(t_1 - t_2)$$

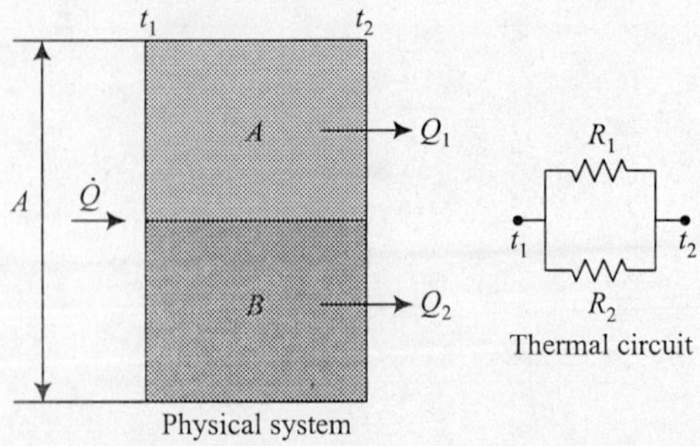

Fig. 14.4 Thermal conduction through parallel walls

where,
$$R_1 = \left(\frac{l}{kA}\right)_A$$

$$R_2 = \left(\frac{l}{kA}\right)_B$$

$$\therefore \qquad \dot{Q} = \frac{(t_1 - t_2)}{\dfrac{R_1 R_2}{(R_1 + R_2)}}$$

Therefore, there is analogy to electrical parallel circuit.

14.6.4 Conduction Through a Wall Consisting of Series and Parallel Thermal Paths

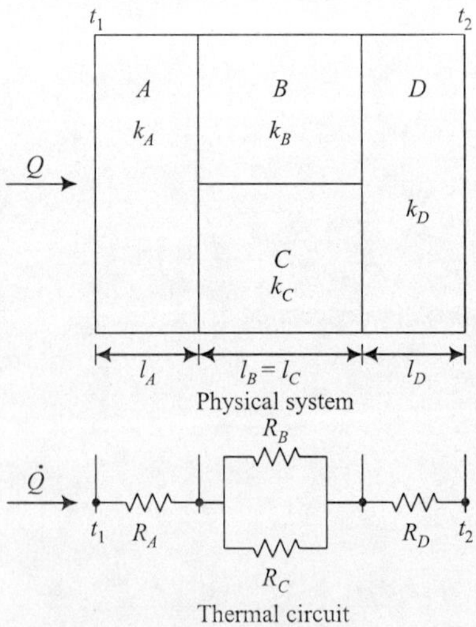

Fig. 14.5 Complex thermal circuit

$$R_1 = R_A = \frac{l_A}{k_A A_A}$$

$$R_B = \frac{l_B}{k_B A_B}$$

$$R_C = \frac{l_C}{k_C A_C}$$

$$R_D = \frac{l_D}{k_D A_D}$$

$$R_3 = R_D$$

$$\frac{1}{R_2} = \frac{1}{R_B} + \frac{1}{R_C} = \frac{R_B + R_C}{R_B R_C}$$

$$R_2 = \frac{R_B R_C}{R_B + R_C}$$

$$\dot{Q} = \frac{(\Delta t)_{\text{overall}}}{\sum_{n=1}^{3} R_n} = \frac{(t_1 - t_2)}{R_1 + R_2 + R_3}$$

$$\dot{Q} = \frac{(t_1 - t_2)}{\frac{l_A}{k_A L_A} + \frac{R_B R_C}{R_B + R_C} + \frac{l_D}{k_D A_D}}$$

14.7 CONDUCTION OF HEAT THROUGH HOLLOW CYLINDER

The heat conduction is considered in radial direction only. The Fourier's equation for a cylinder can be:

$$d\dot{Q} = -kA\frac{dt}{dr}$$

$$= k\,(2\pi r l)\frac{dt}{dr}$$

where, l = length of pipe

$A = 2\pi r l$

Separating the variables and integrating,

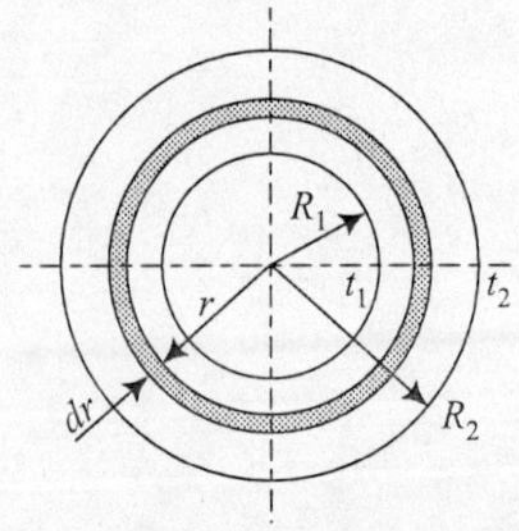

Fig. 14.6 Heat conduction through hollow cylinder

$$\int_{R_1}^{R_2} d\dot{Q}\frac{dr}{r} = \int_{t_1}^{t_2} -k2\pi l\, dt$$

$$\dot{Q}\ln\left(\frac{R_2}{R_1}\right) = -2\pi lk\,(t_2 - t_1) = 2\pi lk\,(t_1 - t_2)$$

$$\therefore \qquad \dot{Q} = \frac{(t_1 - t_2)}{\dfrac{1}{2\pi lk}\ln\left(\dfrac{R_2}{R_1}\right)} = \frac{(t_1 - t_2)}{R_t}$$

The thermal resistance of the pipe of length l,

$$R_t = \frac{\ln\dfrac{R_2}{R_1}}{2\pi lk}$$

The problems on multi-layered pipes can be solved using analogy between electricity and heat flow.

$$\dot{Q} = \frac{(\Delta t)_{\text{overall}}}{\sum_{n=1}^{N} R_n}$$

14.8 CONDUCTION OF HEAT THROUGH HOLLOW SPHERE

The heat flow through a sphere is considered only in radical direction.

The Fourier's equation can be written as:

$$\dot{Q} = -kA\frac{dt}{dr}$$

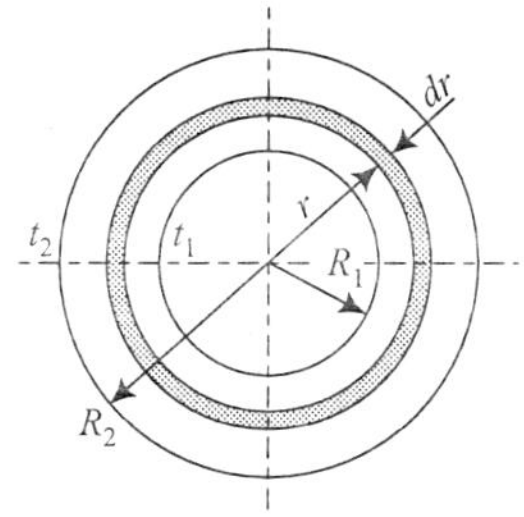

Fig. 14.7 Heat conduction through a hollow sphere

$$\dot{Q} = -k(4\pi r^2)\frac{dt}{dr}$$

where $A = 4\pi r^2$

Separating the variables and integrating,

$$\int_{R_1}^{R_2} \dot{Q}\frac{dr}{r^2} = -4\pi k \int_{t_1}^{t_2} dt$$

$$\dot{Q}\left|-\frac{1}{r}\right|_{R_1}^{R_2} = -4\pi k\,(t_2 - t_1)$$

$\therefore$
$$Q\left[\frac{1}{R_1} - \frac{1}{R_2}\right] = 4\pi k\,(t_1 - t_2)$$

$$\dot{Q} = \frac{4\pi k\,(t_1 - t_2)}{\dfrac{(R_2 - R_1)}{R_1 R_2}} = \frac{(t_1 - t_2)}{\dfrac{(R_2 - R_1)}{4\pi k\, R_1 R_2}}$$

$$= \frac{(t_1 - t_2)}{R_t}$$

The thermal resistance of the sphere,

$$R_t = \frac{(R_2 - R_1)}{4\pi k\, R_1 R_2}$$

The problems on multiwalled sphere can be solved by using the analogy between electricity and heat flow.

$$\dot{Q} = \frac{(\Delta t)_{\text{overall}}}{\sum_{n=1}^{N} R_n}$$

14.9 THERMAL INSULATION

According to Fourier's law, thermal conductivity is defined as

$$k \equiv \frac{\dot{Q}/A}{dt/dx}$$

For engineering calculations, experimental values of thermal conductivity of materials are used. Although for gases at moderate temperature, the kinetic theory of gases may be used but in case of liquids and solids, theories are not adequate to predict the thermal conductivity with satisfactory accuracy. In engineering practice, it is required to decrease heat transfer from a surface for the following reasons:

1. saving fuel
2. ensure temperatures for processes
3. safety

It is attained by applying a layer of heat insulating material upon the wall. A heat insulating material is a material of low thermal conductivity which helps to increase thermal resistance to heat flow. A good insulating material has thermal conductivity below 0.25 W/m-K at 50-100°C. Insulating materials may be used in the following states:

1. Natural state:

   Asbestos, mica, wood, cork, saw duct, peat, earth, etc.

2. With binders or special treatments:

   Glass wool, mineral wool, slag wool, etc.

Thermal conductivity of these materials to a great extent depends upon the porosity. The more porous the material, the lower its thermal conductivity. Heat insulation thickness is calculated with the aid of Fourier's law.

14.10 CRITICAL INSULATION THICKNESS FOR PIPES

The addition of insulation on a plane wall does not affect the face area of heat transfer and rate of heat transfer decreases with increase of insulation thickness. The outer area due to application of heat insulation increases in case of cylindrical pipes and spherical walls. Therefore, rate of heat transfer will decrease due to thickness of insulation but will increase as the outer area increases with increase of insulation thickness. Hence, there will be a critical insulation thickness for cylindrical and spherical walls which will give minimum rate of heat transfer.

An insulated pipe of length l is shown. If the length of the pipe is very long, the heat flow may be assumed radial only. In the steady state, the rate of heat flow is same through each section.

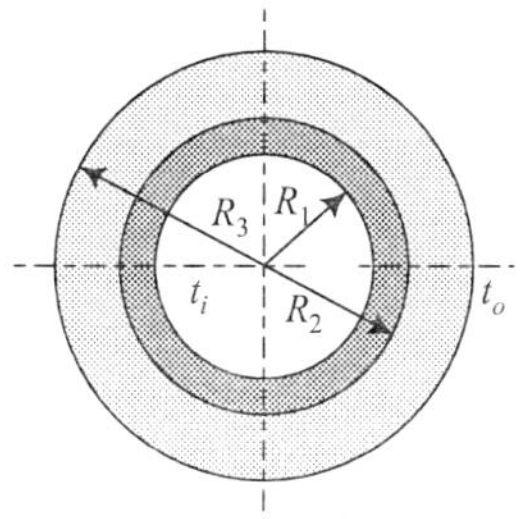

Fig. 14.8 Critical insulation thickness

1. For inner surface, the heat flow through the film by convection,

$$\dot{Q} = 2\pi R_1 l\, h_i\, (t_i - t_1)$$

2. For metal wall,

$$\dot{Q} = -k_1 A \frac{dt}{dr} = -k_1 (2\pi r l) \frac{dt}{dr}$$

Separating the variable and integrating,

$$\dot{Q} \frac{dr}{r} = -k_1 2\pi l\, dt$$

$$\dot{Q} \int_{R_1}^{R_2} \frac{dr}{r} = -k_1 (2\pi l) \int_{t_1}^{t_2} dt$$

$$\dot{Q} \ln \frac{R_2}{R_1} = 2\pi l\, k_1 (t_1 - t_2)$$

$$\dot{Q} = \frac{2\pi l\, k_1 (t_1 - t_2)}{\ln \frac{R_2}{R_1}}$$

3. For insulation,

$$\dot{Q} = \frac{2\pi\, k_2\, l (t_2 - t_3)}{\ln \frac{R_3}{R_2}}$$

4. For outer surface,

$$\dot{Q} = 2\pi R_3 l h_0 (t_3 - t_0)$$

Adding the expressions for temperature differences,

$$(t_i - t_1) + (t_1 - t_2) + (t_2 - t_3) + (t_3 - t_0) = (t_i - t_0)$$

$$= \frac{\dot{Q}}{2\pi R_1 l h_i} + \frac{\dot{Q}\ln\frac{R_2}{R_1}}{2\pi k_1 l} + \frac{\dot{Q}\ln\frac{R_3}{R_2}}{2\pi k_2 l} + \frac{\dot{Q}}{2\pi R_3 h_0 l}$$

$$\therefore \qquad \dot{Q} = \frac{(t_i - t_0)}{\frac{1}{2\pi l}\left\{\frac{1}{h_i R_1} + \frac{\ln\frac{R_2}{R_1}}{k_1} + \frac{\ln\frac{R_3}{R_2}}{k_2} + \frac{1}{h_0 R_3}\right\}}$$

Keeping R_1 and R_2 fixed, maximum heat flow will occur at a particular value of R_3 so that $\frac{d\dot{Q}}{dR_3} = 0$.

$$\therefore \qquad \frac{d\dot{Q}}{dR_3} = \frac{(-2\pi l)(t_i - t_0)\left\{\frac{1}{k_2 R_3} - \frac{1}{h_0 R_3^2}\right\}}{\left\{\frac{1}{h_i R_1} + \frac{\ln R_2/R_1}{k_1} + \frac{\ln R_3/R_2}{k_2} + \frac{1}{h_0 R_3}\right\}^2} = 0$$

$$\therefore \qquad R_3 = \frac{k_2}{h_0} = R_C$$

where R_C is the critical radius.

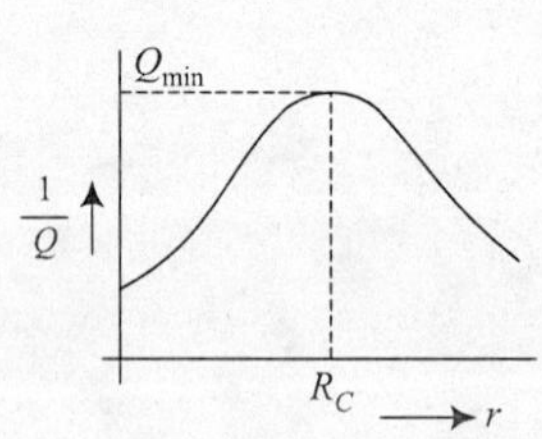

Fig. 14.9 Critical radius

Note:

1. *If* $R_C < R_2$: This is an imaginary case, as outer radius can never be less than inside radius of insulation, but theoretically it is possible to get such a result. The addition of insulation in such a case will always reduce heat transfer rate.
2. *Electrical conductors*: For electrical conductors, the inner radius of insulation or outer radius of wire should always be chosen less than R_C. This will eliminate the anomaly of the necessity of application of insulation against electrical current on one hand and decrease in heat dissipation rate on the other hand, if inside radius of insulation is more than critical radius. The critical radius concept is used to calculate the insulation of electrical wires for which maximum heat removal must be ensured.

14.11 CONVECTIVE HEAT TRANSFER COEFFICIENT

Convection is the transmission of heat from one part of a body to another part by actual motion of the heated particles. As seen in section 14.5.2, the rate of convective heat transfer is given by Newton-Rickman law.

The convective heat transfer coefficient may be defined as the amount of heat transmitted for a temperature difference of 1° between the fluid and unit surface area in one hour. The value of h varies widely with types of fluids, their velocities, temperature involved, dimensions of the pipe, etc. There is no single equation which may express this value under all conditions.

Evaluation of h is therefore difficult because convection is a complex phenomenon. The methods and techniques available can be grouped as:

1. Dimensional analysis and experimental correlation.
2. Hydrodynamic concept of velocity boundary layer and analogous concept of temperature boundary layer.
3. Reynolds similarity between the mechanism of fluid friction in the boundary layer and the transfer of heat by convection.
4. Numerical analysis.

14.11.1 Significance of Dimensionless Groups

(i) Reynolds Number (R_e)

It is the dimensionless parameter that quantitatively relates the viscous and inertia forces and whose value determines the transition from laminar flow to turbulent flow.

$$R_{e_x} = \frac{U_\infty x}{\nu_f}$$

where, U_∞ = free-stream velocity

x = distance from the leading edge

$\nu_f = \dfrac{\mu_f}{\rho_f}$ = kinematic viscosity of the fluid

The transition from laminar to turbulent flow begins at $R_e = 10^5$.

(ii) Prandtl Number (P_r)

It is the dimensionless group that relates the temperature distribution to the velocity distribution. It is the ratio of two molecular transport properties, the kinematic viscosity $\nu = \dfrac{\mu}{\rho}$, which affects the velocity distribution, and thermal diffusibility, $\alpha = k/\rho C_p$, which affects the temperature profile,

$$P_r = \frac{C_p \mu}{k} = \frac{\nu}{\alpha} = \frac{\text{Kinematic viscosity}}{\text{Thermal diffusibility}}$$

In geometrically similar systems having same R_e and P_r, the temperature distribution will be similar.

(iii) Nusselt Number (N_U)

It is the ratio of temperature gradient at a fluid-to-surface interface to a reference temperature gradient.

$$N_{U_L} = \frac{h_c L}{k_f} = \frac{\text{Convection heat transfer in a fluid layer of thickness, } L}{\text{Conduction in the same fluid layer}}$$

This is the dimensionless heat transfer coefficient.

(iv) Stanton Number (S_t)

It is the ratio of heat transfer coefficient to the flow of heat per unit temperature rise due to velocity of fluid.

$$S_t = \frac{h_c}{\rho U_\infty C_p} = \frac{N_{U_L}}{R_{e_L} P_r}$$

Stanton Number can be expressed in terms of other dimensionless numbers. It can be used only in the forced convection. This illustrates the interrelationship between heat transfer and fluid friction process.

(v) Grashoff Number (G_r)

It represents the ratio of buoyant forces to viscous forces.

$$G_{r_L} = \frac{g\beta(T_s - T_\infty)L^3}{\nu^2}$$

where, g = Acceleration due to gravity

β = Coefficient of expansion

$(T_S - T_\infty)$ = Temperature difference

ν = Kinematic viscosity

L = Characteristic length

$$G_{r_L} = \text{Buoyant force} \times \frac{\text{Inertia force}}{(\text{Viscous force})^2}$$

Grashoff number has a role in free convection similar to that played by R_e in forced convection.

14.11.2 Free and Forced Convection

(a) Whenever the flow of fluid over a hot or cold surface takes place due to temperature difference only, the heat transfer between the solid surface and fluid takes place by natural or free convection. The force acting on the fluid to produce its motion is called buoyancy force. It is directly proportional to temperature difference. The heat transfer coefficient is directly proportional to temperature difference. The heat transfer from the condenser of a domestic refrigerator to the surrounding air takes place by natural convection.

$$Nu = \phi(Gr),\ \psi(Pr)$$

The basic equation for free convection is,

$$Nu = C\,(Gr)^a\,(Pr)^b$$

It has been found from rigorous experiments on liquids and gases that $a = b$.

$$\therefore \qquad Nu = C\,(Gr\,Pr)^m$$

Constants C and m depend upon laminar or turbulent flow conditions, geometrical shape (plate or cylinder), and its orientation (vertical or horizontal). The physical properties of the fluid are evaluated at mean film temperature.

For laminar flow, $m = 0.5$ and for turbulent flow $m = 0.33$.

(b) *Forced convection*: When a fluid flows on a hot or cold surface under external pressure, heat transfer takes place by forced convection. The motion of fluid is due to a fan or a blower. Forced convection is used in many engineering heat transfer applications. If the heat transfer coefficient for natural convection with air lies between 12 and 24 W/m^{2}K, its corresponding value for forced convection lies between 60 and 240 W/m^{2}K.

$$Nu = \phi(Re),\ \psi(Pr)$$

$$St = f\,(Re,\ P_r)$$

The basic equations for forced convection are:

$$Nu = C_1(Re)^m\,(Pr)^n$$

$$St = C_2(Re)^a\,(Pr)^b$$

The fluid properties are calculated at bulk temperature.

The average values of convection coefficient,

$$h_c = 0.664 \frac{k}{L}\left(\frac{U_\infty L}{\nu}\right)^{0.5} (\text{Pr})^{0.33}$$

14.12 OVERALL HEAT TRANSFER COEFFICIENT

There is flow of heat between the two fluids separated by a wall. The heat will flow through hot fluid film and cold fluid film by convection and through the wall by conduction.

$$\dot{Q} = h_a A(t_h - t_1)$$

$$\dot{Q} = kA \frac{(t_1 - t_2)}{L}$$

$$\dot{Q} = h_b A (t_2 - t_c)$$

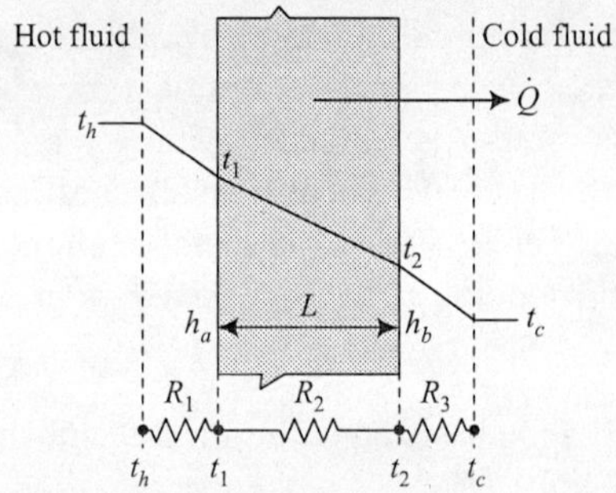

Fig. 14.10

$$(t_h - t_1) = \frac{\dot{Q}}{h_a A}$$

$$(t_1 - t_2) = \frac{\dot{Q}L}{kA}$$

$$(t_2 - t_c) = \frac{\dot{Q}}{h_b A}$$

Adding,

$$(t_h - t_c) = \dot{Q}\left[\frac{1}{h_a A} + \frac{L}{kA} + \frac{1}{h_b A}\right]$$

$$\dot{Q} = \frac{A(t_h - t_c)}{\frac{1}{h_a} + \frac{L}{k} + \frac{1}{h_b}}$$

Let $$U = \frac{1}{\frac{1}{h_a} + \frac{L}{k} + \frac{1}{h_b}}$$

$\therefore$ $$\dot{Q} = UA\,(t_h - t_c)$$

where U is called overall heat transfer coefficient.

Using electricity analogy,

$$R_1 = \frac{1}{(h_a A)}$$

$$R_2 = \frac{L}{(kA)}$$

$$R_3 = \frac{1}{(h_b A)}$$

$$\therefore \qquad \dot{Q} = \frac{(T_h - T_c)}{\sum_{i=1}^{3} R_i} = \frac{\Delta T}{R_1 + R_2 + R_3}$$

$$= \frac{(T_h - T_c)}{\frac{1}{h_a A} + \frac{L}{kA} + \frac{1}{h_b A}}$$

14.12.1 Composite Three-layer Wall with Convection Over Both Exterior Surfaces

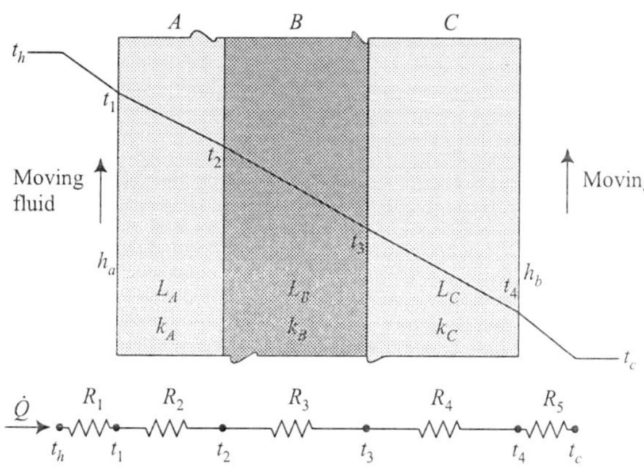

Fig. 14.11

$$R_1 = \frac{1}{h_a A}$$

$$R_2 = \frac{L_A}{k_A A}$$

$$R_3 = \frac{L_B}{k_B A}$$

$$R_4 = \frac{L_C}{k_C A}$$

$$R_5 = \frac{1}{h_b A}$$

$$\dot{Q} = \frac{R(t_h - t_c)}{R_1 + R_2 + R_3 + R_4 + R_5}$$

14.12.2 Fluids Separated by the Walls of a Composite Pipe

Hot fluid flows inside the tube and cold fluid outside.

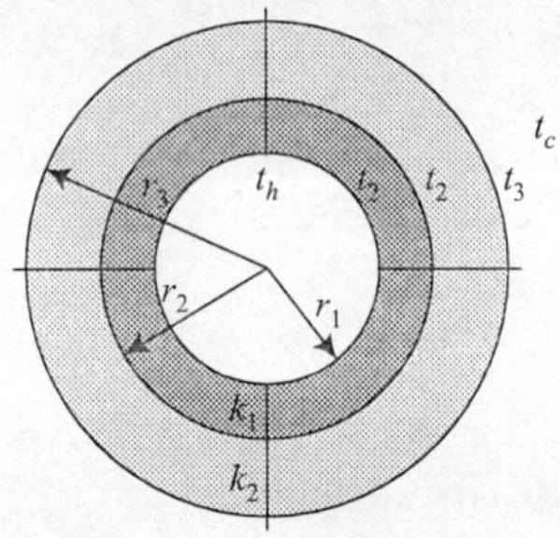

Fig. 14.12

$$\dot{Q} = \frac{(t_h - t_c)}{\dfrac{1}{h_a A_a} + \dfrac{\ln\dfrac{r_2}{r_1}}{2\pi k_1 L} + \dfrac{\ln\dfrac{r_3}{r_2}}{2\pi k_2 L} + \dfrac{1}{A_b h_b}}$$

$$\dot{Q} = U_a A_a (t_h - t_c)$$

$$h = U_b A_b (t_h - t_c)$$

U_a and U_b are overall heat transfer coefficients referred to inside and outside areas respectively.

14.12.3 Fluids Separated by the Walls of a Composite Sphere

$$\dot{Q} = \frac{(t_h - t_c)}{\dfrac{1}{h_a A_a} + \dfrac{r_2 - r_1}{4\pi k_1 r_1 r_2} + \dfrac{r_3 - r_2}{4\pi k_2 r_2 r_3} + \dfrac{1}{h_b A_b}}$$

$$= U_a A_a (t_h - t_c)$$

$$= U_b A_b (t_h - t_c)$$

U_a and U_b are overall heat transfer coefficients referred to inside and outside area respectively.

14.13 RADIATION

Radiation is the transfer of heat between two bodies situated at a distance apart. Thermal radiations are electromagnetic radiations emitted by a body as a result of its temperature. A vacuum is the perfect transmitter of radiant heat. The thermal radiations are propagated at the speed of light.

14.13.1 Radiation Properties

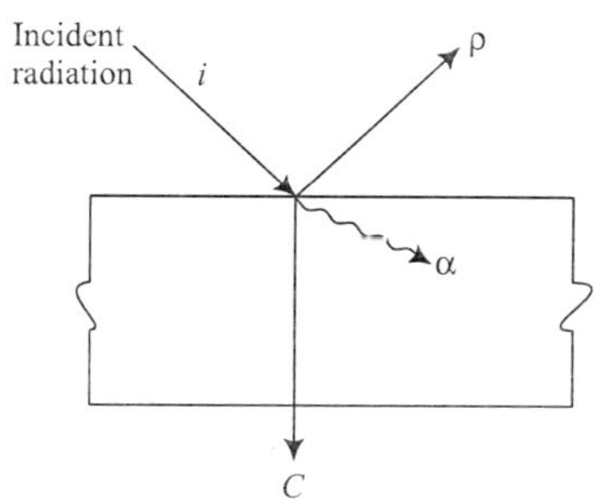

Fig. 14.13

The incident radiation is partly reflected, partly absorbed and partly transmitted

$$\rho + \alpha + \tau = 1$$

where, ρ = reflectivity = fraction of incident radiation reflected

α = absorptivity = fraction of incident radiation absorbed

τ = transmissivity = fraction of incident radiation transmitted

For opaque body, $\tau = 0$

$$\therefore \quad \rho + \alpha = 1$$

For rough surface, incident beam is distributed uniformly in all directions after reflection.

Emission power of a body (E) is the amount of energy emitted per unit area and per unit time.

Emissivity (ε): The ratio of emissive power of a body to emission power of a black body at the same temperature is equal to absorptivity and is called emissivity (ε) of the body.

$$\alpha = 1 \text{ for a black body}$$

$$\varepsilon = 1 \text{ for a black body}$$

$$\alpha = \frac{E}{E_b} = \varepsilon$$

$$\varepsilon = \alpha \text{ Kirchhoff's identity}$$

$$\varepsilon = \varepsilon(T, \lambda)$$

$$\varepsilon \neq \varepsilon(\lambda) \text{ for grey body}$$

Monochromatic Emissivity (ε_λ): It is the ratio of the monochromatic-emissive power of a body to the monochromatic emissive power of a black body.

$$\varepsilon_\lambda = \frac{E_\lambda}{E_{b_\lambda}}$$

where E_{b_λ} is the emissive power of a black body per unit wavelength.

For a grey body, ε_λ = constant

$\therefore$ $\varepsilon = \varepsilon_\lambda$

14.13.2 Black Body Concept

A perfect black body does not exist as all surfaces reflect radiation to some extent. Therefore, concept of a black body is an idealization.

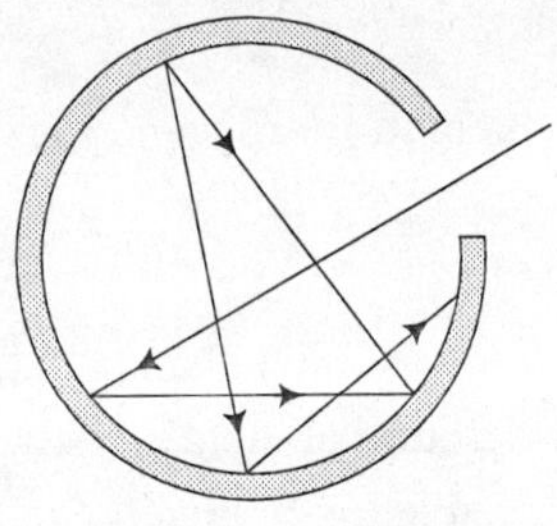

Fig. 14.14 Black body concept

A cavity with a very small aperture behaves approximately as a black body emitter as well as absorber. An incident ray of energy is reflected repeatedly and a fraction of it is absorbed corresponding to the absorptivity of the inside cavity. After many absorptions, practically all incident radiation is absorbed.

14.14 LAWS OF RADIATIONS

14.14.1 Planck's Distribution Law

The emissivities of various substances vary widely with wavelength, temperature and surface conditions. For a given wavelength, a body radiates much more energy at elevated temperatures. Similarly, emitted radiation is strongly influenced by the wavelength even if the temperature of the body remains constant.

The functional relationship for E_{b_λ} was derived by Planck using quantum concept for electromagnetic energy. He suggested the following law for the spectral distribution of emission power.

$$E_{b_\lambda} = \frac{U_A C}{4}$$

$$= \frac{C_1 \lambda^{-5}}{e^{C_2/\lambda T} - 1}$$

where,

U_A = energy density

λ = wavelength

T = temperature

$C_1 = 2\pi C^2 h = 0.374 \times 10^{-15}$ Jm^2/s

$C_2 = \frac{Ch}{K} = 1.4388 \times 10^{-2}$ mK

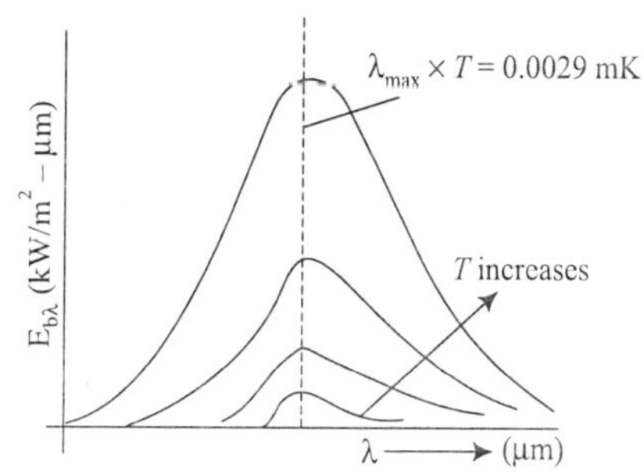

Fig. 14.15

14.14.2 Wien's Displacement Law

The wavelength associated with maximum rate of emission depends upon the absolute temperature of radiating surface.

$$\lambda_{max} T = 28976 \ \mu\text{m-K}$$

The magnitude of the maximum monochromatic emission power varies proportionally with fifth power of the absolute temperature of black body.

$$\left(E_{b_\lambda}\right)_{max} = 1.285 \times 10^{-5} \ T^5 \text{W/m}^2 \text{ per m wavelength}$$

14.14.3 Kirchhoff's Law

It establishes the emissivity and absorptivity of a body

$$\frac{E}{\alpha} = E_b = E(T)$$

The ratio of emissivity power and absorptivity is the same for all materials and is equal to the emissive power of a black body at the same temperature and is a function of temperature only.

The absorptivity and emissivity of a body are numerically equal.

$$\alpha = \varepsilon$$

14.15 HEAT TRANSFER BY RADIATION BETWEEN TWO SURFACES

The heat transfer by radiation between two surfaces depends upon the following factors.

1. The absolute temperatures of the two bodies
2. The emissivity of two bodies
3. The relative position of two bodies

$$\dot{Q} = \sigma\, F_{1-2} \cdot f_{1-2}\, A_1\, (T_1^4 - T_2^4)$$

where, σ = Stefen Boltzmann's constant

$= 5.67 \times 10^{-8}$ W/m^{2}K^4

$T_1 > T_2$

F_{1-2} is shape factor or configuration factor and depends upon geometrical positions of the two bodies.

For example,

(i) Infinite parallel plates $F_{1-2} = 1$

(ii) A small body enclosed in a big body, $F_{1-2} = 1$

(iii) Concentric spheres or cylinders, $F_{1-2} = 1$

f_{1-2} is energy interchange factor and depends upon emissivities of the two bodies.

For example,

(i) Infinite parallel plate $f_{1-2} = \dfrac{1}{\dfrac{1}{\varepsilon_1} + \dfrac{1}{\varepsilon_2} - 1}$

(ii) A small body enclosed in a big body, $f_{1-2} = \varepsilon_1$

14.15.1 Net Radiation Heat Transfer from Surface 1 to Surface 2

$$\dot{Q} = A_1\,\varepsilon_1\,F_{1-2}\,\sigma\,(T_1^4 - T_2^4)$$

$$R_r = \frac{(T_1 - T_2)}{A_1\varepsilon_1F_{1-2}\sigma\left(T_1^4 - T_2^4\right)}$$

∴ $$\dot{Q} = \frac{(T_1 - T_2)}{R_r} = \frac{\Delta T}{R_r}$$

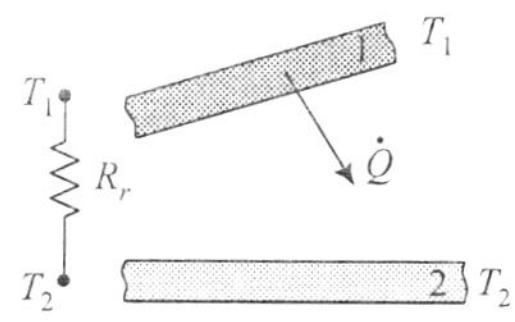

Fig. 14.16

14.15.2 Convection and Radiation in Parallel

Heat transfer,

$$\dot{Q} = \dot{Q}_c + \dot{Q}_r$$
$$= h_cA\,(T_1 - T_2) + h_rA\,(T_1 - T_2)$$
$$= (h_c + h_r)A\,(T_1 - T_2)$$

$$R_C = \frac{1}{h_cA_1}$$

$$R_r = \frac{1}{h_rA_1}$$

∴ $$\dot{Q} = \frac{T_1 - T_2}{R_c} + \frac{T_1 - T_2}{R_r}$$

$$= \frac{T_1 - T_2}{R} = hA\,(T_1 - T_2)$$

$$h = h_r + h_c$$

$$h_r = \frac{\varepsilon_1\sigma(T_1^4 - T_2^4)}{T_1 - T_2} \quad \text{For } F_{1-2} = 1$$

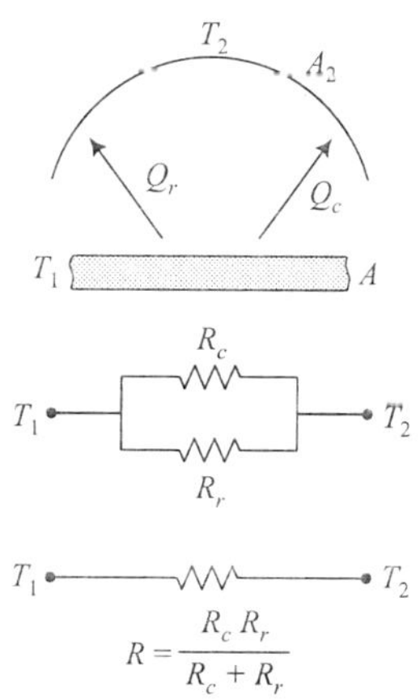

Fig. 14.17

14.16 HEAT EXCHANGERS

A heat exchanger is a device in which heat is transferred between a warmer and a colder substance, usually fluids.

14.16.1 Plane Wall Heat Exchanger

$$\dot{Q} = UA\,\Delta T$$

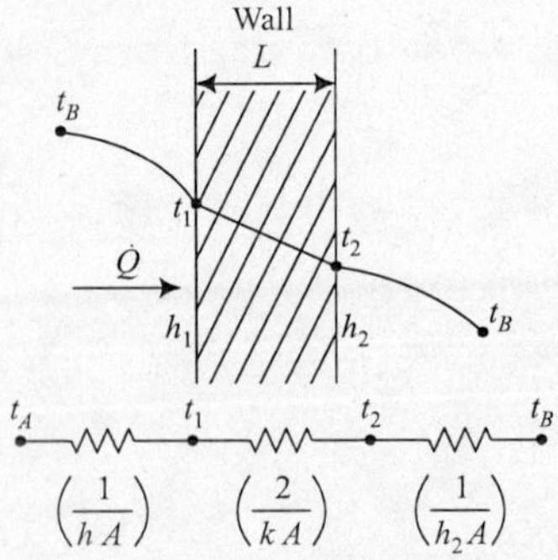

Fig. 14.17

$$\dot{Q} = \frac{T_A - T_B}{\frac{1}{h_1 A} + \frac{L}{kA} + \frac{1}{h_2 A}}$$

14.16.2 Double Pipe Heat Exchanger

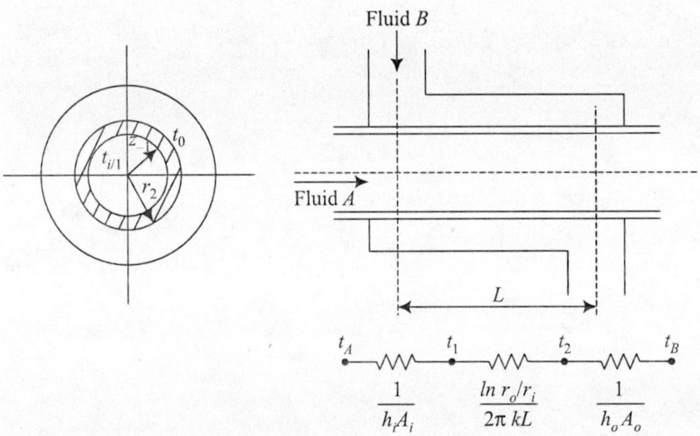

Fig. 14.19 Double pipe heat exchanger

$$\dot{Q} = \frac{T_A - T_B}{\frac{1}{h_i A_i} + \frac{A_i \ln(r_0/r_i)}{2\pi kL} + \frac{1}{h_0 A_0}}$$

$$\dot{Q} = UA\ \Delta T$$

$$U_i = \frac{1}{\dfrac{1}{h_i A_i} + \dfrac{A_i \ln(r_0/r_i)}{2\pi k L} + \dfrac{1}{h_0 A_0}} \quad \text{based on inside area}$$

$$U_o = \frac{1}{\dfrac{A_o}{A_i}\dfrac{1}{h_i} + \dfrac{A_o \ln(r_0/r_i)}{2\pi k L} + \dfrac{1}{h_0}} \quad \text{based on outside area}$$

14.16.3 The Log Mean Temperature Difference

The temperatures of the fluids in a heat exchanger vary from point to point as heat flows from the hotter to the colder fluids.

Temperature Distribution in Heat exchanger

1. Single pass condenser

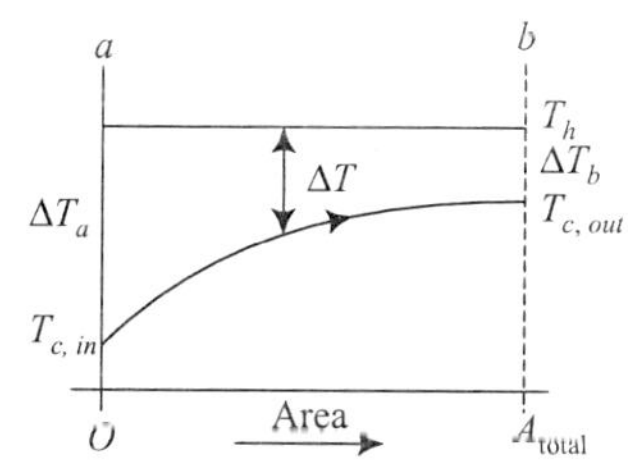

Fig. 14.20(a)

2. Single pass evaporator

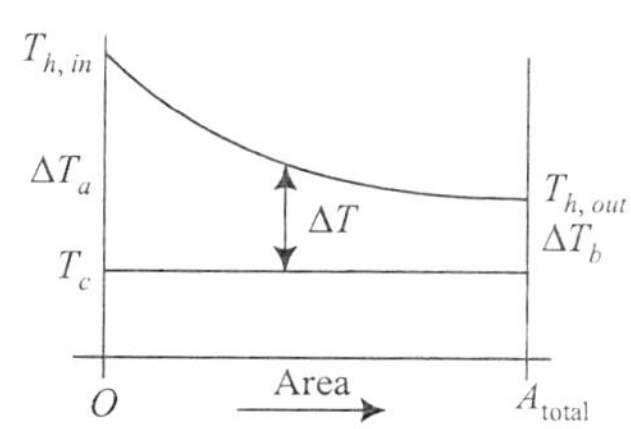

Fig. 14.20(b)

3. Single pass parallel flow heat exchanger

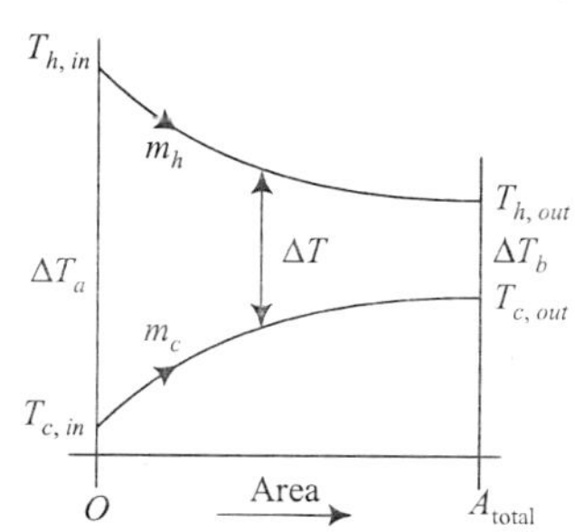

Fig. 14.20(c)

4. Single pass counterflow heat exchanger

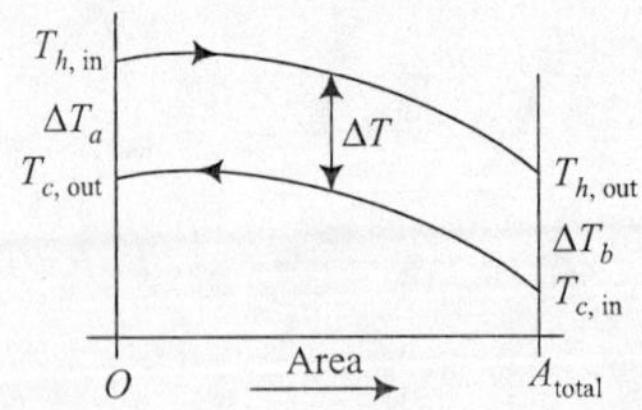

Fig. 14.20(d)

The rate of heat transfer

$$dq = UdA\Delta T$$

Let $$T_h - T_C = \Delta T$$

$$\dot{Q} = UA\frac{\Delta T_a - \Delta T_b}{\ln\left(\frac{\Delta T_a}{\Delta T_b}\right)} = UA\Delta T_m$$

$$\Delta T_m = \frac{\Delta T_a - \Delta T_b}{\ln\left(\frac{\Delta T_a}{\Delta T_b}\right)}$$

ΔT_m is called logarithmic mean temperature difference or LMTD.

If $$\dot{m}_h C_{p_h} = \dot{m}_c C_{p_c}$$

$$\Delta T_m = \Delta T_a = \Delta T_b$$

14.16.4 Effectiveness of Heat Exchanger

The effectiveness of a heat exchanger is the ratio of actual heat transfer and maximum possible heat transfer.

Actual heat transfer for a parallel flow heat exchanger,

$$\dot{Q} = \dot{m}_h C_h\left(T_{h_1} - T_{h_2}\right)$$

$$= \dot{m}_c C_c\left(T_{c_1} - T_{c_2}\right)$$

Actual heat transfer for a counterflow heat exchanger,

$$\dot{Q} = \dot{m}_h C_h\left(T_{h_1} - T_{h_2}\right)$$

$$= \dot{m}_c C_c\left(T_{c_1} - T_{c_2}\right)$$

Maximum possible flow heat transfer,

$$\dot{Q}_{\max} = (\dot{m}_c)_{\min}\left[T_{h_{in}} - T_{c_{in}}\right]$$

For parallel flow heat exchangers,

$$\varepsilon = \frac{T_{h_1} - T_{h_2}}{T_{h_1} - T_{c_1}} = \frac{T_{c_2} - T_{c_1}}{T_{h_1} - T_{c_1}}$$

For counterflow heat exchangers,

$$\varepsilon = \frac{T_{h_1} - T_{h_2}}{T_{h_1} - T_{c_2}} = \frac{T_{c_1} - T_{c_2}}{T_{h_1} - T_{c_2}}$$

TUTORIAL SHEET NO. 14.1 (THEORY)

1. Explain the engineering importance of heat transfer.
2. Explain the following laws of heat transfer
   (a) Fourier's law
   (b) Newton-Rikhman law
   (c) Stefan Boltzmann law
3. What is critical radius of insulation? Derive an expression for critical radius insulation for a cylindrical conductor.
4. Explain the concept of driving potential and resistance as applied to heat transfer problem.
5. What are Grashoff and Prandtl numbers? Explain their significance in heat transfer.
6. Write a note on LMTD correction factors in heat exchanger analysis.
7. Distinguish between:
   (a) A black body and a grey body
   (b) Specular and diffuse surfaces
   (c) Absorptivity and emissivity of a surface
8. Write short note on thermal insulation.
9. Distinguish between free and forced convection.
10. What are various laws of radiation? Explain briefly their significance.
11. What is effectiveness of a heat exchanger?

TUTORIAL SHEET NO. 14.2 (CONDUCTION)

1. The walls of a room consist of parallel layers of concrete, brick and wood of thickness 2 cm, 30 cm and 1 cm, respectively. Find the rate of heat flow by conduction per square metre area and per minute, if the temperature of inside wall surface is 30°C and outside is 5°C. The values of k for concrete, brick and wood are 0.29, 0.46 and 0.167 W/mK respectively.
2. The furnace wall is made of 250 mm fire brick ($k = 1.05$ W/mK) 120 mm of insulation brick ($k = 0.15$ W/mK) and 200 mm of red brick ($k = 0.85$ W/mK). The temperature of inner and outer wall surfaces are 850°C and 65°C respectively. Compute the temperatures at the contact joint surfaces.
3. A pipe of 15 cm outside diameter is covered with 5 cm thick insulation ($k = 0.035$ W/m-K). The pipe is 40 m long. The inside and outside surface temperatures of insulation are 320°C and 15°C respectively.
   (a) Calculate radial heat flow is kJ/hr.
   (b) Thickness of insulation required to reduce heat loss of 5000 kJ/hr.
4. Calculate the critical radius of insulation for asbestos ($k = 0.17$ W/mK) surrounding a pipe and exposed to room temperature at 20°C with $h = 3$ W/m^2-K. Calculate the heat loss from a 200°C, 5 cm diameter pipe when covered with the critical radius of insulation and without insulation.
5. A 10 cm *OD* pipe exposed to atmosphere is to be covered with two layers of insulation each 3 cm thick. The average thermal conductivity of asbestos is 20 times that of magnesia. Work out your preferred arrangement for same temperature difference between steam and atmosphere and give the per cent savings in heat loss by your arrangement.
6. Compare the heat loss from an insulated and uninsulated copper pipe under the following conditions:
   The pipe ($k = 400$ W/m-K) has an insulation diameter of 10 cm and external diameter of 12 cm. Saturated steam flows inside the pipe at 110°C. The pipe is located in a space at 30°C and the heat transfer coefficient on its outer surface is estimated to be 15 W/m^2-K. The insulation available to reduce heat loss is 5 cm thick and its conductivity is 0.20 W/m-K.
7. A current of 200A is passed through a stainless steel wire ($k = 19$ W/m-K) 3 mm in diameter. The resistivity of steel may be taken as 30 μΩ and the length of the wire is 1 m. The wire is submerged in a liquid at 110°C and experiences a convective heat transfer coefficient of 4 kW/m^2. Calculate the centre temperature of wire.
8. A flat plate fuel element for a nuclear reactor is 6 mm thick and is clad on each face with aluminum 2 mm thick. The rate of heat generation is uniform within the element and has a magnitude of 5.6×10^8 W/m^3. If the

coolant temperature is 140°C, make calculations for the temperature at free surface of aluminum, aluminum/uranium interface and at the centre of the fuel element.

Thermal conductivity for uranium and aluminum are 24.5 W/m-K and 200 W/m-K respectively and heat transfer coefficient at the aluminum/coolant interface is 2.85×10^4 W/m^2-K.

SOLUTION OF TUTORIAL SHEET NO. 14.2

1. $$\dot{Q} = \frac{t_i - t_0}{\sum \frac{L}{kA}}$$

$$\Sigma \frac{L}{kA} = \frac{1}{A}\left[\frac{L_1}{k_1} + \frac{L_2}{k_2} + \frac{L_3}{k_3}\right]$$

$$= \frac{0.02}{0.29} + \frac{0.30}{0.46} + \frac{0.01}{0.167}$$

$$= 0.7811$$

$$\therefore \qquad \dot{Q} = \frac{30.5}{0.7811} = \mathbf{32\ W} \quad \textbf{Ans.}$$

2. The heat flow rate per m^2 of wall is

$$\dot{Q} = \frac{1 \times (850 - 65)}{\frac{0.250}{1.05} + \frac{0.120}{0.15} + \frac{0.200}{0.85}} = 616.5 \text{ W/m}^2$$

$$\frac{\dot{Q}}{A} = \frac{t_i - t_1}{\frac{L_1}{k_1}}$$

$$616.5 = \frac{850 - t_1}{\frac{0.250}{1.05}}$$

$$\therefore \qquad t_1 = 703°\text{C}$$

Also, $$\frac{\dot{Q}}{A} = \frac{t_1 - t_2}{\frac{L_2}{k_2}}$$

$$t_2 = 209.8\ °\text{C}$$

The temperature of joint between fire brick and insulation brick is 703°C and that between insulation brick and red brick is 209.8°C.

3. The insulation has inside diameter of 15 cm and outside diameter of $(15 + 2 \times 5) = 25$ cm

$$\frac{\dot{Q}}{L} = \frac{2\pi k(t_1 - t_2)}{\ln r_2 / r_1}$$

For insulation layer

$$\frac{\dot{Q}}{L} = \frac{2\pi \times 0.035\,(320 - 15)}{\ln\left(\frac{25}{15}\right)} = 131.1 \text{ W/m}$$

$$\dot{Q} = 131.1 \times 40 = 5252 \text{ J/s}$$

$$= \frac{5252 \times 3600}{1000} = \mathbf{18907\ kJ/hr} \quad \textbf{Ans.}$$

(b) Let t cm be the thickness of insulation

$$\dot{Q} = \frac{5000 \times 1000}{3600} = 1389 \text{ W}$$

$$\frac{\dot{Q}}{L} = \frac{2\pi k(t_1 - t_2)}{\ln\left(\frac{r_1 + 2t}{r_1}\right)}$$

$$\therefore \quad 1389 = \frac{2\pi \times 0.035(320 - 15) \times 40}{\ln\left(\frac{d + 2t}{d}\right)}$$

$$\ln\left(\frac{d + 2t}{d}\right) = 1.9315$$

$$\frac{d + 2t}{d} = 6.9$$

$$d = 15 \text{ cm}$$

$$\therefore \quad t = \mathbf{44.25\ cm} \quad \textbf{Ans.}$$

4. For critical insulation radius,

$$r_c = \frac{k}{h} = \frac{0.17}{3} = 0.0566 \text{ m}$$

$r_c = 5.67$ cm

Inside radius of insulation,

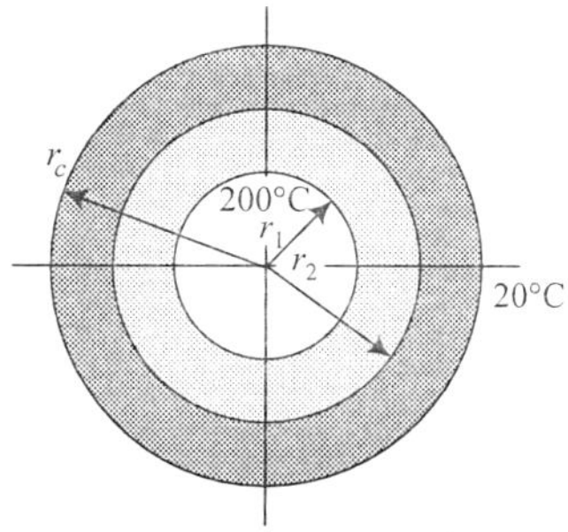

$$r_2 = \frac{5}{2} = 2.5 \text{ cm}$$

$$\frac{\dot{Q}}{L} = \frac{2\pi(T_i - T_\infty)}{\frac{\ln r_c/r_2}{k} + \frac{1}{r_c h}}$$

$$= \frac{2\pi(200-20)}{\frac{\ln(5.67/2.5)}{0.17} + \frac{1}{0.0567 \times 3}} = 105.7 \text{ W/m}$$

Without insulation,

$$\frac{\dot{Q}}{L} = 2\pi rh\,(T_i - T\infty) = 2\pi\,(90.025) \times 3\,(200-20) - 84.8 \text{ W/m}$$

There is an increase in heat transfer by $\frac{105.7 - 84.8}{84.8}$ = **24.6%. Ans.**

5. The thermal circuit is shown

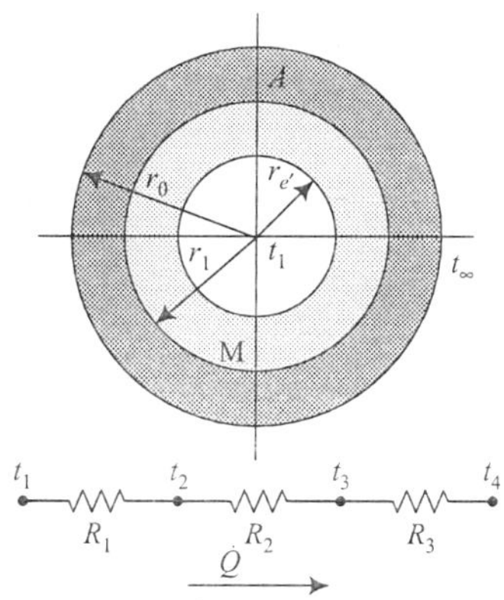

$$R_1 = \frac{\ln r_1/r_i}{2\pi k_1}$$

$$R_2 = \frac{\ln r_0/r_i}{2\pi k_2}$$

$$R_3 = \frac{1}{2\pi r h_0}$$

$$\dot{Q} = \frac{\Delta T}{\Sigma R_i}$$

When magnesia is on the steel pipe

$$|\Sigma R|_m = \frac{\ln 8/5}{2\pi\, k_m} + \frac{\ln 11/8}{2\pi\, 20\, k_m}$$

$$= \frac{0.0748}{k_m} + \frac{0.0025}{k_m} = \frac{0.07733}{k_m}$$

When asbestos is on the steel pipe

$$|\Sigma R|_a = \frac{\ln 8/5}{2\pi \times 20 k_m} + \frac{\ln 11/5}{2\pi k_m}$$

$$= \frac{0.00374}{k_m} + \frac{0.05068}{k_m} = \frac{0.0544}{k_m}$$

$$|\Sigma R|_m > (\Sigma R)_a$$

Therefore, magnesium should be placed on the steel pipe.

$$\dot{Q} = \frac{\Delta T}{\Sigma R}$$

$$|\dot{Q}|_m = \frac{\Delta T \text{ km}}{0.007733} = 12.93\ k_m\ \Delta T$$

$$|\dot{Q}|_a = \frac{\Delta T \text{ km}}{0.0544} = 18.382\ k_m\ \Delta T$$

$$\text{Percentage saving} = \frac{18.382 - 12.93}{12.93} = 42\%$$

6. For uninsulated pipe;

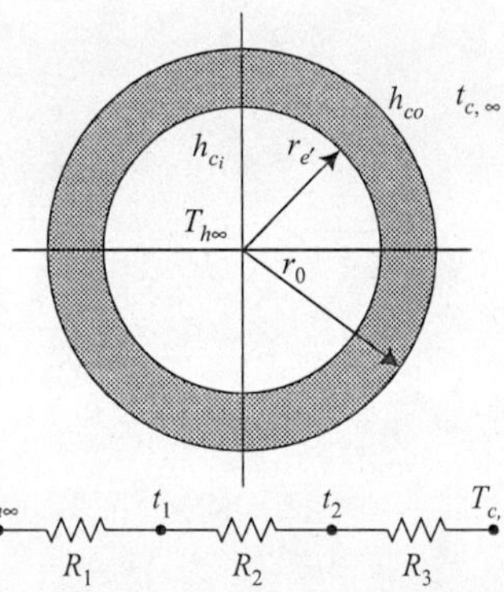

$$\frac{\dot{Q}}{L} = \frac{T_s - T_\infty}{R_1 + R_2 + R_3}$$

$$R_1 = \frac{1}{2\pi r_i R_{c_i}}$$

For saturated steam condition,

$$h_{c_i} = 10{,}000 \text{ W/m}^2\text{-K}$$

$$\therefore \quad R_1 = \frac{1}{2\pi (0.05)(10{,}000)}$$

$$= 0.000318 \text{ mK/W}$$

$$R_2 = \frac{\ln r_0/r_i}{2\pi k_{pipe}} = \frac{0.182}{2\pi \times 400} = 0.00007 \text{ mK/W}$$

$$R_3 = \frac{1}{2\pi r_0 h_{c,o}} = \frac{1}{2\pi (0.06)(15)} = 0.177 \text{ m-K/W}$$

$$R_3 >> R_1$$
$$R_3 >> R_2$$

$$\therefore \quad \frac{\dot{Q}}{L} = \frac{(110-30)}{0.177} = 452 \text{ W/m}$$

For insulated pipe,

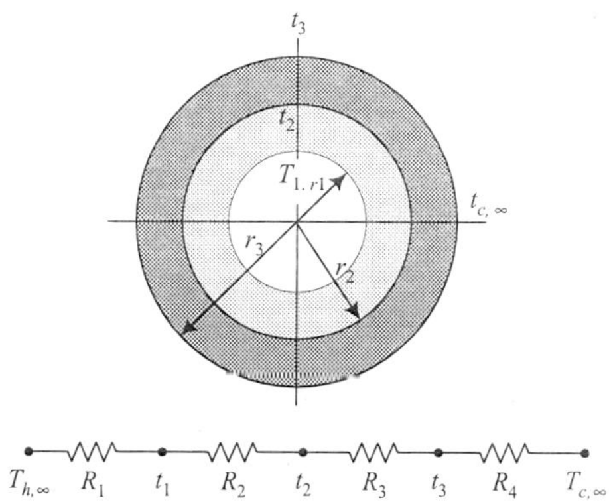

$$R_1 = 0.000318 \text{ mK/W}$$

$$R_2 = \frac{\ln r_2/r_1}{2\pi k_{\text{pipe}}}$$

$$= 0.00007 \text{ mK/W}$$

$$R_3 = \frac{\ln 11/6}{2\pi(0.2)} = 0.482 \text{ mK/W}$$

$$R_4 = \frac{1}{2\pi(0.11)\times 15} = 0.096 \text{ mK/W}$$

$$\Sigma R = R_1 + R_2 + R_3 + R_4 = 0.00007 + 0.000318 + 0.482 + 0.096$$
$$= 0.578 \text{ mK/W}$$

$$\dot{Q} = \frac{80}{0.578} = 138 \text{ W/m}$$

$$\text{Reduction in heat loss} = \left(\frac{452-138}{452}\right) \times 100 = 70\%$$

7. $$P = I^2R = qv = hA\,(T_W - T_\infty)$$

$$R = \rho \frac{L}{A} = \frac{(70 \times 10^{-6}) \times 100}{\pi (0.15)^2} = 0.099\ \Omega$$

$$A = \pi\, dL$$

$$P = (200)^2\ (0.099) = 4000 \times \pi \times 3 \times 10^{-3} \times 1 \times (T_W - 110)$$

$$= 3960 \text{ W}$$

$$\therefore \quad T_W = 215°\text{C}$$

$$P = \dot{Q}\, v = \dot{Q}\, \pi r^2 L$$

$$\dot{Q} = \frac{3960}{\pi (1.5 \times 10^{-3})^2 \times (1)} = 560.2 \text{ MW/m}^3$$

$$T_o = \frac{\dot{Q} r_0^2}{4k} + T_W = \frac{(560.2 \times 10^8)\,(1.5 \times 10^{-3})^2}{4 \times 19} + 45$$

$$= \mathbf{231.6°C} \quad \textbf{Ans.}$$

8. $$\dot{Q}_G = 5.6 \times 10^8 \times 1 \times 6 \times 10^{-3}$$

$$= 33.6 \times 10^5 \text{ W/unit area}$$

In each direction,

$$\frac{33.6 \times 10^5}{2} = \frac{kA}{L}(t_1 - t_2) = h_c A (t_2 - t_\infty)$$

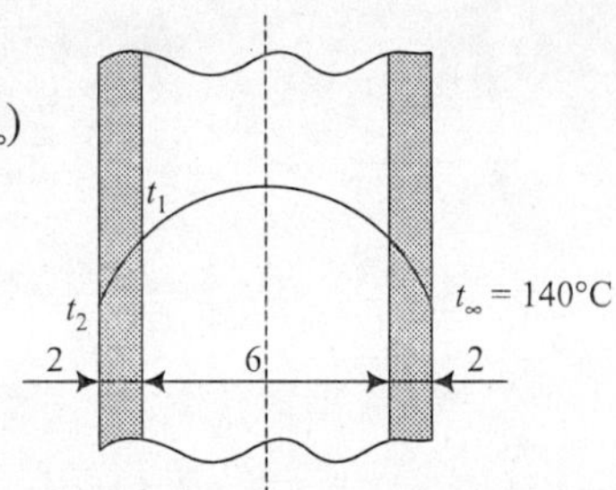

For $A = 1 \text{ m}^2$

$$t_2 = \frac{16.85 \times 10^5}{2.85 \times 10^4 \times 1} + 140$$

$$= 199.12\ °\text{C}$$

$$t_1 = \frac{16.85 \times 10^5 \times 0.2 \times 10^{-2}}{200 \times 1} + 199.12 = 215.97°\text{C}$$

$$t_{\max} = \frac{\dot{Q}_G}{8k} L^2 + t_1$$

$$= \frac{5.6 \times 10^8}{8 \times 24.5} (6 \times 10^{-3})^2 + 215.97$$

$$= \mathbf{318.8°C} \quad \textbf{Ans.}$$

TUTORIAL SHEET NO. 14.3 (CONVECTION)

1. The average Reynold number for air passing in turbulent flow over a 2 m long flat plate is 2.4×10^6. Under these conditions, the average Nusselt number was found to be equal to 4150. Determine the average heat transfer coefficient for an oil having the following thermal properties at 30°C at the same Reynold number in flow over the same plate.

   Thermal conductivity, $k = 0.11$ W/m-K

   Viscosity, $\mu = 0.0134$ N-s/m^2

   Specific heat, $C_p = 1780$ J/kg-K

2. A square channel with a side 10 mm and length 1.5 m carries water with a velocity of 5 m/s. Measurements indicate that lengthwise mean temperature of water is 30°C, whilst the inner surface of channel is at 80°C. Calculate the convective coefficient of heat transfer from the channel wall to water. Use the correlation.

$$Nu = 0.021\, R_e^{0.8}\, P_r^{0.43} \left(\frac{P_r}{P_{r_w}}\right)^{0.25}$$

   where thermo-physical properties pertain to those at the mean bulk temperature of water, P_{r_w} corresponds to the value of Prandtl number at the channel surface temperature and equivalent diameter is the reference dimension.

   The physical properties of water at 30°C are:

   $\rho = 995.07$ kg/m^3; $C_p = 4174$ J/kg-K; $k = 0.6172$ W/m-K;

   $\mu = 2.88$ kg/m-hr; $\nu = 0.805 \times 10^{-6}$ m^2/s; $P_r = 5.42$

   At wall temperature, $t_w = 80$°C, $P_{r_w} = 2.21$.

3. Liquid mercury flows through a copper tube of 2 cm inner diameter at the rate of 1.25 kg/s. The mercury enters at 15°C and is heated to 25°C as it passes through the tube. Determine the tube length which would satisfy the condition of a constant heat flux at the wall which is at an average temperature of 40°C.

   For the liquid metals, the following correlation is presumed to agree well with experimental results.

$$N_U = 7 + 0.025\,(P_e)^{0.8}$$

   where P_e is Peclet No. $P_e = P_r \times R_e$

   Mean bulk temperature, $T_b = \dfrac{15 + 25}{2} = 20$°C.

The physico-thermal properties of liquid mercury are:

$$\rho = 13580 \text{ kg/m}^3;\ k = 8.685 \text{ W/m-K};\ C_p = 139.35 \text{ J/kg-K}$$

$$\nu = 1.148 \times 10^{-7} \text{ m}^2\text{/s and } P_r = 0.0249$$

4. The following correlation applies to cooling of a car engine with water

$$N_U = 0.031\ (R_e)^{0.75}\ (P_r)^{0.30}$$

A special rust preventor is added to water which increases the viscosity of mixture by 20%; other fluid properties remain the same as for water. Calculate the per cent change in the rate of cooling per degree temperature difference.

SOLUTION TO TUTORIAL SHEET NO. 14.3

1. $P_r = \dfrac{C_p\mu}{k} = \dfrac{1780 \times 0.0134}{0.11} = 217$

For $R_{e_L} > 5 \times 10^5$ (turbulent flow and $P_r > 0.5$)

$$\overline{N}_{U_L} = 0.036\, P_r^{\,0.33}\, [R_{e_L}^{\,0.8} - 23{,}200] \quad \text{...Given}$$

$$= 0.036\, (217)^{0.33}\, [(2.4 \times 10^6)^{0.8} - 23{,}200]$$

$$= 22080$$

$$\therefore \quad \bar{h}_c = \frac{\overline{N}_{U_L} k}{L} = \frac{22080 \times 0.11}{2} = \mathbf{1214\ W/m^2\text{-}K} \quad \textbf{Ans.}$$

2. $$D_H = 4\frac{A}{P} = 4\frac{0.01 \times 0.01}{4 \times 0.01} = 0.01 \text{ m}$$

$$R_{e_{D_H}} = \frac{VD_H}{\nu} = \frac{5 \times 0.01}{0.805 \times 10^{-6}} = 62100$$

$$N_U = 0.021\, (62100)^{0.8}\, (5.42)^{0.43} \left(\frac{5.42}{2.21}\right)^{0.35} = 371.4$$

$$h = \frac{Nuk}{D_H} = \frac{371.4 \times 0.6172}{0.01} = \mathbf{22923\ W/m^2\text{-}K} \quad \textbf{Ans.}$$

3. Heat gained by mercury is

$$\dot{Q} = \dot{m}\, C_p\, \Delta T = 1.25 \times 139.35 \times (25 - 15)$$

$$= 1741.87 \text{ W}$$

$$R_e = \frac{UD\rho}{\mu} = \frac{md}{A\mu} = \frac{4m}{\pi d\mu} = \frac{4m}{\pi d\rho v}$$

$$= \frac{4 \times 1.25}{\pi \times 0.02 \times 13580 \times 1.145 \times 10^{-3}} = 51115$$

$$\therefore \quad N_U = 7 + 0.025\,(5115 \times 0.0249)^{0.8} = 14.616$$

$$h = \frac{Nuk}{D} = \frac{14.616 \times 8.685}{0.02}$$

$$= 6347 \text{ W/m}^2\text{-K}$$

Heat gained, $\dot{Q} = hA\,\Delta T$

$$1741.87 = 6347\,(\pi \times 0.02 \times L)(25 - 15)$$

$$\therefore \quad L = \frac{1741.87}{6347 \times \pi \times 0.02 \times 10} = \mathbf{0.4370\ m} \quad \textbf{Ans.}$$

4. $$Nu_l = 0.031 \left(\frac{UL\rho}{\mu_1}\right)^{0.75} \left(\frac{\mu_1 C_p}{k}\right)^{0.30}$$

$$= A\,\mu^{-0.45}$$

$$Nu_2 = A\,\mu_2^{-0.45} = A\,(1.2\,\mu_1)^{-0.45}$$

$$\frac{Nu_1 - Nu_2}{Nu_1} = 1 - \frac{Nu_2}{Nu_1} = 1 - \left(\frac{1.2\mu_1}{\mu_1}\right)^{-0.45}$$

$$= 1 - \left(\frac{1}{1.2}\right)^{0.45} = 7.87\% \text{ decrease}$$

TUTORIAL SHEET NO. 14.4 (RADIATION)

Q1. The sun emits maximum radiation at $\lambda = 0.52$ μm. Assuming the sun to be a black body, calculate the surface temperature of the sun and the emissive ability of the sun's surface at that temperature. Also determine the maximum monochromatic emissive power of the sun's surface.

Q2. Determine the radiation heat flux between two closely spaced, black parallel plates, radiating only to each other if their temperatures are 850 K and 425 K respectively. Recalculate the heat flux presuming that each of the parallel plates has an emissibility of 0.5. In each case, the plates have an area of 4 m^2.

Q3. A pipe with a surface temperature of 480 K is kept within a large enclosure whose walls are at 380 K. Presuming the pipe surface to be black, calculate the coefficient of radiant heat transfer. If the heat transfer coefficient including the effect of radiation and convection is 34.9 W/m^2-K, fluid the convective heat transfer coefficient.

Q4. A hot water radiator of overall dimensions 2.5 m × 1.25 m × 0.25 m and having surface temperature of 335 K is being used to maintain the room temperature at 290 K. Calculate the coefficient of radiant heat transfer, convective heat transfer coefficient and the heat loss from the radiator due to both convection and radiation. The radiations are considered close to that of a black body and the convective heat transfer coefficient is prescribed by the relation.

$$h_c = 1.32(\Delta T)^{0.33} \text{ [W/m}^2\text{-K]}$$

For convection, the actual surface area of the radiator is twice the area of the envelope.

Q5. Two very large parallel planes with emissivities 0.3 and 0.8 exchange heat. Find the percentage reduction in heat transfer when a polished aluminum radiation shield ($\varepsilon = 0.04$) is placed between them.

Q6. A thermocouple is used to measure the temperature of a flame in a combustion chamber. If the thermocouple temperature is 1033 K and the walls of the chamber are at 700 K, what is the error in the thermocouple reading due to radiation to the walls? Assume all surfaces are black and the convection coefficient is 568 W/m^2-°C on the thermocouple.

Q7. Calculate the equilibrium temperature of a thermocouple in a large air duct if the air temperature is 1367 K, the duct wall temperature is 533 K, the emittance of the thermocouple is 0.5 and convective heat transfer coefficient, h_c is 114 W/m^2-K. State clearly the assumptions made and draw the thermal circuit.

SOLUTIONS OF TUTORIAL SHEET NO. 14.4

1. (a) $\lambda_{max} T = 2.898 \times 10^{-3}$ [Wien's Displacement Law]

$$\therefore \qquad T = \frac{2.898 \times 10^{-3}}{0.52 \times 10^{-6}}$$

$$= 5573 \text{ K}$$

(b) $E = \sigma_b T^4$ [Stefan-Boltzmann Law]

$$= 5.67 \times 10^{-8} (5573)^4$$

$$= 5.47 \times 10^7 \text{ W/m}^2$$

(c) $(E_\lambda)_{max} = 1.285 \times 10^{-5}\, T^5$

$= 1.285 \times 10^{-5}\, (5573)^5$

$= 6908 \times 10^{13}$ W/m^2 per metre wavelength

2. For black parallel plates radiating only to each other, $F_{12} = 1$

$$\dot{Q} = F_{12}\, A_1\, \sigma\, (T_1^4 - T_2^4)$$
$$= 1 \times 4 \times 5.67 \times 10^{-8}\, [(850)^4 - (425)^4]$$
$$= 110 \times 10^{-3} \text{ W}$$

For greay surfaces,

$$\dot{Q} = (F_{12})_g\, A_1\, \sigma\, (T_1^4 - T_2^4)$$
$$F_{12} = 1\; A_1 = A_2$$

$$(F_{12})_g = \frac{1}{\dfrac{1-\varepsilon_1}{\varepsilon_1} + \dfrac{1}{F_{12}} + \dfrac{1-\varepsilon_2}{\varepsilon_2} \times \dfrac{A_1}{A_2}}$$

$$= \frac{1}{\dfrac{1-\varepsilon_1}{\varepsilon_1} + \dfrac{1-\varepsilon_2}{\varepsilon_2} - 1}$$

$$= \frac{1}{\dfrac{1-0.5}{0.5} + \dfrac{1-0.5}{0.5} - 1} = 0.333$$

$$\dot{Q} = 0.333 \times 4 \times 5.67 \times 10^{-8}\, [(850)^4 - (425)^4]$$
$$= \mathbf{37 \times 10^{-3}\ W} \quad \textbf{Ans.}$$

3. $$\dot{Q} = \sigma A\, (T_1^4 - T_2^4)$$
$$= 5.67 \times 10^{-8} \times 1\, [(480)^4 - (380)^4]$$
$$= 1827.6 \text{ W}$$

Also, $\dot{Q} = h_r A\, (T_1 - T_2)$

$$1827.6 = h_r \times 1\, (480 - 380)$$
$$h_r = 18.23 \text{ W/m}^2\text{-K}$$
$$h_{overall} = h_r + h_c$$
$$\therefore \quad h_c = 34.9 - 18.23$$
$$= \mathbf{16.62\ W/m^2\text{-}K} \quad \textbf{Ans.}$$

4. Area of radiator = 2[2.5 × 1.25 + 1.25 × 0.25 + 0.25 × 2.5]

$$= 8.105 \text{ m}^2.$$

$$\dot{Q}_r = \sigma A (T_1^4 - T_2^4)$$

$$= h_r A (T_1 - T_2)$$

$$h_r = \sigma (T_1 + T_2)(T_1^2 - T_2^2)$$

$$= 5.67 \times 10^{-8} (335 + 290)(335^2 - 290^2)$$

$$= 6.96 \text{ W/m-k}$$

$$\dot{Q}_r = \sigma A (T_1^4 - T_2^4) = h_r A (T_1 - T_2)$$

$$= 6.96 \times 8.105 (335 - 290)$$

$$= 2538 \text{ W}$$

$$h_c = 1.32 (\Delta T)^{0.33}$$

$$= 1.32 (335 - 290)^{0.33}$$

$$= 4.63 \text{ W/m}^2\text{-K}$$

$$\therefore \quad \dot{Q}_c = h_c A \Delta T$$

$$= 4.63 \times 2 \times 8.105 \times (335 - 290)$$

$$= 3377 \text{ W}$$

$$\dot{Q}_{total} = \dot{Q}_r + \dot{Q}_c$$

$$= 3377 + 2538 = \mathbf{5914\ W} \quad \textbf{Ans.}$$

5. $$\left(\frac{\dot{Q}}{A}\right)_{\text{without shield}} = \frac{\sigma(T_1^4 - T_2^4)}{\dfrac{1}{\varepsilon_1} + \dfrac{1}{\varepsilon_2} - 1}$$

$$= \frac{\sigma(T_1^4 - T_2^4)}{\dfrac{1}{0.3} + \dfrac{1}{0.8} - 1} = 0.279\, \sigma (T_1^4 - T_2^4)$$

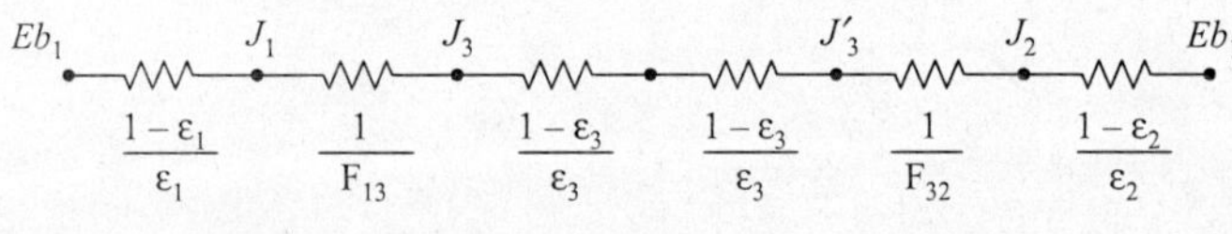

$$\frac{1-\varepsilon_1}{\varepsilon_1} = \frac{1-0.3}{0.3} = 2.333$$

$$\frac{1-\varepsilon_3}{\varepsilon_3} = \frac{1-0.04}{0.04} = 24.0$$

$$\frac{1-\varepsilon_2}{\varepsilon_2} = \frac{1-0.8}{0.8} = 0.25$$

Total resistance = 2.333 + 1 + 24.0 + 24.0 + 1 + 0.25 = 52.583

$$\therefore \quad \left(\frac{\dot{Q}}{k}\right)_{\text{with shield}} = \frac{\sigma(T_1^4 - T_2^4)}{52.583} = 0.01902\ \sigma\ (T_1^4 - T_2^4)$$

% reduction in heat transfer due to shield

$$= \frac{0.279 - 0.01902}{0.279} = \mathbf{93.18\%} \quad \textbf{Ans.}$$

6. For thermocouple,

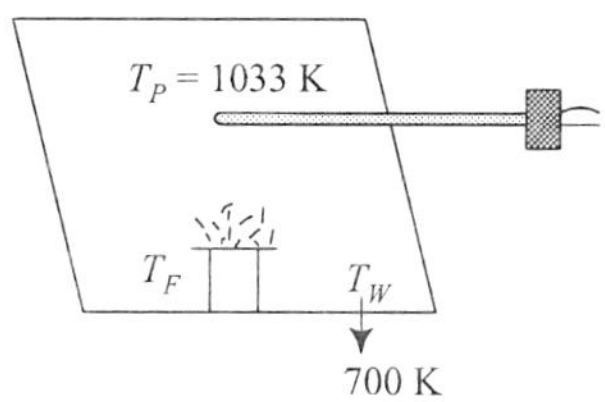

$$\varepsilon = 1$$

$$h_c = 568\ \text{W/m}^2\text{–°C}.$$

For walls, $\varepsilon = 1.$

Balance of convection and radiation,

$$h_c(T_f - T_p) = \sigma(T_p^4 - T_w^4)$$

$$568\ (T_f - 1033) = 5.67 \times 10^{-8}\ [(1033)^4 - (700)^4]$$

$$= 50949$$

$$\therefore \quad T_f = 1123\ \text{K}$$

$$\text{Error} = 1123 - 1033 = 90\ \text{K}$$

7. Assumption:
   1. Neglect probe conduction error
   2. Walls are black
   3. Air contains no H_2 or CO_2

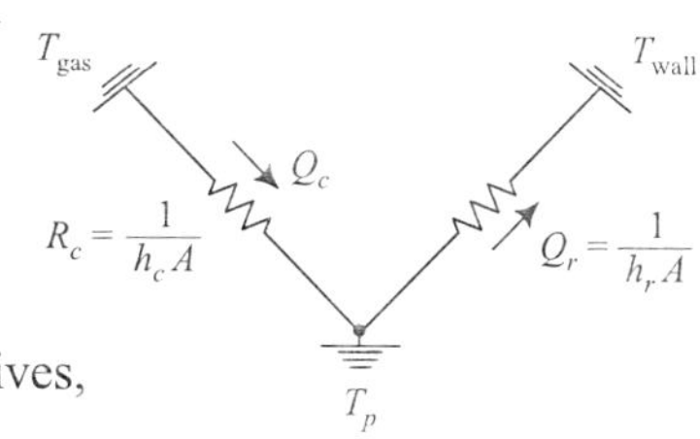

Balance of convection and radiation gives,

$$\therefore \quad h_C(T_a - T_p) = \varepsilon_{TC}\ \sigma\ (T_p^4 - T_w^4)$$

$$114\ (1367 - T_p) = (0.5)\ (5.67) \left[\left(\frac{T_p}{1000}\right)^4 - 5.33^4\right]$$

$$T_p = \mathbf{1065\ K} \quad \textbf{Ans.}$$

TUTORIAL SHEET NO. 14.5 (HEAT EXCHANGERS)

1. Water at the rate of 68 kg/min is heated from 35°C to 75°C by an oil having a specific heat of 1.9 kJ/kg-K. The fluids are used in a counterflow double-pipe heat exchanger, and the oil enters the exchanger at 110°C and leaves at 75°C. The overall heat transfer coefficient is 320 W/m^2–°C. Calculate the heat exchanger area.
2. Water at the rate of 3.783 kg/sec is heated from 37.78°C to 54.44°C in a shell-and-tube heat exchanger. On the shell side, one pass is used with water as the working fluid, 1.892 kg/s entering the exchanger at 93.33°C. The overall heat transfer coefficient is 1414 W/m^2-°C and the average water velocity in the 1.905 cm diameter tube is 0.366 m/s. Because of space limitation, the length of the tube must not be longer than 2.438 m. Calculate the number of tube passes, the number of tubes per pass, and the length of the tubes, consistent with this restriction.
3. The engine oil at 150°C is cooled to 80°C in a parallel flow heat exchanger by water entering at 25°C and leaving at 60°C. Estimate the exchanger effectiveness. If the fluid flow rates and the inlet conditions remain unchanged work out the lowest temperature to which the oil may be cooled by increasing the length of the exchanger.
4. In a surface condenser, the water flowing through a series of tubes at the rate of 200 kg/hr is heated from 15°C to 75°C. The steam condenses on the outside surface of tubes at atmospheric pressure and overall heat transfer coefficient is estimated at 0.25 W/m^2-K. Use NTU method to work out the length of the tube and steam consumption rate. Presume that the tube is 25 mm in diameter. At the condensing pressure, steam has a saturation temperature of 100°C and h_{fg} = 2257 kJ/kg. Further, the steam is initially just saturated and condensate leaves the exchanger without sub-cooling.

SOLUTION OF TUTORIAL SHEET NO. 14.5

1. $T_{h1} = 110°C$

$T_{h2} = 75°C$

$T_{c1} = 35°C$

$T_{c2} = 75°C$

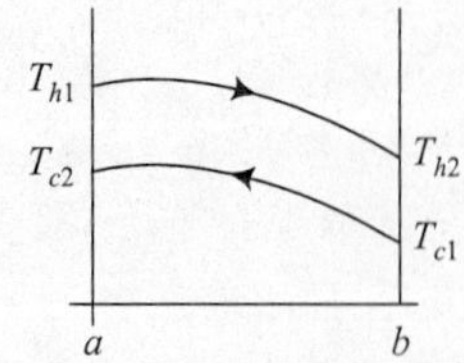

Fig. (Problem 1)

$$\dot{Q} = \dot{m}_w C_w \Delta T_w$$

$$= 68\ (4180)\ (75 - 35)$$

$$= 11.37 \text{ MJ/min}$$

$$= 189.5 \text{ kW}$$

$$\text{LMTD} = \frac{\Delta T_a - \Delta T_b}{\ln\left(\dfrac{\Delta T_a}{\Delta T_b}\right)}$$

$$\Delta T_m = \frac{(110 - 75) - (75 - 35)}{\ln \dfrac{(110 - 75)}{(75 - 35)}} = 37.44°\text{C}$$

$$\dot{Q} = UA\,\Delta T_m$$

$$A = \frac{\dot{Q}}{U\,\Delta T_m} = \frac{1.89 \times 10^5}{(320)\,(37.44)}$$

$$= \mathbf{15.82\ m^2}. \quad \textbf{Ans.}$$

2. $$\dot{Q} = \dot{m}_C\, C_C\, \Delta T_C$$

$$= \dot{m}_h C_h \Delta T_h$$

$\therefore$ $$\Delta T_h = \frac{3.783 \times 4.19 \times (54.44 - 37.78)}{1.892 \times 4.19}$$

$$= 33.33°\text{C}$$

$$T_{h_{out}} = 93.33 \quad 33.33$$

$$= 60°\text{C}$$

$$\dot{Q} = (3.783)(4182)(54.44 - 37.78)$$

$$= 263.6\ \text{kW}$$

For counterflow heat exchanger,

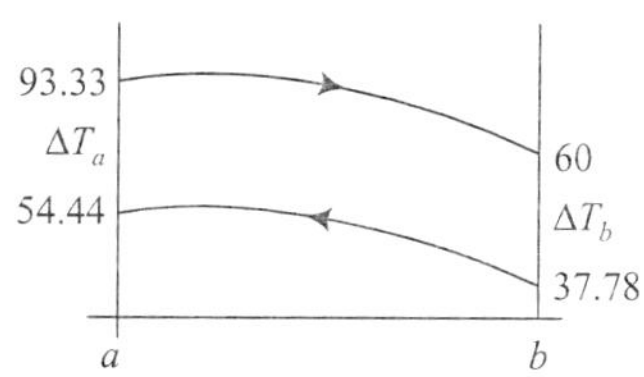

Fig. (Problem 2)

$$\text{LMTD} = \Delta T_m = \frac{\Delta T_a - \Delta T_b}{\ln \dfrac{\Delta T_a}{\Delta T_b}}$$

$$= \frac{(93.33 - 54.44) - (60 - 37.78)}{\ln \dfrac{(93.33 - 54.44)}{(60 - 37.78)}}$$

$$= 29.78°\text{C}$$

$$\dot{Q} = UA\,\Delta T_m$$

$$A = \frac{\dot{Q}}{U\Delta T_m} = \frac{2.636 \times 10^5}{(1419)\,(2978)} = 6.238\ \text{m}^3$$

Now, $\dot{m}_C = \rho A_f V$

Cross-sectional area of flow,

$$A_f = \frac{\dot{m}_C}{\rho U} = \frac{3.783}{(1000)(0.366)} = 0.01034 \text{ m}^2$$

$$\frac{\pi d^2}{4} n = A_f$$

No. of tubes, $n = \dfrac{4A_f}{\pi d^2} = \dfrac{(0.01034)(4)}{\pi.(0.01905)^2} = 36.3 \approx 36.$

Surface area, $NdLn = 6.238$

$$\therefore \qquad L = \frac{6.238}{(36)(\pi)(0.01905)} = 2.898 \text{ m} > 2.438 \text{ m}$$

$\therefore$ Use 2 passes,

$$A_{\text{total}} = \frac{\dot{Q}}{U\Delta T_m} = \frac{2.636 \times 10^5}{(1419)(29.38)} = 6.24 \text{ m}^2$$

No. of tubes per pass = 36 because of velocity consideration

$$A_{\text{total}} = 2\pi dLn$$

$$L = \frac{6.24}{(2)(\pi)(36)(0.01905)} = 1.447 \text{ m} < 2.438 \text{ m}$$

No. of tubes per pass = 36
No. of passes = 2
Length of tubes per pass = 1.447 m

3. (a) $m_c\, C_c\left(T_{c_2} - T_{c_1}\right) = m_h\, C_h\left(T_{h_1} - T_{h_2}\right)$

$$\therefore \qquad \frac{m_h\, C_h}{m_c\, C_c} = \frac{\left(T_{c_2} - T_{c_1}\right)}{\left(T_{h_1} - T_{h_2}\right)} = \frac{60 - 25}{150 - 80} = 0.5$$

The hot fluid has minimum thermal capacity,

$$\frac{C_{\min}}{C_{\max}} = 0.5$$

The exchanger effectiveness,

$$\varepsilon = \frac{C_h\left(T_{h_1} - T_{h_2}\right)}{C_{\min}\left(T_{h_1} - T_{c_1}\right)} = \frac{\left(T_{h_1} - T_{h_2}\right)}{\left(T_{h_1} - T_{c_1}\right)} = \frac{150 - 80}{150 - 25}$$

$$= 0.56$$

PART V

ANNEXURES

Annexure I: Introduction to SI Units

International System of Units (SI units) have seven fundamental units and two supplementary units, which cover the entire field of science and engineering. These basic units are given in Table 1.1.

Table I.1 SI System: Basic Units

S. No.	*Physical Quantity*	*Unit*	*Symbol*
	Fundamental Units		
1.	Length (L)	Metre	m
2.	Mass (M)	Kilogram	kg
3.	Amount of substance (n)	Mole	mol
4.	Time (t)	Second	s
5.	Temperature (T)	Kelvin	K
6.	Electric current (I)	Ampere	A
7.	Luminous intensity (Iv)	Candela	cd
	Supplementary Units		
1.	Plane angle (α, β, θ, ϕ)	Radian	rad
2.	Solid angle (Ω)	Steradian	sr

The dimensions of all other quantities are derived from the basic units. Some of the derived units are given in Table 1.2.

Table I.2 SI System: Derived Units

S. No.	*Quantity*	*Unit*	*Symbol*	*Alternative unit*	*In basic units*
1	*2*	*3*	*4*	*5*	*6*
1.	Force, Weight (F, W)	newton	N	–	kg m/s^2
2.	Work, Energy, Enthalpy (W, E, H)	joule	J	N-m	kg m^2/s^2
3.	Power (P)	watt	W	J/s	kg m^2/s^2
4.	Pressure (p)	pascal	Pa	N/m^2	kg/m^2
5.	Gas constant (R)	-	J/kg K	-	m^2/Ks^2
6.	Molar mass or Molecular mass (M)	-	kg/mol	-	kg/mol
7.	Specific heat (c)	-	J/kg K	-	m/s^{2}K
8.	Thermal conductivity (k)	-	w/mK	-	kg m/s^{2}K
9.	Mass density (ρ)	-	kg/m^3	-	kg/m^3
10.	Linear velocity (V)	-	m/s	-	m/s
11.	Linear acceleration (a)	-	m/s^2	-	m/s^2
12.	Angular velocity (ω)	-	rad/s	-	rad/s
13.	Angular acceleration (α)	-	rad/s^2	-	rad/s^2
14.	Dynamic viscosity (μ)	-	N-s/m^2	-	kg/sm
15.	Kinematic viscosity (μ)	-	m^2/s	-	m^2/s
16.	Frequency (*f*)	hertz	Hz	-	s^{-1}

It is often convenient and desirable to use multiples of various units, the standard list of which is given in Table 1.3.

Table I.3 SI System: Standard Multipliers

Factor	*Prefix*	*Abbreviation*
10^{12}	tera	T
10^{9}	giga	G
10^{6}	mega	M
10^{3}	kilo	k
10^{-3}	milli	m
10^{-6}	micro	m
10^{-9}	nano	n
10^{-12}	pico	P

Rules for SI Units

There are certain styles and usage of SI units.

1. For large numerals use groups of three without a coma. Grouping should start both sides from decimal. No grouping is done if figures are only 4.

Correct	*Incorrect*
234 869.739 45	23 4869.7 394 5
40 000	40000
3456.72	3456.7 2
0.123 457	0.123457

2. A dash should be used to separate units that are multiplied together.

Correct	*Incorrect*
N-m	Nm or mN which stand for millinewton

3. Plurals are never used with symbols or full names. For example, metre or metres are written as *m*.
4. Full names of units or prefixes shall be written in upright roman, lower case letters. No capitals are used even if unit is named after a person.

Correct	*Incorrect*
kelvin	Kelvin
newton	Newton
joule	Joule
watt	Watt
volt	Volt

5. All symbols are written in small letters except those derived from the proper names.

Correct	*Incorrect*
m	M
s	S
kg	Kg
W	w
J	j
N	n
K	k

6. No abbreviation should be used for units. Write either as a symbol or full name.

Correct	*Incorrect*
N or newton	newt
kg	kgram

7. No full stop or a punctuation mark be used.

Correct	*Incorrect*
3 m	3m,
100.25 kg	100.25 kg.

8. If a prefix is used with a symbol no space be left in between.

Correct	*Incorrect*
MJ	M J

9. Not more than one prefix be used. For 10^{-9} m.

Correct	*Incorrect*
nm	m μ m

10. Power to base unit applies to whole and not to base alone.

Correct	*Incorrect*
$km^2 = (km)^2 = 10^6\ m^2$	$km^2 \neq k(m)^2 = 10^3\ m^2$

11. Leave a space between numeral and its unit.

Correct	*Incorrect*
64 kW	64kW

12. A solidus or oblique may be used or a negative power may be used.

Correct	
$\frac{m}{s}$, m/s or ms^{-1}	

Annexure II: Steam Tables in SI Units

Table II.1 Saturated water and steam (temperatures) tables

Temperature in °C	*Absolute pressure in bar*	*Specific volume in m²/kg*		*Specific enthalpy in kJ/kg*			*Specific entropy in kJ/kg K*			*Temperature in °C*
(t)	(p)	*Water* (v_f)	*Steam* (v_g)	*Water* (h_f)	*Evaporation* (h_{fg})	*Steam* (h_g)	*Water* (s_f)	*Evaporation* (s_{fg})	*Steam* (s_g)	(t)
0	0.006 11	0.001 000	206.31	0.0	2 501.6	2 501.6	0.000	9.158	9.158	**0**
1	0.006 57	0.001 000	192.61	4.2	2 499.2	2 503.4	0.015	9.116	9.131	**1**
2	0.007 06	0.001 000	179.92	8.4	2 496.8	2 505.2	0.031	9.074	9.105	**2**
3	0.007 58	0.001 000	168.17	12.6	2 494.5	2 507.1	0.046	9.033	9.079	**3**
4	0.008 13	0.001 000	157.27	16.8	2 492.1	2 508.9	0.061	8.992	9.053	**4**
5	0.008 72	0.001 000	147.16	21.0	2 489.7	2 510.7	0.076	8.951	9.027	**5**
6	0.009 35	0.001 000	137.78	25.2	2 487.4	2 512.6	0.091	8.911	9.002	**6**
7	0.010 01	0.001 000	129.06	29.4	2 485.0	2 514.4	0.106	8.870	8.976	**7**
8	0.010 72	0.001 000	120.97	33.6	2 482.6	2 516.2	0.121	8.830	8.951	**8**
9	0.011 47	0.001 000	113.44	37.8	2 480.3	2 518.1	0.136	8.791	8.927	**9**
10	0.012 27	0.001 000	106.43	42.0	2 477.9	2 519.9	0.151	8.751	8.902	**10**
11	0.013 12	0.001 000	99.909	46.2	2 475.5	2 521.7	0.166	8.712	8.878	**11**
12	0.014 01	0.001 000	93.835	50.4	2 473.2	2 523.6	0.181	8.673	8.854	**12**
13	0.014 97	0.001 001	88.176	54.6	2 470.8	2 525.4	0.195	8.635	8.830	**13**
14	0.015 97	0.001 001	82.900	58.7	2 468.5	2 527.2	0.210	8.597	8.806	**14**
15	0.017 04	0.001 001	77.978	62.9	2 466.1	2 529.1	0.224	8.559	8.783	**15**
16	0.018 17	0.001 001	73.384	67.1	2 463.8	2 530.9	0.239	8.520	8.759	**16**
17	0.019 36	0.001 001	69.095	71.3	2 461.4	2 532.7	0.253	8.483	8.736	**17**

(t)	(p)	(v_f)	(v_g)	(h_f)	(h_{fg})	(h_g)	(s_f)	(s_{fg})	(s_g)	(t)
18	0.020 62	0.001 001	65.087	75.5	2 459.0	2 534.5	0.268	8.446	8.714	**18**
19	0.021 96	0.001 002	61.341	79.7	2 456.7	2 536.4	0.282	8.409	8.691	**19**
20	0.023 37	0.001 002	57.838	83.9	2 454.3	2 538.2	0.296	8.372	8.668	**20**
21	0.024 85	0.001 002	54.561	88.0	2 452.0	2 540.0	0.310	8.336	8.646	**21**
22	0.026 42	0.001 002	51.492	92.2	2449.6	2 541.8	0.325	8.299	8.624	**22**
23	0.28 08	0.001 002	48.619	96.4	2 447.2	2 543.6	0.339	8.263	8.602	**23**
24	0.029 82	0.001 002	45.926	100.6	2 444.9	2 545.5	0.353	8.228	8.581	**24**
25	0.031 66	0.001 003	43.402	104.8	2 442.5	2 547.3	0.367	8.192	8.559	**25**
26	0.033 60	0.001 003	41.034	108.9	2 440.2	2 549.1	0.381	8.157	8.538	**26**
27	0.035 64	0.001 003	38.813	113.1	2 437.8	2 550.9	0.395	8.122	8.517	**27**
28	0.037 78	0.001 004	36.728	117.3	2 435.4	2 552.7	0.409	8.087	8.496	**28**
29	0.040 04	0.001 004	34.769	121.5	2 433.1	2 554.5	0.423	8.052	8.475	**29**
30	0.042 42	0.001 004	32.929	125.7	2 430.7	2 556.4	0.437	8.018	8.455	**30**
31	0.044 91	0.001 005	31.199	129.8	2.428.3	2 558.2	0.450	8.984	8.434	**31**
32	0.047 53	0.001 005	29.572	134.0	2 425.9	2 560.0	0.464	7.950	8.414	**32**
33	0.050 29	0.001 005	28.042	138.2	2 423.6	2 561.8	0.478	7.916	8.394	**33**
34	0.053 18	0.001 006	26.601	142.4	2 421.2	2 563.6	0.491	7.883	8.374	**34**
35	0.056 22	0.001 006	25.245	146.6	2 418.8	2 565.4	0.505	7.849	8.354	**35**
36	0.059 40	0.001 006	23.967	150.7	2 416.4	2 567.2	0.518	7.817	8.335	**36**
37	0.062 74	0.001 007	22.763	154.9	2 414.1	2 569.0	0.532	7.783	8.315	**37**
38	0.066 24	0.001 007	21.627	159.1	2 411.7	2 570.8	0.545	7.751	8.296	**38**
39	0.069 91	0.001 007	20.557	163.3	2 409.3	2 572.6	0.559	7.718	8.277	**39**
40	0.073 75	0.001 008	19.546	167.5	2 406.9	2 574.4	0.572	7.686	8.258	**40**
41	0.077 77	0.001 008	18.592	171.6	2 404.5	2 576.2	0.585	7.654	8.239	**41**
42	0.081 99	0.001 009	17.692	175.8	2 402.1	2 577.9	0.599	7.622	8.221	**42**
43	0.086 39	0.001 009	16.841	180.0	2 399.7	2 579.7	0.612	7.591	8.203	**43**
44	0.091 00	0.001 009	16.036	184.2	2 397.3	2.581.5	0.625	7.559	8.184	**44**

(t)	(p)	(v_f)	(v_g)	(h_f)	(h_{fg})	(h_g)	(s_f)	(s_{fg})	(s_g)	(t)
45	0.095 82	0.001 010	15.276	188.4	2 394.9	2 583.3	0.638	7.528	8.166	**45**
46	0.100 80	0.001 010	14.557	192.5	2 392.5	2 585.1	0.651	7.497	8.148	**46**
47	0.106 12	0.001 011	13.877	196.7	2 390.1	2 586.9	0.664	7.466	8.130	**47**
48	0.111 62	0.001 011	13.233	200.9	2 387.7	2 588.6	0.678	7.435	8.113	**48**
49	0.117 36	0.001 012	12.623	205.1	2 385.3	2 590.4	0.691	7.404	8.095	**49**
50	0.123 35	0.001 012	12.046	209.3	2382.9	2 592.2	0.704	7.374	8.078	**50**
51	0.129 61	0.001 013	11.499	213.4	2380.5	2 593.9	0.716	7.344	8.060	**51**
52	0.136 11	0.001 013	10.980	217.6	2378.1	2 595.7	0.729	7.314	8.043	**52**
53	0.142 93	0.001 014	10.488	221.8	2375.7	2 597.5	0.742	7.284	8.026	**53**
54	0.150 02	0.001 014	10.022	226.0	2373.2	2 599.2	0.755	7.254	8.009	**54**
55	0.157 41	0.001 015	9.5789	230.2	2 370.8	2 601.0	2.768	7.225	7.993	**55**
56	0.165 11	0.001 015	9.1587	234.3	2368.4	2602.7	0.780	7.196	7.976	**56**
57	0.173 13	0.001 016	8.759 8	238.5	2366.0	2 604.5	0.793	7.166	7.959	**57**
58	0.181 47	0.001 016	8.380 8	242.7	2 363.5	2 606.2	0.806	7.137	7.943	**58**
59	0.190 16	0.001 017	8.020 8	246.9	2 361.1	2 608.0	0.818	7.109	7.927	**59**
60	0.199 20	0.001 017	7.678 5	251.1	2 358.6	2 609.7	0.831	7.080	7.911	**60**
61	0.208 61	0.001 018	7.353 2	255.3	2 356.1	2 611.4	0.844	7.051	7.895	**61**
62	0.218 38	0.001 018	7.043 7	259.5	2 353.7	2 613.2	0.856	7.023	7.879	**62**
63	0.228 55	0.001 019	6.749 3	263.6	2 351.3	2 614.9	0.868	6.995	7.863	**63**
64	0.239 12	0.001 019	6.469 0	267.8	2 348.8	2 616.6	0.881	6.967	7.848	**64**
65	0.250 09	0.001 020	6.202 3	272.0	2 346.4	2 618.4	0.893	6.939	7.832	**65**
66	0.261 50	0.001 020	5.948 2	276.2	2 343.9	2 620.1	0.906	6.911	7.817	**66**
67	0.273 34	0.001 021	5.706 2	280.4	2 341.4	2 621.8	0.918	6.884	7.802	**67**
68	0.285 63	0.001 022	5.475 6	284.6	2 338.9	2 623.5	0.930	6.856	7.786	**68**
69	0.298 38	0.001 022	5.255 8	288.8	2 336.4	2 625.2	0.943	6.828	7.771	**69**
70	0.311 62	0.001 023	5.046 3	293.0	2 333.9	2 626.9	0.955	6.802	7.757	**70**
71	0.325 35	0.001 024	4.846 4	297.2	2 331.4	2 628.6	0.967	6.775	7.742	**71**

(t)	(p)	(v_f)	(v_g)	(h_f)	(h_{fg})	(h_g)	(s_f)	(s_{fg})	(s_g)	(t)
72	0.339 58	0.001 024	4.655 7	301.3	2 329.0	2 630.3	0.979	6.748	7.727	**72**
73	0.354 34	0.001 025	4.473 7	305.5	2 326.5	2 632.0	0.991	6.721	7.712	**73**
74	0.369 64	0.001 025	4.300 0	309.7	2 324.0	2 633.7	1.003	6.695	7.698	**74**
75	0.385 49	0.001 026	4.134 1	313.9	2 321.5	2 635.4	1.015	6.668	7.683	**75**
76	0.401 91	0.001 027	3.975 7	318.1	2 318.9	2 637.0	1.027	6.642	7.669	**76**
77	0.418 91	0.001 027	3.824 3	322.3	2 316.4	2 638.7	1.039	6.616	7.655	**77**
78	0.436 52	0.001 028	3.679 6	326.5	2 313.9	2 640.4	1.051	6.590	7.641	**78**
79	0.454 74	0.001 029	3.541 3	330.7	2 311.4	2 642.1	1.063	6.564	7.627	**79**
80	0.473 60	0.001 029	3.409 1	334.9	2 308.9	2 643.8	1.075	6.538	7.613	**80**
81	0.493 60	0.001 030	3.282 6	339.1	2 306 .3	2 645.4	1.087	6.512	7.599	**81**
82	0.513 29	0.001031	3.161 6	343.3	2 303.8	2 647.1	1.099	0.407	7.686	**82**
83	0.534 16	0.001 031	3.045 8	347.5	2 301.2	2 648.7	1.111	6.461	7.572	**83**
84	0.555 73	0.001 032	2.935 0	351.7	2 298.7	2 650.4	1.123	6.436	7.559	**84**
85	0.578 03	0.001 033	2.828 8	355.9	2 296.1	2 652.0	1.134	6.411	7.545	**85**
86	0.601 08	0.001 033	2.727 2	360.1	2 293.5	2 653.6	1.146	6.386	7.532	**86**
87	0.624 89	0.001 034	2.629 8	364.3	2 291.0	2 655.3	1.158	6.361	7.519	**87**
88	0.649 48	0.001 035	2.536 5	368.5	2 288.4	2 656.9	1.169	6.337	7.506	**88**
89	0.674 87	0.001 035	2.447 0	372.7	2 285.8	2 658.5	1.181	6.312	7.493	**89**
90	0.701 09	0.001 036	2.361 3	376.9	2 283.2	2 660.1	1.193	6.287	7.480	**90**
91	0.728 15	0.001 037	2.279 1	381.1	2 280.6	2 661.7	1.204	6.263	7.467	**91**
92	0.756 06	0.001 038	2.200 2	385.4	2 278.0	2 663.4	1.216	6.238	7.454	**92**
93	0.784 89	0.001 038	2.124 5	389.6	2 275.4	2 665.0	1.227	6.215	7.442	**93**
94	0.814 61	0.001 039	2.051 9	393.8	2 272.8	2 666.6	1.239	6.190	7.429	**94**
95	0.845 26	0.001 040	1.982 2	398.0	2 270.1	2 668.1	1.250	6.167	7.417	**95**
96	0.876 86	0.001 041	1.915 2	402.2	2 267.5	2 669.7	1.261	6.143	7.404	**96**
97	0.909 44	0.001 041	1.851 0	406.4	2 264.9	2 671.3	1.273	6.119	7.392	**97**

(t)	(p)	(v_f)	(v_g)	(h_f)	(h_{fg})	(h_g)	(s_f)	(s_{fg})	(s_g)	(t)
98	0.943 01	0.001 042	1.789 2	410.6	2 262.3	2 672.9	1.284	6.096	7.380	**98**
99	0.977 61	0.001 043	1.730 0	414.8	2 259.6	2 674.4	1.296	6.072	7.368	**99**
100	1.013 3	0.001 044	1.673 0	419.1	2 256.9	2 676.0	1.307	6.048	7.355	**100**
102	1.087 6	0.001 045	1.565 5	427.5	2 251.6	2 679.1	1.329	6.002	7.331	**102**
104	1.166 8	0.001 047	1.466 2	335.9	2 246.3	2 682.2	1.352	5.956	7.308	**104**
106	1.250 4	0.001 048	1.347 2	444.4	2 240.9	2 685.3	1.374	5.910	7.284	**106**
108	1.339 0	0.001 050	1.288 9	452.9	2 235.4	2 688.3	1.396	5.865	7.261	**108**
110	1.432 7	0.001 052	1.209 9	461.3	2 230.0	2 691.3	1.418	5.821	7.239	**110**
112	1.531 6	0.001 054	1.136 6	469.8	2 224.5	2 694.3	1.440	5.776	7.216	**112**
114	1.636 2	0.001 055	1.068 5	478.3	2 218.9	2 697.2	1.462	5.732	7.194	**114**
116	1.746 5	0.001 057	1.005 2	486.7	2 213.5	2 700.2	1.484	5.688	7.172	**116**
118	1.862 8	0.001 059	0.946 34	495.2	2 207.9	2 703.1	1.506	5.645	7.151	**118**
120	1.985 4	0.001 061	0.891 52	503.7	2 202.3	2 706.0	1.528	5.601	7.129	**120**
122	2.114 5	0.001 063	0.840 45	512.2	2 196.6	2 708.8	1.549	5.559	7.108	**122**
124	2.250 4	0.001 064	0.792 83	520.7	2 190.9	2 711.6	1.570	5.517	7.087	**124**
126	2.393 3	0.001 066	0.748 40	529.2	2 185.2	2 714.4	1.592	5.475	7.067	**126**
128	2.543 5	0.001 068	0.706 91	537.8	2 179.4	2 717.2	1.613	5.433	7.046	**128**
130	2.701 3	0.001 070	0.668 14	546.3	2 173.6	2 719.9	1.634	5.392	7.026	**130**
132	2.867 0	0.001 072	0.631 88	554.8	2 167.8	2 722.6	1.655	5.351	7.006	**132**
134	3.040 7	0.001 074	0.597 95	563.4	2 161.9	2 725.3	1.676	5.310	6.986	**134**
136	3.222 9	0.001 076	0.566 18	572.0	2 155.9	2 727.9	1.697	5.270	6.967	**136**
138	3.413 8	0.001 078	0.536 41	580.5	2 150.0	2 730.5	1.718	5.229	6.947	**138**
140	3.613 9	0.001 080	0.508 49	589.1	2 144.0	2 733.1	1.739	5.189	6.928	**140**
142	3.823 1	0.001 082	0.482 30	597.7	2 137.9	2 735.6	1.760	5.150	6.910	**142**
144	4.042 0	0.001 084	0.457 71	606.3	2 131.8	2 738.1	1.780	5.111	6.891	**144**
146	4.270 9	0.001 086	0.434 60	614.9	2 125.7	2 740.6	1.801	5.071	6.872	**146**
148	4.510 1	0.001 089	0.412 88	623.5	2 119.5	2 743.0	1.821	5.033	6.854	**148**

(t)	(p)	(v_f)	(v_g)	(h_f)	(h_{fg})	(h_g)	(s_f)	(s_{fg})	(s_g)	(t)
150	4.760 0	0.001 091	0.392 45	632.2	2 113.2	2 745.4	1.842	4.994	6.836	**150**
155	5.433 3	0.001 096	0.346 44	653.8	2 097.4	2 751.2	1.892	4.899	6.791	**155**
160	6.180 6	0.001 102	0.306 76	675.5	2 081.2	2 756.7	1.943	4.805	6.748	**160**
165	7.007 7	0.001 108	0.272 40	697.2	2 064.8	2 762.0	1.992	4.713	6.705	**165**
170	7.920 2	0.001 114	0.242 55	719.1	2 048.0	2 767.1	2.042	4.621	6.663	**170**
175	8.924 4	0.001 121	0.216 54	741.1	2 030.7	2 771.8	2.091	4.531	6.622	**175**
180	10.027	0.001 128	0.193 80	763.1	2 013.2	2 776.3	2.139	4.443	6.582	**180**
185	11.233	0.001 135	0.173 86	785.3	1 995.1	2 780.4	2.187	4.355	6.542	**185**
190	12.551	0.001 142	0.156 32	807.5	1 976.8	2 784.3	2.236	4.268	6.504	**190**
195	13.087	0.001 149	0.140 84	829.9	1 957.9	2 287.8	2.283	4.182	6.465	**195**
200	15.549	0.001 156	0.127 16	852.4	1 938.5	2 790.9	2.331	4.097	6.428	**200**
205	17.243	0.001 164	0.115 03	875.0	1 918.8	2 793.8	2.378	4.013	6.391	**205**
210	19.077	0.001 172	0.104 24	897.7	1 898.5	2 796.2	2.425	3.929	6.354	**210**
215	21.060	0.001 181	0.094 625	920.6	1 877.7	2 798.3	2.471	3.846	6.317	**215**
220	23.198	0.001 190	0.086 038	943.7	1 856.2	2 799.9	2.518	3.764	6.282	**220**
225	25.501	0.001 199	0.078 349	966.9	1 834.3	2 801.2	2.564	3.682	6.246	**225**
230	27.976	0.001 209	0.071 450	990.3	1 811.7	2 802.0	2.610	3.601	6.211	**230**
235	30.632	0.001 219	0.065 245	1 013.8	1788.5	2 802.3	2.656	3.519	6.175	**235**
240	33.478	0.001 229	0.059 654	1 037.6	1 764.6	2 802.2	2.702	3.439	6.141	**240**
245	36.523	0.001 240	0.054 606	1 061.6	1 740.0	2 801.6	2.748	3.358	6.106	**245**
250	39.776	0.001 251	0.050 037	1 085.8	1 714.6	2 800.4	2.794	3.277	6.071	**250**
255	43.246	0.001 263	0.045 896	1 110.2	1 688.5	2 798.7	2.839	3.197	6.036	**255**
260	46.943	.001 276	0.042 134	1 134.9	1 661.5	2 796.4	2.885	3.116	6.001	**260**
265	50.877	0.001 289	0.038 710	1 1.59.9	1 633.6	2 793.5	2.931	3.035	5.966	**265**
270	55.058	0.001 303	0.035 588	1 185.2	1604.7	2 789.9	2.976	2.954	5.930	**270**

(t)	(p)	(v_f)	(v_g)	(h_f)	(h_{fg})	(h_g)	(s_f)	(s_{fg})	(s_g)	(t)
275	59.496	0.001 317	0.032 746	1 210.8	1 574.7	2 785.5	3.022	2.873	5.895	**275**
280	64.202	0.001 332	0.030 126	1 236.8	1 543.6	2 780.4	3.068	2.790	5.858	**280**
285	69.186	0.001 349	0.027 733	1 263.2	1 511.3	2 774.5	3.115	2.707	5. 822	285
290	74.461	0.001 366	0.025 535	1 290.0	1 477.6	2 767.6	3.161	2.624	5.785	290
295	80.037	0.001 384	0.023 513	1 317.3	1 442.5	2 759.8	3.208	2.539	5.447	295
300	85.927	0.001 404	0.021 649	1 345.0	1 406.0	2 751.0	3.255	2.453	5.708	**300**
305	92.144	0.001 425	0.019 927	1 373.4	1 367.7	2 741.1	3.303	2.366	5.669	**305**
310	98.700	0.001 448	0.018 334	1 402.4	1 327.6	2 730.0	3.351	2.277	5.628	**310**
315	105.61	0.001 473	0.016 856	1 432.1	1 285.5	2 717.6	3.400	2.186	5.586	**315**
320	112.89	0.001 500	0.015 480	1 462.6	1 241.1	2 703.7	3.450	2.092	5.542	**320**
325	120.56	0.001 529	0.014 195	1 494.0	1 194.0	2 688.0	3.501	1.996	5.497	**325**
330	128.63	0.001 562	0.012 989	1 526.5	1 143.7	2 670.2	3.553	1.896	5.449	**330**
335	137.12	0.001 598	0.011 854	1 560.2	1 089.5	2 649.7	3.606	1.792	5.398	**335**
340	146.05	0.001 639	0.010 780	1 595.5	1 030.7	2 626.2	3.662	1.681	5.343	**340**
345	155.45	0.001 686	0.097 631	1 632.5	966.4	2.598.9	3.719	1.564	5.283	**345**
350	165.35	0.001 741	0.087 991	1 671.9	895.8	2 567.7	3.780	1.438	5.218	**350**
355	175.77	0.001 809	0.078 592	1 716.6	813.8	2 530.4	3.849	1.295	5.144	**355**
360	186.75	0.001 896	0.069 398	1 764.2	721.2	2 485.4	3.921	1.139	5.060	**360**
365	198.33	0.002 016	0.060 116	1 818.0	610.0	2 428.0	4.002	0.956	4.958	**365**
370	210.54	0.002 214	0.049 728	1 890.2	452.6	2 342.8	4.111	0.703	4.814	370
374.15	221.20	0.003 170	0.003 170	2 107.4	0.0	2 107.4	4.443	0.000	4.443	**374.15**

Table II.2 Saturated water and steam (pressure) tables.

Absolute pressure in bar	Temperature in °C	Specific volume in m^2/kg		Specific enthalpy in kJ/kg			Specific entropy in kJ/kg K			Absolute pressure in bar
(p)	(t)	Water (v_f)	Steam (v_g)	Water (h_f)	Evaporation (h_{fg})	Steam (h_g)	Water (s_f)	Evaporation (s_{fg})	Steam (s_g)	(p)
0.0061	0.000	0.001 000	206.31	0.0	2 501.6	2 501.6	0.000	9.158	9.158	**0.006 1**
0.010	6.983	0.001 000	129.21	29.3	2 485.1	2 514.4	0.106	8.871	8.977	**0.010**
0.015	13.04	0.001 001	87.982	54.7	2 470.8	2 525.5	0.196	8.624	8.830	**0.015**
0.020	17.51	0.001 001	67.006	73.5	2 460.1	2 533.0	0.261	8.464	8.725	**0.020**
0.025	21.10	0.001 002	54.256	88.4	2 451.8	2 540.2	0.312	8.333	8.645	**0.025**
0.030	24.10	0.001 003	45.667	101.0	2 444.6	2 545.6	0.354	8.224	8.578	**0.030**
0.035	26.69	0.001 003	39.479	111.8	2 438.6	2 550.4	0.391	8.132	8.523	**0.035**
0.040	28.98	0.001 004	34.802	121.4	2 433.1	2 554.5	0.423	8.053	8.476	**0.040**
0.045	31.03	0.001 005	31.141	130.0	2 428.2	2 558.2	0.451	7.983	8.434	**0.045**
0.050	32.90	0.001 005	28.194	137.8	2 423.8	2 561.6	0.476	7.920	8.396	**0.050**
0.060	36.18	0.001 006	23.741	151.5	2 416.0	2 567.5	0.521	7.810	8.331	**0.06**
0.070	39.03	0.001 007	20.531	163.4	2 409.2	2 572.6	0.559	7.718	8.277	**0.07**
0.080	41.53	0.001 008	18.105	173.9	2 403.2	2 577.1	0.593	7.637	8.230	**0.08**
0.090	43.79	0.001 009	16.204	183.3	2 397.8	2 581.1	0.622	7.566	8.188	**0.09**
0.100	45.83	0.001 010	14.675	191.8	2 392.9	2 584.7	0.649	7.502	8.151	**0.10**
0.11	47.71	0.001 011	13.416	199.7	2 388.4	2 588.1	0.674	7.444	8.118	**0.11**
0.12	49.45	0.001 012	12.362	206.9	2 384.3	2 591.2	0.696	7.391	8.087	**0.12**
0.13	51.06	0.001 013	11.466	213.7	2 380.3	2 594.0	0.717	7.342	8.059	**0.13**

(p)	(t)	(v_f)	(v_g)	(h_f)	(h_{fg})	(h_g)	(s_f)	(s_{fg})	(s_g)	(p)
0.14	52.57	0.001 013	10.694	220.0	2 376.7	2 596.7	0.737	7.296	8.033	**0.14**
0.15	54.00	0.001 014	10.023	226.0	2 373.2	2 599.2	0.755	7.254	8.009	**0.15**
0.16	55.34	0.001 015	9.433 1	231.6	2 370.0	2 601.6	0.772	7.215	7.987	**0.16**
0.17	56.62	0.001 015	8.911 1	236.9	2 366.9	2 603.8	0.788	7.178	7.996	**0.17**
0.18	57.83	0.001 016	8.445 2	242.0	2 363.9	2 605.9	0.804	7.142	7.946	**0.18**
0.19	58.98	0.001 017	8.027 2	246.8	2 361.1	2 607.9	0.818	7.109	7.927	**0.19**
0.20	60.09	0.001 017	8.649 8	251.5	2 358.4	2 609.9	0.832	7.077	7.909	**0.20**
0.21	61.15	0.001 018	7.367 3	255.9	2 355.8	2 611.7	0.845	7.047	7.892	**0.21**
0.22	62.16	0.001 018	6.995 1	260.1	2 353.4	2 613.5	0.858	7.018	7.876	**0.22**
0.23	63.14	0.001 019	6.709 3	264.2	2 351.0	2 615.2	0.870	6.991	7.861	**0.23**
0.24	64.08	0.001 019	6.446 7	268.2	2 348.6	2 616.8	0.882	6.964	7.846	**0.24**
0.25	64.99	0.001 020	6.204 5	272.0	2 346.3	2 618.3	0.893	6.939	7.832	**0.25**
0.26	65.87	0.001 020	5.980 3	275.7	2 344.2	2 619.9	0.904	6.915	7.819	**0.26**
0.27	66.72	0.001 021	5.772 4	279.2	2 342.1	2 621.3	0.915	6.891	7.806	**0.27**
0.28	67.55	0.001 021	5.577 8	282.7	2 340.0	2 622.7	0.925	6.868	7.793	**0.28**
0.29	68.35	0.001 022	5.398 2	286.0	2 338.1	2 624.1	0.935	6.847	7.781	**0.29**
0.30	69.12	0.001 022	5.229 3	289.3	2 336.1	2 625.4	0.944	6.825	7.769	**0.30**
0.32	70.62	0.001 023	4.922 0	295.6	2 332.4	2 628.0	0.962	6.785	7.747	**0.32**
0.34	72.03	0.001 024	4.650 4	301.5	2 328.9	2 630.4	0.980	6.747	7.727	**0.34**
0.36	73.37	0.001 025	4.407 6	307.1	2 325.5	2 632.6	0.996	6.711	7.707	**0.36**
0.38	74.66	0.001 026	4.190 0	312.5	2 322.3	2 634.8	1.011	6.677	7.688	**0.38**
0.40	75.89	0.001 027	3.993 4	317.7	2 319.2	2 636.9	1.026	6.645	7.671	**0.40**
0.42	77.06	0.001 027	3.814 8	322.6	2 316.3	2 638.9	1.040	6.614	7.654	**0.42**
0.44	78.19	0.001 028	3.652 2	327.3	2 313.4	2 640.7	1.054	6.584	7.638	**0.44**
0.46	79.28	0.001 029	3.503 2	331.9	2 310.7	2 642.6	1.067	6.556	7.623	**0.46**
0.48	80.33	0.001 029	3.366 3	336.3	2 308.0	2 644.3	1.079	6.530	7.609	**0.48**
0.50	81.35	0.001 030	3.240 1	340.6	2 305.4	2 646.0	1.091	6.504	7.595	**0.50**

(p)	(t)	(v_f)	(v_g)	(h_f)	(h_{fg})	(h_g)	(s_f)	(s_{fg})	(s_g)	(p)
0.52	82.33	0.001 031	3.123 3	344.7	2 302.9	2 647.6	1.103	6.478	7.581	**0.52**
0.54	83.28	0.001 031	3.014 8	348.7	2 300.5	2 649.2	1.114	6.455	7.569	**0.54**
0.56	84.19	0.001 032	3.913 9	352.5	2 298.2	2 650.7	1.125	6.431	7.556	**0.56**
0.58	85.09	0.001 033	3.819 7	356.3	2 295.8	2 652.1	1.135	6.409	7.544	**0.58**
0.60	85.95	0.001 033	3.731 7	359.9	2 293.7	2 653.6	1.145	6.388	7.533	**0.60**
0.62	86.80	0.001 034	2.649 1	363.5	2 291.4	2 654.9	1.155	6.367	7.552	**0.62**
0.64	87.62	0.001 034	2.571 5	366.9	2 289.4	2 656.3	1.165	6.346	7.511	**0.64**
0.66	88.42	0.001 035	2.498 5	370.3	2 287.3	2 657.6	1.174	6.326	7.500	**0.66**
0.68	89.20	0.001 036	2.429 7	373.6	2 285.2	2 658.8	1.183	6.307	7.490	**0.68**
0.70	89.96	0.001 036	2.364 7	376.8	2 283.3	2 660.1	1.192	6.288	7.480	**0.70**
0.72	90.70	0.001 037	2.303 1	379.9	2 281.4	2 661.3	1.201	6.270	7.471	**0.72**
0.74	91.43	0.001 037	2.244 8	382.9	2 279.5	2 662.4	1.209	6.253	7.462	**0.74**
0.76	92.14	0.001 038	2.189 5	385.9	2 277.7	2 663.4	1.217	6.235	7.452	**0.76**
0.78	92.83	0.001 038	2.136 9	388.9	2 275.8	2 664.7	1.225	6.219	7.444	**0.78**
0.80	93.51	0.001 039	2.086 9	391.7	2 274.1	2 665.8	1.235	6.202	7.435	**0.80**
0.85	95.15	0.001 040	1.972 1	398.6	2 269.8	2 668.4	1.252	6.163	7.415	**0.85**
0.90	96.71	0.001 041	1.869 1	405.2	2 265.7	2 670.9	1.270	6.125	7.395	**0.90**
0.95	98.20	0.001 042	1.777 1	411.5	2 261.7	2 673.2	1.287	6.091	7.378	**0.95**
1.00	99.63	0.001 043	1.693 8	417.5	2 257.9	2 675.4	1.303	6.057	7.360	**1.00**
1.013 25	100.00	0.001 044	1.673 0	419.1	2 256.9	2 676.0	1.307	6.048	7.355	**1.013 25**
1.05	101.0	0.001 045	1.618 1	423.3	2 254.3	2 670.6	1.318	6.025	7.343	**1.05**
1.10	102.3	0.001 046	1.549 2	428.8	2 250.8	2 679.6	1.333	5.995	7.328	**1.10**
1.15	103.6	0.001 047	1.486 1	434.2	2 247.4	2 681.6	1.347	5.966	7.313	**1.15**
1.20	104.8	0.001 048	1.428 1	439.3	2 244.1	2 683.4	1.361	5.937	7.298	**1.20**
1.25	106.0	0.001 049	1.374 6	444.4	2 240.8	2 685.2	1.374	5.911	7.285	**1.25**
1.30	107.1	0.001 050	1.325 0	449.2	2 237.8	2 687.0	1.387	5.885	7.272	**1.30**
1.35	108.2	0.001 050	1.279 1	453.4	2 234.8	2 688.7	1.399	5.860	7.259	**1.35**

(p)	(t)	(v_f)	(v_g)	(h_f)	(h_{fg})	(h_g)	(s_f)	(s_{fg})	(s_g)	(p)
1.40	109.3	0.001 051	1.236 3	458.4	2 231.9	2 690.3	1.411	5.836	7.247	**1.40**
1.45	110.4	0.001 052	1.196 3	462.8	2 229.0	2 691.8	1.423	5.812	7.235	**1.45**
1.50	111.4	0.001 053	1.159 0	467.1	2 226.3	2 693.4	1.433	5.790	7.223	**1.50**
1.60	113.3	0.001 055	1.091 1	475.4	2 220.8	2 696.2	1.455	5.747	7.202	**1.60**
1.70	115.2	0.001 056	1.030 9	483.2	2 215.8	2 699.0	1.475	5.706	7.181	**1.70**
1.80	116.9	0.001 058	0.977 18	490.7	2 210.8	2 701.5	1.494	5.668	7.162	**1.80**
1.90	118.6	0.001 059	0.928 95	497.9	2 206.1	2 704.0	1.513	5.631	7.144	**1.90**
2.00	120.2	0.001 061	0.885 40	504.7	2 201.6	2 706.3	1.530	5.597	7.127	**2.00**
2.1	121.8	0.001 062	0.845 86	511.3	2 197.2	2 708.5	1.547	5.564	7.111	**2.1**
2.2	123.3	0.001 064	0.809 80	517.6	2 193.0	2 710.6	1.563	5.532	7.095	**2.2**
2.3	124.7	0.001 065	0.776 77	523.7	2 188.9	2 712.6	1.578	5.502	7.080	**2.3**
2.4	126.1	0.001 066	0.746 41	529.6	2 184.9	2 714.5	1.593	5.473	7.066	**2.4**
2.5	127.4	0.001 068	0.718 40	535.3	2 181.1	2 716.4	1.607	5.445	7.052	**2.5**
2.6	128.7	0.001 069	0.692 47	540.9	2 177.3	2 718.2	1.621	5.418	7.039	**2.6**
2.7	130.0	0.001 070	0.668 40	546.2	2 173.7	2 719.9	1.634	5.392	7.026	**2.7**
2.8	131.2	0.001 071	0.646 00	551.4	2 170.1	2 721.5	1.647	5.367	7.014	**2.8**
2.9	132.4	0.001 072	0.625 09	556.5	2 166.6	2 723.1	1.660	5.342	7.002	**2.9**
3.0	133.5	0.001 074	0.605 53	561.5	2 163.2	2 724.7	1.672	5.319	6.991	**3.0**
3.1	134.7	0.001 075	0.587 18	566.2	2 159.9	2 726.1	1.683	5.297	6.980	**3.1**
3.2	135.8	0.001 076	0.569 95	570.9	2 156.7	2 727.6	1.695	5.274	6.969	**3.2**
3.3	136.8	0.001 077	0.553 73	575.5	2 153.5	2 729.0	1.706	5.253	6.959	**3.3**
3.4	137.9	0.001 078	0.538 43	579.9	2 150.4	2 730.3	1.717	5.232	6.949	**3.4**
3.5	138.9	0.001 079	0.523 97	584.3	2 147.3	2 731.6	1.727	5.212	6.939	**3.5**
3.6	139.9	0.001 080	0.510 29	588.5	2 144.4	2 732.9	1.738	5.192	6.930	**3.6**
3.7	140.8	0.001 081	0.497 33	592.7	2 141.4	2 734.1	1.748	5.173	6.921	**3.7**
3.8	141.8	0.001 082	0.485 02	596.7	2 138.6	2 735.3	1.758	5.154	9.912	**3.8**
3.9	142.7	0.001 083	0.473 33	600.8	2 135.7	2 736.5	1.767	5.136	9.903	**3.9**

(p)	(t)	(v_f)	(v_g)	(h_f)	(h_{fg})	(h_g)	(s_f)	(s_{fg})	(s_g)	(p)
4.0	143.6	0.001 084	0.462 20	604.7	2 132.9	2 737.6	1.776	5.118	6.894	**4.0**
4.1	144.5	0.001 085	0.451 59	608.5	2 130.2	2 738.7	1.786	5.100	6.886	**4.1**
4.2	145.4	0.001 086	0.441 47	612.3	2 127.5	2 739.8	1.795	5.083	6.878	**4.2**
4.3	146.3	0.001 087	0.431 81	616.0	2 124.9	2 740.9	1.803	5.067	6.870	**4.3**
4.4	147.1	0.001 088	0.422 57	619.6	2 122.3	2 741.9	1.812	5.050	6.862	**4.4**
4.5	147.9	0.001 089	0.413 73	623.2	2 119.7	2 742.9	1.820	5.035	6.855	**4.5**
4.6	148.7	0.001 090	0.405 26	626.7	2 117.2	2 743.9	1.829	5.018	6.847	**4.6**
4.7	149.5	0.001 090	0.397 14	630.1	2 114.7	2 744.8	1.837	5.003	6.840	**4.7**
4.8	150.3	0.001 091	0.389 34	633.5	2 112.2	2 745.7	1.845	4.988	6.833	**4.8**
4.9	151.1	0.001 092	0.381 86	636.8	2 109.8	2 746.6	1.853	4.973	6.826	**4.9**
5.0	151.8	0.001 093	0.374 66	640.1	2 107.4	2 747.5	1.860	4.959	6.819	**5.0**
5.2	153.3	0.001 095	0.361 06	646.5	2 102.7	2 749.2	1.875	4.931	6.806	**5.2**
5.4	154.8	0.001 096	0.348 44	652.8	2 098.1	2 750.9	1.890	4.903	6.793	**5.4**
5.6	156.2	0.001 098	0.336 69	658.8	2 093.7	2 752.5	1.904	4.877	6.781	**5.6**
5.8	157.5	0.001 099	0.325 72	664.7	2 089.3	2 754.0	1.918	4.851	6.769	**5.8**
6.0	158.8	0.001 101	0.315 46	670.4	2 085.1	2 755.5	1.931	4.827	6.758	**6.0**
6.2	160.1	0.001 102	0.305 84	676.1	2 080.8	2 756.9	1.944	4.803	6.747	**6.2**
6.4	161.4	0.001 104	0.296 80	681.5	2 076.7	2 758.2	1.956	4.780	6.736	**6.4**
6.6	162.6	0.001 105	0.288 29	686.8	2 072.7	2 759.5	1.968	4.757	6.725	**6.6**
6.8	163.8	0.001 107	0.280 26	692.0	2 068.8	2 760.8	1.980	4.735	6.715	**6.8**
7.0	165.0	0.001 108	0.272 68	697.1	2 064.9	2 762.0	1.992	4.713	6.705	**7.0**
7.2	166.1	0.001 110	0.265 50	702.0	2 061.2	2 763.2	2.003	4.693	6.696	**7.2**
7.4	167.2	0.001 111	0.258 70	706.9	2 057.4	2 764.3	2.014	4.672	6.686	**7.4**
7.6	168.3	0.001 112	0.252 24	711.7	2 053.7	2 765.4	2.025	4.652	6.677	**7.6**
7.8	169.4	0.001 114	0.246 10	716.3	2 050.1	2 766.4	2.035	4.633	6.668	**7.8**
8.0	170.4	0.001 115	0.240 26	720.9	2 046.5	2 767.4	2.046	4.614	6.660	**8.0**

(p)	(t)	(v_f)	(v_g)	(h_f)	(h_{fg})	(h_g)	(s_f)	(s_{fg})	(s_g)	(p)
8.2	171.4	0.001 116	0.234 69	725.4	2 043.0	2 768.4	2.056	4.595	6.651	**8.2**
8.4	172.4	0.001 118	0.229 38	729.9	2 039.6	2 769.4	2.066	4.577	6.643	**8.4**
8.6	173.4	0.001 119	0.224 31	734.2	2 036.2	2 770.4	2.075	4.560	6.635	**8.6**
8.8	174.4	0.001 120	0.219 46	738.5	2 032.8	2 771.3	2.085	4.542	6.627	**8.8**
9.0	175.4	0.001 121	0.214 82	742.6	2 029.5	2 772.1	2.094	4.525	6.619	**9.0**
9.2	176.3	0.001 123	0.210 37	746.8	2 026.2	2 773.0	2.103	4.509	6.612	**9.2**
9.4	177.2	0.001 124	0.206 10	750.8	2 023.0	2 773.8	2.112	4.492	6.604	**9.4**
9.6	178.1	0.001 125	0.202 01	754.8	2 019.8	2 774.6	2.121	4.476	6.597	**9.6**
9.8	179.0	0.001 126	0.198 08	758.7	2 016.7	2 775.4	2.130	4.460	6.590	**9.8**
10.0	179.9	0.001 127	0.194 30	762.6	2 013.6	2 776.2	2.138	4.445	6.583	**10.0**
10.5	182.0	0.001 130	0.185 48	772.0	2 006.0	2 778.0	2.159	4.407	6.566	**10.5**
11.0	184.1	0.001 133	0.177 39	781.1	1 998.6	2 779.7	2.179	4.371	6.550	**11.0**
11.5	186.0	0.001 136	0.170 02	789.9	1 991.4	2 781.3	2.198	4.336	6.534	**11.5**
12.0	188.0	0.001 139	0.163 21	798.4	1 984.3	2 782.7	2.216	4.303	6.519	**12.0**
12.5	189.8	0.001 141	0.156 96	806.7	1 977.5	2 784.2	2.234	4.271	6.505	**12.5**
13.0	191.6	0.001 144	0.151 14	814.7	1 970.7	2 785.4	2.251	4.240	6.491	**13.0**
13.5	193.3	0.001 146	0.145 76	822.5	1 964.2	2 786.7	2.267	4.211	6.478	**13.5**
14.0	195.0	0.001 149	0.140 73	830.1	1 957.7	2 787.8	2.284	4.181	6.465	**14.0**
14.5	196.7	0.001 151	0.136 06	837.5	1 951.2	2 788.9	2.299	4.154	6.453	**14.5**
15.0	198.3	0.001 154	0.131 67	844.6	1 945.3	2 789.9	2.314	4.127	6.441	**15.0**
15.5	199.8	0.001 156	0.127 56	851.6	1 939.2	2 790.8	2.329	4.100	6.429	**15.5**
16.0	201.4	0.001 159	0.123 70	858.5	1 933.2	2 791.7	2.344	4.074	6.418	**16.0**
16.5	202.9	0.001 161	0.120 06	865.3	1 952.3	2 792.6	2.358	4.049	6.407	**16.5**
17.0	204.3	0.001 163	0.116 64	871.8	1 921.6	2 793.4	2.371	4.025	6.396	**17.0**
17.5	205.7	0.001 166	0.113 40	878.2	1 915.9	2 794.1	2.384	4.001	6.385	**17.5**
18.0	207.1	0.001 168	0.110 33	884.5	1 910.3	2 794.8	2.398	3.977	6.375	**18.0**
18.5	208.5	0.001 170	0.107 42	890.7	1 904.8	2 795.5	2.410	3.955	6.365	**18.5**

(p)	(t)	(v_f)	(v_g)	(h_f)	(h_{fg})	(h_g)	(s_f)	(s_{fg})	(s_g)	(p)
19.0	209.8	0.001 172	0.104 67	896.8	1 899.3	2 796.1	2.423	3.933	6.356	**19.0**
19.5	211.1	0.001 174	0.102 04	902.7	1 894.0	2 796.7	2.435	3.911	6.346	**19.5**
20.0	212.4	0.001 177	0.099 55	908.6	1 888.7	2 797.2	2.447	3.890	6.337	**20.0**
21.0	214.8	0.001 181	0.094 902	919.9	1 878.3	2 798.2	2.470	3.849	6.319	**21.0**
22.0	217.2	0.001 185	0.090 663	930.9	1 868.1	2 799.1	2.492	3.809	6.301	**22.0**
23.0	219.6	0.001 189	0.086 780	941.6	1 858.2	2 799.8	2.514	3.771	6.285	**23.0**
24.0	221.8	0.001 193	0.083 209	951.9	1 848.5	2 800.4	2.534	3.735	6.269	**24.0**
25.0	223.9	0.001 197	0.079 915	961.9	1 839.1	2 801.0	2.554	3.699	6.253	**25.0**
26.0	226.0	0.001 201	0.076 865	971.7	1 829.7	2 801.4	2.574	3.665	6.239	**26.0**
27.0	228.1	0.001 205	0.074 033	981.2	1 820.5	2 801.7	2.592	3.632	6.224	**27.0**
28.0	230.0	0.001 209	0.071 396	990.5	1 811.5	2 802.0	2.611	3.600	6.211	**28.0**
29.0	232.0	0.001 213	0.068 036	999.5	1 802.7	2 802.2	2.628	3.569	6.197	**29.0**
30.0	233.8	0.001 216	0.066 632	1008.3	1 794.0	2 802.3	2.646	3.538	6.184	**30.0**
31.0	235.7	0.001 220	0.064 473	1017.1	1 785.4	2 802.3	2.662	3.509	6.171	**31.0**
32.0	237.4	0.001 224	0.062 443	1025.4	1 776.9	2 802.3	2.679	3.480	6.159	**32.0**
33.0	239.2	0.001 227	0.060 533	1033.7	1 768.6	2 802.3	2.694	3.452	6.146	**33.0**
34.0	240.9	0.001 231	0.058 731	1041.8	1 760.3	2 802.1	2.710	3.424	6.134	**34.0**
35.0	242.5	0.001 235	0.057 028	1049.7	1 752.3	2 802.0	2.725	3.398	6.123	**35.0**
36.0	244.2	0.001 238	0.055 417	1057.5	1 744.2	2 801.7	2.740	3.371	6.111	**36.0**
37.0	245.8	0.001 242	0.053 889	1065.2	1 736.2	2 801.4	2.755	3.345	6.100	**37.0**
38.0	247.3	0.001 245	0.052 439	1072.7	1 728.4	2 801.1	2.769	3.321	6.090	**38.0**
39.0	248.8	0.001 249	0.051 061	1080.1	1 720.7	2 800.8	2.783	3.296	6.079	**39.0**
40.0	250.3	0.001 252	0.049 749	1087.4	1 712.9	2 800.3	2.797	3.272	6.069	**40.0**
42.0	253.2	0.001 259	0.047 306	1101.6	1 697.8	2 799.4	2.823	3.225	6.048	**42.0**
44.0	256.1	0.001 266	0.045 078	1115.4	1 682.9	2 798.3	2.849	3.180	6.029	**44.0**
46.0	258.8	0.001 273	0.043 036	1128.8	1 668.2	2 797.0	2.874	3.136	6.010	**46.0**
48.0	261.4	0.001 279	0.041 158	1141.8	1 653.9	2 795.7	2.897	3.094	5.991	**48.0**
50.0	263.9	0.001 286	0.039 425	1154.5	1 639.7	2 794.2	2.921	3.053	5.974	**50.0**

(p)	(t)	(v_f)	(v_g)	(h_f)	(h_{fg})	(h_g)	(s_f)	(s_{fg})	(s_g)	(p)
52.0	266.4	0.001 293	0.037 820	1166.9	1 625.7	2 792.6	2.943	3.013	5.956	**52.0**
54.0	268.8	0.001 299	0.036 330	1179.0	1 611.8	2 790.8	2.965	2.974	5.939	**54.0**
56.0	271.1	0.001 306	0.034 942	1190.8	1 598.2	2 789.0	2.986	2.937	5.923	**56.0**
58.0	273.4	0.001 312	0.033 646	1202.4	1 584.6	2 787.0	3.007	2.899	5.906	**58.0**
60.0	275.5	0.001 319	0.032 433	1213.7	1 571.3	2 785.0	3.027	2.863	5.890	**60.0**
62.0	277.7	0.001 325	0.031 295	1224.9	1 558.0	2 782.9	3.047	2.828	5.875	**62.0**
64.0	279.8	0.001 332	0.030 225	1235.8	1 544.8	2 780.6	3.066	2.794	5.860	**64.0**
66.0	281.9	0.001 338	0.029 218	1246.5	1 531.8	2 778.3	3.085	2.760	5.845	**66.0**
68.0	283.9	0.001 345	0.028 267	1257.1	1 518.8	2 775.9	3.104	2.727	5.831	**68.0**
70.0	285.8	0.001 351	0.027 368	1267.4	1 506.0	2 773.4	3.122	2.694	5.816	**70.0**
72.0	287.7	0.001 358	0.026 517	1277.7	1 493.2	2 770.9	3.140	2.662	5.802	**72.0**
74.0	289.6	0.001 365	0.025 711	1287.8	1 480.4	2 768.2	3.157	2.631	5.788	**74.0**
76.0	291.4	0.001 371	0.024 944	1297.8	1 467.8	2 765.5	3.174	2.600	5.774	**76.0**
78.0	293.2	0.001 378	0.024 215	1307.5	1 455.4	2 762.7	3.191	2.569	5.760	**78.0**
80.0	295.0	0.001 384	0.023 521	1317.5	1 442.7	2 759.9	3.208	2.539	5.747	**80.0**
82.0	296.7	0.001 391	0.022 860	1326.7	1 430.3	2 757.0	3.224	2.510	5.734	**82.0**
84.0	298.4	0.001 398	0.022 228	1336.2	1 417.8	2 754.0	3.240	2.481	5.721	**84.0**
86.0	300.1	0.001 404	0.021 624	1345.4	1 405.5	2 750.9	3.256	2.452	5.708	**86.0**
88.0	301.7	0.001 411	0.021 046	1354.7	1 393.1	2 747.8	3.271	2.424	5.695	**88.0**
90.0	303.3	0.001 418	0.020 493	1363.8	1 380.8	2 744.6	3.287	2.395	5.682	**90.0**
92	304.9	0.001 425	0.019 962	1372.8	1 368.5	2 741.3	3.302	2.367	5.669	**92**
94	306.5	0.001 432	0.019 453	1381.7	1 356.3	2 738.0	3.317	2.340	5.657	**94**
96	308.0	0.001 439	0.018 964	1390.6	1 344.1	2 734.7	3.332	2.313	5.644	**96**
98	309.5	0.001 446	0.018 493	1396.4	1 331.9	2 731.2	3.346	2.286	5.632	**98**
100	311.0	0.001 453	0.018 041	1408.0	1 319.7	2 727.7	3.361	2.259	5.620	**100**
105	314.6	0.001 470	0.016 981	1429.5	1 289.9	2 718.7	3.396	2.194	5.590	**105**
110	318.0	0.001 489	0.016 007	1450.5	1 258.8	2 709.3	3.430	2.129	5.560	**110**

(p)	(t)	(v_f)	(v_g)	(h_f)	(h_{fg})	(h_g)	(s_f)	(s_{fg})	(s_g)	(p)
115	321.4	0.001 508	0.015 114	1471.3	1 228.2	2 699.5	3.464	2.066	5.530	**115**
120	324.6	0.001 527	0.014 285	1491.7	1 197.5	2 698.2	3.497	2.003	5.500	**120**
125	327.8	0.001 547	0.013 518	1511.9	1 166.5	2 678.4	3.530	1.941	5.471	**125**
130	330.8	0.001 567	0.012 800	1531.9	1 135.1	2 667.0	3.561	1.880	5.441	**130**
135	333.8	0.001 588	0.012 130	1551.8	1 103.3	2 655.1	3.593	1.818	5.411	**135**
140	336.6	0.001 611	0.011 498	1571.5	1 070.9	2 642.4	3.624	1.756	5.380	**140**
145	339.4	0.001 634	0.010 905	1591.3	1 037.9	2 629.2	3.655	1.694	5.349	**145**
150	342.1	0.001 658	0.010 343	1610.9	1 004.2	2 615.1	3.686	1.632	5.318	**150**
155	344.8	0.001 683	0.009 813	1630.7	969.7	2 600.4	3.716	1.570	5.286	**155**
160	347.3	0.001 710	0.009 310	1650.4	934.5	2 584.9	3.747	1.506	5.253	**160**
165	349.7	0.001 739	0.008 833	1670.4	898.5	2 568.9	3.778	1.442	5.220	**165**
170	352.3	0.001 770	0.008 372	1691.6	860.0	2 551.6	3.811	1.375	5.186	**170**
175	354.6	0.001 803	0.007 927	1713.3	820.0	2 533.3	3.844	1.306	5.150	**175**
180	357.0	0.001 840	0.007 497	1734.8	779.1	2 513.9	3.877	1.236	5.113	**180**
185	359.0	0.001 881	0.007 082	1756.5	736.5	2 493.0	3.910	1.164	5.074	**185**
190	361.4	0.001 926	0.006 676	1778.7	691.8	2 470.5	3.943	1.090	5.033	**190**
195	363.6	0.001 978	0.006 276	1801.9	643.9	2 445.8	3.978	1.011	4.989	**195**
200	365.7	0.002 037	0.005 875	1826.6	591.6	2 418.2	4.015	0.926	4.941	**200**
205	367.8	0.002 110	0.005 462	1854.2	532.0	2 386.2	4.056	0.830	4.886	**205**
210	369.8	0.002 202	0.005 023	1886.3	461.2	2 347.5	4.105	0.717	4.822	**210**
215	371.8	0.002 342	0.004 509	1929.4	365.2	2 294.6	4.170	0.566	4.736	**215**
220	373.7	0.002 668	0.003 735	2010.3	186.3	2 196.6	4.293	0.288	4.581	**220**
221.2	374.15	0.003 170	0.003 170	2107.4	000.0	2 107.4	4.443	0.000	4.443	**221.2**

Table II.3 Specific volume of superheated steam

Absolute pressure in bar	*Saturation Temperature in °C*	*Specific volume (v) in m³/kg at various temperature in °C*										
(p)	*(t_s)*	*100*	*150*	*200*	*250*	*300*	*350*	*400*	*500*	*600*	*700*	*800*
0.02	17.5	86.08	97.63	109.2	120.7	132.2	143.8	155.3	178.4	201.5	224.6	247.6
0.04	29.0	43.03	48.81	54.58	60.35	66.12	71.89	77.66	89.20	100.7	112.3	123.8
0.06	36.2	28.68	32.53	36.38	40.23	44.08	47.93	51.77	59.47	67.16	74.85	82.54
0.08	41.5	21.50	24.40	27.28	30.17	33.06	35.94	38.83	44.60	50.37	56.14	61.91
0.10	45.8	17.20	19.51	21.83	24.14	26.45	28.75	31.06	35.68	40.30	44.91	49.53
0.15	54.0	11.51	13.06	14.61	16.16	17.71	19.25	20.80	23.89	26.98	30.07	33.16
0.20	60.1	8.585	9.748	10.91	12.07	13.22	14.37	15.53	17.84	20.15	22.45	24.76
0.25	65.0	6.874	7.808	8.737	9.665	10.59	11.52	12.44	14.29	16.14	17.99	19.84
0.30	69.1	5.714	6.493	7.268	8.040	8.811	9.581	10.35	11.89	13.43	14.70	16.51
0.35	72.7	4.898	5.568	6.233	6.896	7.557	8.218	8.879	10.20	11.52	12.84	14.16
0.40	75.9	4.279	4.866	5.448	6.028	6.607	7.185	7.763	8.918	10.07	11.23	12.38
0.45	78.7	3.803	4.325	4.844	5.360	5.875	6.389	6.903	7.930	8.957	9.984	10.99
0.50	81.3	3.418	3.889	4.356	4.821	5.284	5.747	6.209	7.134	8.057	8.981	9.904
0.60	86.0	2.844	3.238	3.628	4.016	4.402	4.788	5.174	5.944	6.714	7.484	8.254
0.70	90.0	2.434	2.773	3.108	3.441	3.772	4.103	4.434	5.095	5.755	6.415	7.074
0.80	93.5	2.126	2.425	2.718	3.010	3.300	3.590	3.879	4.457	5.035	5.613	6.190
0.90	96.7	1.887	2.153	2.415	2.674	2.933	3.190	3.448	3.962	4.475	4.989	5.502
1.00	99.6	1.696	1.936	2.172	2.406	2.639	2.871	3.103	3.565	4.028	4.490	4.952
1.50	111.4	..	1.285	1.444	1.601	1.757	1.912	2.067	2.376	2.685	2.993	3.301
2.00	120.2	..	0.959 5	1.080	1.199	1.316	1.433	1.549	1.781	2.013	2.244	2.475
2.5	127.4	..	0.764 1	0.862 0	0.957 4	1.052	1.145	1.239	1.424	1.610	1.795	1.980
3.0	133.5	..	0.633 7	0.716 4	0.796 4	0.875 3	0.953 5	1.031	1.187	1.341	1.496	1.650
3.5	138.9	..	0.540 6	0.612 3	0.681 4	0.749 3	0.816 6	0.883 5	1.017	1.149	1.282	1.414

(p)	(t_s)	100	150	200	250	300	350	400	500	600	700	800
4.0	143.6	..	0.470 7	0.534 3	0.595 2	0.654 9	0.713 9	0.772 5	0.889 2	1.005	1.121	1.237
4.5	147.0	..	0.416 5	0.473 8	0.528 4	0.581 7	0.634 3	0.686 5	0.790 5	0.893 9	0.997 1	1.100
5.0	151.8	..	..	0.425 0	0.474 4	0.522 6	0.570 1	0.617 2	0.710 8	0.804 0	0.896 9	0.989 6
6.0	158.8	..	..	0.352 0	0.393 9	0.434 4	0.474 2	0.513 6	0.591 8	0.669 6	0.747 1	0.824 5
7.0	165.0	..	..	0.299 9	0.336 4	0.371 4	0.405 7	0.439 6	0.506 9	0.573 7	0.640 2	0.706 6
8.0	170.4	.	..	0.260 8	0.293 2	0.324 1	0.354 3	0.384 2	0.443 2	0.501 7	0.560 0	0.618 1
9.0	175.4	..	..	0.230 3	0.259 6	0.287 4	0.314 4	0.341 0	0.393 6	0.445 8	0.497 6	0.549 3
10.0	179.9	..	..	0.205 9	0.232 8	0.258 0	0.282 4	0.306 5	0.354 0	0.401 0	0.447 7	0.494 3
11.0	184.1	..	..	0.185 9	0.210 8	0.233 9	0.256 3	0.278 2	0.321 5	0.364 4	0.406 9	0.449 2
12.0	188.0	..	..	0.169 2	0.192 4	0.213 9	0.234 5	0.254 7	0.294 5	0.333 8	0.372 9	0.411 8
13.0	191.6	..	..	0.155 1	0.176 9	0.196 9	0.216 1	0.234 8	0.271 6	0.308 0	0.344 1	0.380 0
14.0	195.0	..	..	0.142 9	0.163 6	0.182 3	0.200 2	0.217 7	0.252 0	0.285 9	0.319 4	0.352 8
15.0	198.3	..	..	0.132 4	0.152 0	0.169 7	0.186 5	0.202 9	0.235 0	0.266 7	0.298 0	0.329 2
16.0	201.4	..	..	..	0.141 9	0.158 7	0.174 5	0.190 0	0.220 2	0.249 9	0.279 3	0.308 6
17.0	204.3	..	..	..	0.132 9	0.148 9	0.164 0	0.178 6	0.207 0	0.235 1	0.262 8	0.290 4
18.0	207.1	..	..	..	0.125 0	0.140 2	0.154 6	0.168 4	0.195 4	0.221 9	0.248 1	0.274 2
19.0	209.8	..	..	..	0.117 9	0.132 5	0.146 1	0.159 3	0.184 9	0.210 1	0.235 0	0.259 7
20.0	212.4	..	..	..	0.111 5	0.125 5	0.138 6	0.151 1	0.175 6	0.199 5	0.223 2	0.246 7
22.0	217.2	..	..	..	0.100 4	0.113 4	0.125 5	0.137 0	0.159 3	0.181 2	0.202 8	0.224 2
24.0	221.8	..	..	..	0.091 08	0.103 4	0.114 6	0.125 2	0.145 8	0.165 9	0.185 8	0.205 4
26.0	226.0	..	..	..	0.083 21	0.094 83	0.105 3	0.115 3	0.134 4	0.153 0	0.171 4	0.189 5
28.0	230.0	..	..	..	0.076 44	0.087 51	0.097 40	0.106 7	0.124 6	0.141 9	0.159 0	0.175 9
30.0	233.8	..	..	..	0.070 55	0.081 16	0.090 53	0.099 31	0.116 1	0.132 3	0.148 3	0.164 1
32.0	237.4	..	..	..	0.065 38	0.075 59	0.084 51	0.092 83	0.108 7	0.123 9	0.139 0	0.153 8
34.0	240.9	..	..	..	0.060 80	0.070 68	0.079 20	0.087 11	0.102 1	0.116 5	0.130 7	0.144 7
36.0	244.2	..	..	..	00.56 70	0.066 30	0.074 48	0.082 02	0.096 26	0.110 0	0.123 4	0.136 6
38.0	247.3	..	..	..	0.053 02	0.062 37	0.070 25	0.077 47	0.091 04	0.104 1	0.116 8	0.129 4
40.0	250.3	..	..	..	..	0.058 83	0.066 45	0.073 38	0.086 34	0.098 76	0.110 9	0.122 9

(*p*)	(t_s)	*100*	*150*	*200*	*250*	*300*	*350*	*400*	*500*	*600*	*700*	*800*
42.0	253.2	..	..	..	..	0.055 63	0.063 00	0.069 67	0.082 09	0.093 97	0.105 6	0.117 0
44.0	256.0	..	..	..	..	0.052 70	0.059 86	0.066 30	0.078 23	0.089 61	0.100 7	0.111 6
46.0	258.8	..	..	..	..	0.050 03	0.056 99	0.063 22	0.074 70	0.085 62	0.096 26	0.106 7
48.0	261.4	..	..	..	..	0.047 57	0.054 36	0.060 39	0.071 47	0.081 97	0.092 19	0.102 2
50.0	263.9	..	..	..	..	0.045 30	0.051 94	0.057 79	0.068 49	0.078 62	0.088 45	0.098 09
55.0	269.9	..	..	..	..	0.043 43	0.046 66	0.052 13	0.062 02	0.071 31	0.080 31	0.089 12
60.0	275.6	..	..	..	..	0.036 15	0.042 22	0.047 38	0.056 59	0.065 18	0.073 48	0.081 59
65.0	280.8	..	..	..	..	0.032 58	0.038 48	0.043 38	0.052 03	0.060 03	0.067 74	0.075 26
70.0	285.8	..	..	..	..	0.029 46	0.035 23	0.039 92	0.048 09	0.055 59	0.062 79	0.069 80
75.0	290.5	..	..	..	..	0.026 72	0.032 43	0.036 94	0.044 69	0.051 76	0.058 52	0.065 09
80.0	295.0	..	..	..	..	0.024 26	0.029 95	0.034 31	0.041 70	0.048 39	0.054 77	0.060 96
85.0	299.2	..	..	..	..	0.021 91	0.027 76	0.032 00	0.039 08	0.045 44	0.051 48	0.057 32
90.0	303.3	..	..	..	..	..	0.025 79	0.029 93	0.036 74	0.042 80	0.048 53	0.054 08
95.0	307.2	..	..	..	..	..	0.024 03	0.028 08	0.034 65	0.040 45	0.045 91	0.051 19
100.0	311.0	..	..	..	..	..	0.022 42	0.026 41	0.032 76	0.038 32	0.043 55	0.048 58
110.0	318.0	..	..	..	..	..	0.019 61	0.023 51	0.029 50	0.034 66	0.039 47	0.044 08
120.0	324.6	..	..	..	..	..	0.017 21	0.021 08	0.026 79	0.031 60	0.036 07	0.040 33
130.0	330.8	..	..	..	..	..	0.015 10	0.019 02	0.024 49	0.029 02	0.033 19	0.037 16
140.0	336.6	..	..	..	..	..	0.013 21	0.017 23	0.022 51	0.026 80	0.030 72	0.034 44
150.0	342.1	..	..	..	..	..	0.011 46	0.015 66	0.020 80	0.024 88	0.028 59	0.032 09
160.0	347.3	..	..	..	..	..	0.009 76	0.014 28	0.019 29	0.023 20	0.026 72	0.030 03
170.0	352.3	..	..	..	..	..	..	0.013 03	0.017 97	0.021 72	0.025 07	0.028 21
180.0	357.0	..	..	..	..	..	..	0.011 91	0.016 79	0.020 40	0.023 00	0.026 59
190.0	361.4	..	..	..	..	..	..	0.010 89	0.015 73	0.019 22	0.022 29	0.025 15
200.0	365.7	..	..	..	..	..	..	0.009 95	0.014 77	0.018 16	0.021 11	0.023 85
210.0	369.8	..	..	..	..	..	..	0.009 07	0.013 91	0.017 20	0.020 04	0.022 67
220.0	373.7	..	..	..	..	..	..	0.008 25	0.013 12	0.016 33	0.019 07	0.021 60
221.2	374.15	..	..	..	..	..	..	0.008 16	0.013 03	0.016 22	0.018 95	0.021 35

Table II.4 Specific enthalpy of superheated steam

Absolute pressure in bar	*Saturation Temperature in °C*	*Specific enthalpy (h) in kJ/kg at various temperature in °C*										
(p)	*(t_s)*	*100*	*150*	*200*	*250*	*300*	*350*	*400*	*500*	*600*	*700*	*800*
0.02	17.5	2 688.5	2 783.7	2 880.0	2 977.7	3 076.8	3 177.5	3 279.7	3 489.2	3 705.6	3 928.8	4 158.7
0.04	29.0	2 688.3	2 783.5	2 879.9	2 977.6	3 076.8	3 177.4	3 279.7	3 489.2	3 705.6	3 928.8	4 158.7
0.06	36.2	2 688.0	2 783.4	2 879.8	2 977.6	3 076.7	3 177.4	3 279.6	3 489.2	3 705.6	3 928.8	4 158.7
0.08	41.5	2 687.8	2 783.2	2 879.7	2 977.5	3 076.7	3 177.3	3 279.6	3 489.1	3 705.5	3 928.8	4 158.7
0.10	45.8	2 687.5	2 783.1	2 879.6	2 977.4	3 076.6	3 177.3	3 279.6	3 489.1	3 705.5	3 928.8	4 158.7
0.15	54.0	2 686.9	2 782.4	2 879.5	2 977.3	3 076.5	3 177.7	3 279.5	3 489.1	3 705.5	3 928.7	4 158.7
0.20	60.1	2 686.3	2 782.3	2 879.2	2 977.1	3 076.1	3 177.1	3 279.4	3489.0	3 705.4	3 928.7	4 158.7
0.25	65.0	2 685.7	2 782.0	2 879.0	2 977.0	3 076.3	3 177.0	3 279.3	3 489.0	3 705.4	3 928.7	4 158.0
0.30	69.1	2 685.1	2 781.6	2 878.7	2 976.8	3 076.1	3 176.9	3 279.3	3 488.9	3 705.4	3 928.7	4 158.6
0.35	72.7	2 684.5	2 781.2	2 878.5	2 976.7	3 076.0	3 176.8	3 279.2	3 488.9	3 705.3	3 928.7	4 158.6
0.40	75.9	2 683.8	2 780.9	2 878.2	2 976.5	3 075.9	3 176.8	3 279.1	3 488.8	3 705.3	3 928.6	4 158.6
0.45	78.7	2 683.2	2 780.5	2 878.0	2 976.3	3 075.8	3 176.7	3 279.1	3 488.8	3 705.2	3 928.6	4 158.5
0.50	81.3	2 682.6	2 780.1	2 877.7	2 976.1	3 075.7	3 176.6	3 279.0	3 488.7	3 705.2	3 928.6	4 158.5
0.60	86.0	2 681.3	2 779.4	2 877.3	2 975.8	3 075.4	3 176.4	3 278.8	3 488.6	3 705.1	3 928.5	4 158.5
0.70	90.0	2 680.0	2 778.6	2 876.8	2 975.5	3 075.2	3 176.2	3 278.7	3 488.5	3 705.0	3 928.4	4 158.4
0.80	93.5	2 678.8	2 777.8	2 876.3	2 975.2	3 075.0	3 176.0	3 278.5	3 488.4	3 705.0	3 928.4	4 158.4
0.90	96.7	2 677.5	2 777.1	2 875.8	2 974.8	3 074.7	3 175.8	3 278.4	3 488.3	3 704.9	3 928.3	4 158.3
1.00	99.6	2 676.2	2 776.3	2 875.4	2 974.5	3 074.5	3 175.6	3 278.2	3 488.1	3 704.8	3 928.2	4 158.3
1.50	111.4	..	2 772.5	2 872.9	2 972.9	3 073.3	3 174.7	3 277.5	3 487.6	3 704.4	3 927.9	4 158.0
2.00	120.2	..	2 768.5	2 870.5	2 971.2	3 072.1	3 173.8	3 276.7	3 487.0	3 704.0	3 927.6	4 157.8
2.5	127.4	..	2 764.5	2 868.0	2 969.6	3 070.9	3 172.8	3 275.9	3 486.5	3 703.6	3 927.3	4 157.6
3.0	133.5	..	2 760.4	2 865.5	2 967.9	3 069.7	3 171.9	3 275.2	3 486.0	3 703.2	3 927.0	4 157.3
3.5	138.9	..	2 756.3	2 863.0	2 966.2	3 068.4	3 170.9	3 274.4	3 485.4	3 702.7	3 926.7	4 157.1

(p)	(t_s)	100	150	200	250	300	350	400	500	600	700	800
4.0	143.6	..	2 752.0	2 860.4	2 964.5	3 067.2	3 170.0	3 273.6	3 484.9	3 702.3	3 926.4	4 156.9
4.5	147.9	..	2 746.7	2 857.8	2 962.8	3 066.0	3 169.1	3 272.9	3 484.3	3 701.9	3 926.1	4 156.7
5.0	151.8	..	..	2 855.1	2 961.1	3 064.8	3 168.1	3 272.1	3 483.8	3 701.5	3 925.8	4 156.4
6.0	158.8	..	..	2 849.7	2 957.6	3 062.3	3 166.2	3 270.6	3 482.7	3 700.7	3 925.1	4 155.9
7.0	165.0	..	..	2 844.2	2 954.0	3 059.8	3 164.3	3 269.0	3 481.6	3 699.9	3 924.5	4 155.5
8.0	170.4	..	..	2 838.6	2 950.4	3 057.3	3 162.4	3 267.5	3 480.5	3 699.1	3 923.9	4 155.0
9.O	175.4	..	..	2 832.7	2 946.8	3 054.7	3 160.5	3 266.0	3 479.4	3 698.2	3 923.3	4 154.5
10.0	179.9	..	..	2 826.8	2 943.0	3 052.1	3 158.5	3 264.4	3 478.3	3 697.4	3 922.7	4 154.1
11.0	184.1	..	..	2 820.7	2 939.3	3 049.6	3 156.6	3 262.9	3 477.2	3 696.6	3 922.0	4 153.6
12.0	188.0	..	..	2 814.4	2 935.4	3 046.9	3 154.6	3 261.3	3 476.1	3 695.8	3 921.4	4 153.1
13.0	191.6	..	..	2 808.0	2 931.5	3 044.3	3 152.7	3 259.7	3 475.0	3 695.0	3 920.8	4 152.7
14.0	195.0	..	..	2 801.4	2 927.6	3 041.6	3 150.7	3 258.2	3 473.9	3 694.1	3 920.2	4 152.2
15.0	198.3	..	..	2 794.7	2 923.5	3 038.9	3 148.7	3 256.6	3 472.8	3 693.3	3 919.6	4 151.7
16.0	201.4	..	..	..	2 919.4	3 036.2	3 146.7	3 255.0	3 471.7	3 692.5	3 918.9	4 151.3
17.0	204.3	..	..	..	2 915.3	3 033.5	3 144.7	3 253.5	3 470.6	3 691.7	3 918.3	4 150.8
18.0	207.1	..	..	..	2 911.0	3 030.7	3 142.7	3 251.9	3 469.5	3 690.9	3 917.7	4 150.3
19.0	209.8	..	..	..	2 906.7	3 027.9	3 140.7	3 250.3	3 468.4	3 690.0	3 917.1	4 149.8
20.0	212.4	..	..	..	2 902.4	3 025.0	3 138.6	3 248.7	3 467.3	3 689.2	3 916.5	4 149.4
22.0	217.2	..	..	..	2 893.4	3 019.3	3 134.5	3 245.5	3 465.1	3 687.6	3 915.2	4 148.4
24.0	221.8	..	..	..	2 884.2	3 013.4	3 130.4	3 242.3	3 462.9	3 685.9	3 914.0	4 147.5
26.0	226.0	..	..	..	2 874.7	3 007.4	3 126.1	3 239.0	3 460.6	3 684.3	3 912.7	4 146.6
28.0	230.0	..	..	..	2 864.9	3 001.3	3 121.9	3 235.8	3 458.4	3 682.6	3 911.5	4 145.6
30.0	233.8	..	..	..	2 854.8	2 995.1	3 117.5	3 232.5	3 456.2	3 681.0	3 910.3	4 144.7
32.0	237.4	..	..	..	2 844.4	2 988.7	3 113.2	3 229.2	3 454.0	3 679.3	3 909.0	4 143.8
34.0	240.9	..	..	..	2 833.6	2 982.2	3 108.7	3 225.9	3 451.7	3 677.7	3 907.8	4 142.8
36.0	244.2	..	..	..	2 822.5	2 975.6	3 104.2	3 222.5	3 449.5	3 676.1	3 906.5	4 141.9
38.0	247.3	..	..	..	2 811.0	2 968.9	3 099.7	3 219.1	3 447.2	3 674.4	3 905.3	4 141.0
40.0	250.3	..	..	..	..	2 962.0	3 095.1	3 215.7	3 445.0	3 672.8	3 904.1	4 140.0

(p)	(t_s)	100	150	200	250	300	350	400	500	600	700	800
42.0	253.2	..		..	..	2 955.0	3 090.4	3 212.3	3 442.7	3 671.1	3 902.8	4 139.1
44.0	256.0	..	..	..	..	2 947.8	3 085.7	3 208.8	3 440.5	3 669.5	3 901.6	4 138.2
46.0	258.8	..	..	..	..	2 940.8	3 080.9	3 205.3	3 438.2	3 667.8	3 900.3	4 137.2
48.0	261.4	..	..	..	..	2 933.1	3 076.1	3 201.8	3 435.9	3 666.2	3 899.1	4 136.3
50.0	263.9	..	..	..	..	2 925.5	3 071.2	3 198.3	3 433.7	3 664.5	3 897.9	4 135.3
55.0	269.9	..	..	..	..	2 905.6	3 058.7	3 189.3	3 427.9	3 660.4	3 894.8	4 133.0
60.0	275.6	..	..	..	..	2 885.0	3 045.8	3 180.1	3 422.2	3 656.2	3 891.7	4 130.7
65.0	280.8	..	..	..	..	2 863.0	3 032.4	3. 170.8	3 416.4	3 652.1	3 888.6	4 128.8
70.0	285.8	..	..	..	..	2 839.0	3 018.7	3 161.2	3 410.6	3 647.9	3 885.4	4 126.0
75.0	290.5	..		..	..	2 814.1	3 004.5	3 151.6	3 404.7	3 643.7	3 882.4	4 123.7
80.0	295.0	..	..	..	..	2 780.0	2 089.9	3 141.6	3 398.8	3 639.5	3 879.2	4 121.3
85.0	299.2	..	..	..	..	2 757.1	2 974.7	3 131.5	3 392.8	3 635.4	3 876.1	4 119.0
90.0	303.3	..	..	..	..	..	2 959.0	3 121.2	3 386.8	3 631.1	3 873.0	4 116.7
95.0	307.2	..	..	..	..	..	2 942.7	3 110.7	3 380.7	3 627.0	3 869.9	4 114.4
100.0	311.0	..	..	..	..	..	2 925.8	3 099.9	3 374.6	3 622.7	3 866.8	4 112.0
110.0	318.0	..	..	..	..	..	2 889.6	3 077.8	3 362.2	3 614.2	3 860.5	4 107.0
120.0	324.6	..	..	..	..	..	2 849.7	3 054.8	3 349.6	3 605.7	3 854.3	4 102.7
130.0	330.8	..	..	..	..	..	2 805.0	3 030.7	3 336.8	3 597.1	3 848.0	4 098.0
140.0	336.6	..	..	..	..	..	2 754.2	3 005.6	3 323.8	3 588.5	3 841.7	4 093.0
150.0	342.1	..	..	..	..	..	2 694.8	2 979.1	3 310.6	3 579.8	3 835.4	4 088.6
160.0	347.3	..	..	..	..	..	2 620.8	2 951.3	3 297.1	3 571.0	3 829.1	4 084.0
170.0	352.3	..	..	..	..	..	..	2 921.7	3 283.5	3 562.2	3 822.8	4 079.3
180.0	357.0	..	..	..	..	..	..	2 890.3	2 269.6	3 553.4	3 816.5	4 074.0
190.0	361.4	..	..	..	..	..	..	2 856.7	3 255.4	3 544.5	3 810.2	4 070.0
200.0	365.7	..	..	..	..	..	..	2 820.5	3 241.1	3 535.5	3 803.8	4 065.3
210.0	369.8	..	..	..	..	..	..	2 781.3	3 226.5	3 526.5	3 797.5	4 060.6
220.0	373.7	..	..	..	..	..	..	2 738.8	3 211.7	3 517.4	3 791.1	4 055.9
221.2	374.15	..	..	..	..	..	..	2 734.5	32 10.7	3 516.4	3 789.1	4 054.7

Table II.5 Specific entropy of superheated steam

Absolute Pressure in bar	Saturation Temperature in °C	*Specific entropy (s) in kJ/kg-K at various temperature in °C*										
(p)	*(t_s)*	*100*	*150*	*200*	*250*	*300*	*350*	*400*	*500*	*600*	*700*	*800*
0.02	17.5	9.193	9.433	9.648	9.844	10.025	10.193	10.351	10.641	10.904	11.146	11.371
0.04	29.0	8.873	9.113	9.328	9.524	9.705	9.874	10.031	10.321	10.585	10.827	11.051
0.06	36.2	8.685	8.925	9.141	9.337	9.518	9.686	9.844	10.134	10.397	10.639	10.864
0.08	41.5	8.552	8.792	9.008	9.204	9.385	9.554	9.711	10.001	10.265	10.507	10.731
0.10	45.8	8.449	8.689	8.905	9.101	9.282	9.450	9.608	9.898	10.162	10.404	10.628
0.15	54.0	8.261	8.502	8.718	8.915	9.096	9.264	9.422	9.712	9.975	10.217	10.442
0.20	60.1	8.126	8.368	8.584	8.781	8.962	9.130	9.288	9.578	9.842	10.084	10.309
0.25	65.0	8.022	8.264	8.481	8.678	8.859	9.028	9.186	9.476	9.739	9.981	10.206
0.30	69.1	7.936	8.179	8.396	8.593	8.774	8.943	9.101	9.391	9.654	9.897	10.121
0.35	72.7	7.864	8.107	8.325	8.522	8.703	8.872	9.030	9.320	9.583	9.826	10.050
0.40	75.9	7.801	8.045	8.263	8.460	8.641	8.810	8.968	9.258	9.522	9.764	9.989
0.45	78.7	7.745	7.990	8.208	8.405	8.587	8.755	8.914	9.204	9.467	9.709	9.934
0.50	81.3	7.695	7.941	8.159	8.356	8.538	8.707	8.865	9.155	9.419	9.661	9.886
0.60	86.0	7.609	7.855	8.074	8.272	8.454	8.622	8.781	9.071	9.334	9.576	9.801
0.70	90.0	7.535	7.783	8.002	8.200	8.382	8.551	8.709	9.000	9.263	9.505	9.730
0.80	93.5	7.470	7.720	7.940	8.138	8.320	8.489	8.648	8.938	9.214	9.444	9.669
0.90	96.7	7.413	7.664	7.884	8.083	8.266	8.435	8.593	8.884	9.147	9.389	9.614
1.00	99.6	7.362	7.614	7.835	8.034	8.217	80386	8.544	8.835	9.098	9.341	9.565
1.50	111.4	..	7.419	7.644	7.845	8.028	8.198	8.356	8.647	8.911	9.153	9.378
2.00	120.2	..	7.279	7.507	7.710	7.894	8.064	8.223	8.514	8.778	9.020	9.245
2.5	127.4	..	7.169	7.400	7.604	7.789	7.960	8.119	8.410	8.674	8.917	9.142
3.0	133.5	..	7.077	7.312	7.518	7.703	7.874	8.034	8.326	8.590	8.833	9.058
3.5	138.5	..	6.998	7.237	7.444	7.631	7.802	7.962	8.254	8.518	8.761	8.986

(p)	(t_s)	100	150	200	250	300	350	400	500	600	700	800
4.0	143.6	..	6.929	7.171	7.380	7.568	7.740	7.899	8.192	8.456	8.699	8.925
4.5	147.9	..	6.866	7.112	7.323	7.512	7.684	1.844	8.137	8.402	8.645	8.870
5.0	151.8	..	..	7.059	7.272	7.461	7.634	7.795	8.088	8.353	8.596	8.821
6.0	158.8	..	..	6.966	7.183	7.374	7.548	7.709	8.003	8.268	8.511	8.737
7.0	165.0	..	..	6.886	7.107	7.300	7.475	7.636	7.931	8.196	8.440	8.665
8.0	170.4	..	..	6.815	7.040	17.235	7.411	7.573	7.868	8.134	8.377	8.603
9.0	175.4	..	..	6.751	6.980	7.177	7.354	7.517	7.812	8.079	8.323	8.549
10.0	179.9	..	..	6.692	6.926	7.125	7.303	7.467	7.763	8.029	8.273	8.500
11.0	184.1	..	..	6.638	6.876	7.078	7.257	7.42]	7.718	7.985	8.229	8.455
12.0	188.0	..	..	6.587	6.831	7.034	7.214	7.379	7.677	7.944	8.188	8.415
13.0	191.6	..	..	6.539	6.788	6.994	7.175	7.340	7.639	7.906	8.151	8.378
14.0	195.0	..	..	6.494	6.748	6.956	7.139	7.305	7.603	7.871	8.116	8.343
15.0	198.3	..	..	6.451	6.710	6.921	7.104	7.271	7.570	7.839	8.084	8.311
16.0	201.4	..	..	..	6.674	6.887	7.072	7.239	7.540	7.808	8.054	8.281
17.0	204.3	..	..	..	6.640	6.856	7.042	7.210	7.511	7.779	8.025	8.252
18.0	207.1	..	..	..	6.607	6.826	7.013	7.182	7.483	7.752	7.998	8.226
19.0	209.8	..	..	..	6.576	6.797	7.986	7.155	7.457	7.727	7.973	8.200
20.0	212.4	..	..	..	6.545	6.770	6.960	7.130	7.432	7.702	7.949	8.176
22.0	217.2	..	..	..	6.488	6.718	6.911	7.082	7.386	7.657	7.904	8.132
24.0	221.8	..	..	..	6.434	6.670	6.866	7.038	7.344	7.615	7.862	8.091
26.0	226.0	..	..	..	6.382	6.625	6.824	6.998	7.305	7.577	7.825	8.053
28.0	230.0	..	..	..	6.333	6.582	6.784	6.960	7.269	7.541	7.789	8.018
30.0	233.8	..	..	..	6.286	6.542	6.747	6.925	7.235	7.508	7.756	7.986
32.0	237.4	..	..	..	6.240	6.504	6.712	6.891	7.203	7.477	7.726	7.955
34.0	240.9	..	..	..	6.195	6.467	6.679	6.860	7.172	7.447	7.697	7.927
36.0	244.2	..	..	..	6.151	6.432	6.647	6.829	7.144	7.420	7.669	7.900
38.0	247.3	..	..	..	6.109	6.397	6.616	6.801	7.117	7.393	7.643	7.874
40.0	250.3	..	..	..	..	6.364	6.587	6.773	7.091	7.368	7.619	7.850

(p)	(t_s)	*100*	*150*	*200*	*250*	*300*	*350*	*400*	*500*	*600*	*700*	*800*
42.0	253.2	..	..	..	..	6.332	6.559	6.747	7.066	7.344	7.595	7.826
44.0	256.0	..	..	..	..	6.301	6.532	6.722	7.043	7.321	7.573	7.804
46.0	258.8	..	..	..	..	6.270	6.505	6.697	7.020	7.299	7.551	7.783
48.0	261.4	..	..	..	..	6.240	6.479	6.674	6.998	7.278	7.531	7.763
50.0	263.9	..	..	..	..	6.211	6.455	6.651	6.977	7.258	7.511	7.743
55.0	269.9	..	..	..	..	6.139	6.395	6.597	6.928	7.210	7.464	7.697
60.0	275.6	..	..	..	..	6.069	6.339	6.546	6.882	7.166	7.422	7.655
65.0	280.8	..	..	..	..	6.001	6.285	6.499	6.839	7.126	7.382	7.617
70.0	285.8	..	..	..	..	5.933	6.233	6.454	6.799	7.088	7.346	7.581
75.0	290.5	..	..	..	..	5.864	6.184	6.411	6.762	7.053	7.311	7.547
80.0	295.0	..	..	..	..	5.794	6.135	6.369	6.726	7.019	7.279	7.516
85.0	299.2	..	..	..	..	5.744	6.088	6.330	6.692	6.987	7.249	7.486
90.0	303.3	..	..	..	..	..	6.041	6.292	6.660	6.957	7.220	7.458
95.0	307.2	..	..	..	..	..	5.995	6.254	6.629	6.929	7.192	7.431
100.0	311.0	..	..	..	..	..	5.949	6.218	6.599	6.901	7.166	7.406
110.0	318.0	..	..	..	..	..	5.857	6.148	6.543	6.850	7.117	7.358
120.0	324.6	..	..	..	..	..	5.764	6.081	6.491	6.802	7.072	1.315
130.0	330.8	..	..	..	..	..	5.666	6.016	6.441	6.758	7.030	7.274
140.0	336.6	..	..	..	..	..	5.562	5.951	6.394	6.716	6.991	7.237
150.0	342.1	..	..	..	..	..	5.447	5.888	6.349	6.676	6.954	7.201
160.0	347.3	..	..	..	..	..	5.311	5.824	6.305	6.639	6.919	7.168
170.0	352.3	..	..	..	..	..	..	5.760	6.264	6.603	6.886	7.137
180.0	357.0	..	..	..	..	..	..	5.695	6.223	6.569	6.854	7.107
190.0	361.4	..	..	..	..	..	..	5.628	6.184	6.536	6.824	7.078
200.0	365.7	..	..	..	..	..	..	5.559	6.146	6.504	6.795	7.051
210.0	369.8	..	..	..	..	..	..	5.486	6.108	6.474	6.768	7.025
200.0	373.7	..	..	..	..	..	..	5.410	6.072	6.444	6.741	7.000
221.2	374.15	..	..	..	..	..	..	5.399	6.068	6.441	6.738	6.994

Annexure III: Objective Type Questions

SECTION A LAWS AND RELATIONS

Tick (√) the right answer.

1. Which of the following statements is correct?
   - (a) The greater the pressure difference in throttling, the lesser the irreversibility.
   - (b) The primary source of internal irreversibility in power cycle is fluid friction in rotary machines.
   - (c) The greater the irreversibility, the greater the increase in adiabatic process.
   - (d) The entropy of the universe is continually on the increase.
2. Consider the following statements:
   - (a) temperature is due to Zeroth Law of thermodynamics.
   - (b) entropy is due to First Law of thermodynamics.
   - (c) internal energy is due to Second Law of thermodynamics.
   - (d) reversibility is due to Kelvin-Planck's statement.

   Following groups of statements are correct.

   (a) 1,2 and 3 (b) 1, 3 and 4
   (c) 1 alone (d) 2 alone
3. Joule-Thompson Coefficient is defined as:

   (a) $\left[\frac{\partial T}{\partial p}\right]_R$ (b) $\left[\frac{\partial h}{\partial p}\right]_T$

   (c) $\left[\frac{\partial h}{\partial T}\right]_p$ (d) $\left[\frac{\partial p}{\partial T}\right]_h$
4. Heat transfer takes place according to
   - (a) Zeroth Law of thermodynamics
   - (b) First Law of thermodynamics
   - (c) Second Law of thermodynamics
   - (d) Third Law of thermodynamics
5. Zeroth Law of thermodynamics states that
   - (a) two thermodynamic systems are always in thermal equilibrium with each other.
   - (b) if two systems are in thermal equilibrium, then the third system will also be in thermal equilibrium.

(c) two systems not in thermal equilibrium with a third system will also not be in thermal equilibrium with each other.

(d) when two systems are in thermal equilibrium with a third system, they are in thermal equilibrium with each other.

6. Which one of the following statements applicable to a perfect gas will also be true for an irreversible process?

(a) $dQ = du + pdv$ (b) $dQ + Tds$

(c) $Tds = du + pdv$ (d) None of the above

7. Consider the following statements:

1. Availability is generally conserved.
2. Availability can either be negative or positive.
3. Availability is the maximum theoretical work obtained.
4. Availability can be destroyed in irreversibilities.

The following group of statements is correct.

(a) 3 and 4 (b) 1 and 2

(c) 1 and 3 (d) 2 and 4

8. For a given volume of dry saturated steam, Clapeyron's equation is given by:

(a) $(V_g - V_f) = \frac{dTs}{dp} \times \frac{Ts}{h_g - h_f}$ (b) $(V_g - V_f) = \frac{dTs}{dp} \times \frac{h_g - h_f}{T_s}$

(c) $(V_g - V_f) = \frac{dp}{dT_s} \times \frac{h_g - h_f}{T_s}$ (d) $(V_g - V_f) = \frac{dp}{dT_s} \times \frac{T_s}{h_g - h_f}$

9. The Joule-Thomson Coefficient is the

(a) $\left(\frac{\partial T}{\partial p}\right)_h$ of pressure-temperature curve of real gases.

(b) $\left(\frac{\partial T}{\partial S}\right)_v$ of temperature-entropy curve of real gases.

(c) $\left(\frac{\partial h}{\partial S}\right)_T$ of enthalpy-entropy curve of real gases.

(d) $\left(\frac{\partial V}{\partial T}\right)_p$ of pressure-volume curve of real gases.

10. The thermodynamic parameters are:

I. Temperature

II. Specific volume

III. Pressure

IV. Enthalpy

V. Entropy

The Clapeyron Equation of state provides relationship between:

(a) I and II (b) II, III and V

(c) III, IV and V (d) I, II, III and IV

11. For a thermodynamic cycle to be irreversible, it is necessary that

(a) $\oint \frac{\delta Q}{T} = 0$ (b) $\oint \frac{\delta Q}{T} < 0$

(c) $\oint \frac{\delta Q}{T} > 0$ (d) $\oint \frac{\delta Q}{T} \geq 0$

12. Neglecting changes in kinetic energy and potential energy, for unit mass the availability of a non-flow process becomes $a = \phi - \phi_0$ where ϕ is the availability function of the

(a) open system (b) closed system

(c) isolated system (d) steady flow process

13. Consider the following statements:

In an irreversible process,

1. Entropy always increases
2. The sum of the entropy of all the bodies taking part in a process always increases
3. Once created, entropy cannot be destroyed

The following group of statements is correct.

(a) 1 and 2 (b) 1 and 3

(c) 2 and 3 (d) 1, 2 and 3

14. Clausius-Clapeyron equation gives the slope of a curve in

(a) p-v diagram (b) p-h diagram

(c) p-T diagram (d) T-s diagram

15. Gibb's free energy 'G' is defined as:

(a) $G = H\text{-}TS$ (b) $G = U\text{-}TS$

(c) $G = U + pV$ (d) $G = H + TS$

16. Joule-Thomson coefficient is the ratio of

(a) Pressure change to pressure change occurring when a gas undergoes the process of adiabatic throttling

(b) Temperature change to pressure change occurring when a gas undergoes the process of adiabatic throttling

(c) Temperature change to pressure change occurring when a gas undergoes the process of adiabatic compression

(d) Pressure change to temperature change occurring when a gas undergoes the process of adiabatic compressor

17. In which of the following situations the entropy change will be negative?
    (a) Air expands isothermally from 6 bars to 3 bars.
    (b) Air is compressed to half the volume at constant pressure.
    (c) Heat is supplied to air at constant volume till the pressure becomes three folds.
    (d) Air expands isentropically from 6 bar to 3 bar.

18. Consider the following thermodynamic relations:
    1. Tds = du + pdv   2. Tds = du – pdv
    3. Tds = dh + pdv   4. Tds = dh – pdv

    The following statements are correct.
    (a) 1 and 3   (b) 1 and 4
    (c) 2 and 3   (d) 2 and 4

19. The Chapeyron equation with usual notations is given by

    (a) $\left(\frac{dT}{dp}\right)_{\text{sat}} = \frac{h_{fg}}{Tv_{fg}}$ (b) $\left(\frac{dT}{dT}\right)_{\text{sat}} = \frac{h_{fg}}{Tv_{fg}}$

    (c) $\left(\frac{dT}{dp}\right)_{\text{sat}} = \frac{Th_{fg}}{v_{fg}}$ (d) $\left(\frac{dp}{dT}\right)_{\text{sat}} = \frac{Th_{fg}}{v_{fg}}$

20. Availability function for a closed system is expressed as:
    (a) $\phi = u + p_o v - T_o s$ (b) $\phi = du + p_o dv + T_o ds$
    (c) $\phi = du + p_o dv - T_o ds$ (d) $\phi = u + p_o v + T_o s$

21. Tds equation can be expressed as:

    (a) $Tds = CvdT + \frac{T\beta dv}{k}$ (b) $Tds = CvdT + \frac{T}{k}dv$

    (c) $Tds = CvdT + \frac{Tk}{\beta}dv$ (d) $\text{Tds} = CvdT + \frac{T\beta}{k}dp$

22. Which of the following functions represents the Clapeyron equation pertaining to change of phase of a pure substance?
    (a) $f(T, p, h_{fg})$ (b) $f(T, p, h_{fg}, v_{fg})$
    (c) $f(T, p, h_{fg}, s_{fg})$ (d) $f(T, p, h_{fg}, s_{fg}, v_{fg})$

23. Consider the following statements:
    1. Zeroth law of thermodynamics is related to temperature.
    2. Entropy is related to first law of thermodynamics.
    3. Internal energy of an ideal gas is a function of temperature and pressure.

4. Van der Waals equation is related to an ideal gas.

Which of the above statements is/are correct?

(a) 1 only (b) 2, 3 and 4

(c) 1 and 3 (d) 2 and 4

Keys

1 (d); 2 (c); 3 (a); 4 (c); 5 (d); 6 (c); 7 (a); 8 (b); 9 (a);
10 (d); 11 (b); 12 (a); 13 (c); 14 (c); 15 (a); 16 (b); 17 (b); 18 (b);
19 (b); 20 (a); 21 (a); 22 (b); 23 (d)

SECTION B PROPERTIES OF STEAM

Tick (√) the right answer.

1. Mark the correct statement.
   (a) Pressure and temperature are independent during phase change.
   (b) An isothermal line is also a constant pressure like in the wet region.
   (c) Entropy decreases during evaporation.
   (d) The term dryness fraction is used to specify the fraction by mass of liquid in a mixture of liquid and vapour.

2. The conversion of water from 40°C to steam at 200°C at a pressure of 1 bar is best represented by.

(a) 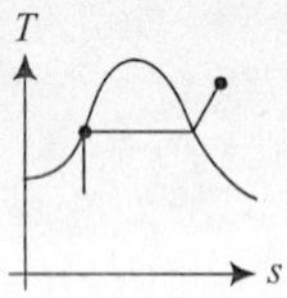

(b) 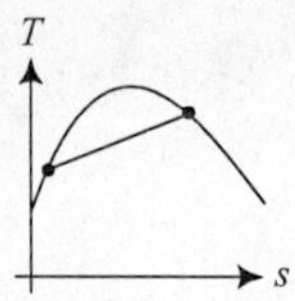

(c) 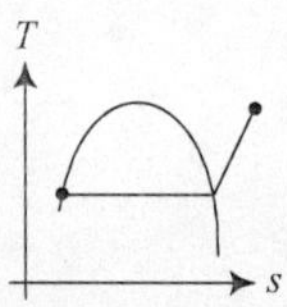

(d) 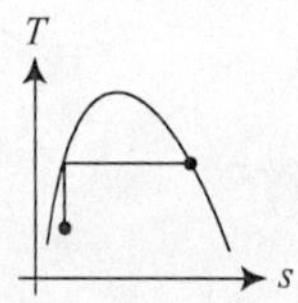

3. Select the correct code from List I and List II for the following *T-s* diagram for steam

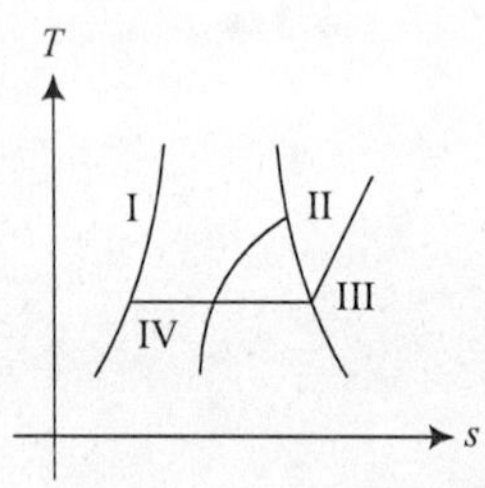

List I	List II
A. Curve I	1. Saturated liquid line
B. Curve II	2. Saturated vapour line
C. Curve III	3. Constant pressure line
D. Curve IV	4. Constant volume line

Codes	A	B	C	D
(a)	2	1	4	3
(b)	2	1	3	4
(c)	1	2	3	4
(d)	1	2	4	3

4. Which one of the following p-v diagrams for steam illustrate the isothermal process undergone by superheated steam till it becomes wet?

(a) 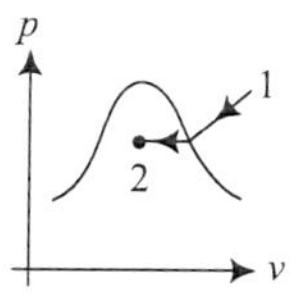

(b) 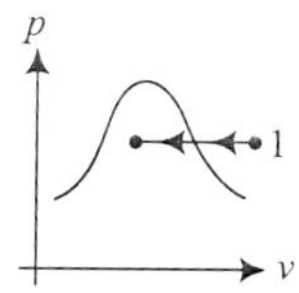

(c) 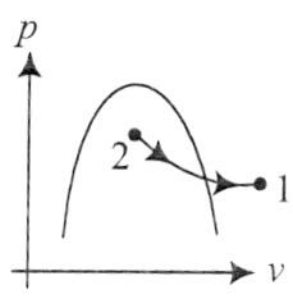

(d) 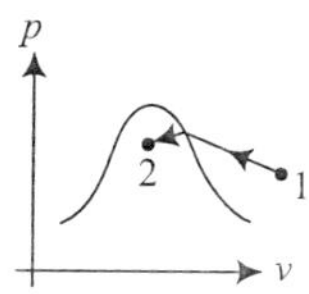

5. Match List I with List II and select the right code.

	List I		List II
A.	As saturation pressure increases	1.	Entropy increases
B.	As saturation temperature increases	2.	Specific volume increases
C.	As saturation pressure decreases	3.	Enthalpy of evaporation decreases
D.	As dryness fraction increases	4.	Saturation temperature increases

Code:	A	B	C	D
(a)	1	3	2	4
(b)	4	3	2	1
(c)	4	3	1	2
(d)	2	4	3	1

6. The given diagram shows the throttling process of a pure substance. The ordinance and abscissa are respectively

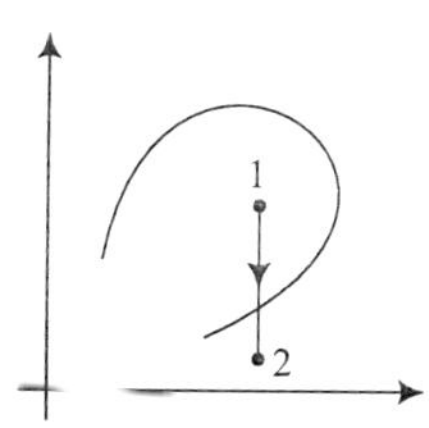

(a) pressure and volume (b) enthalpy and entropy
(c) temperature and entropy (d) pressure and enthalpy

7. The ordinate and abscissa in the given figure showing the saturated liquid and vapour region of a pure substances represents

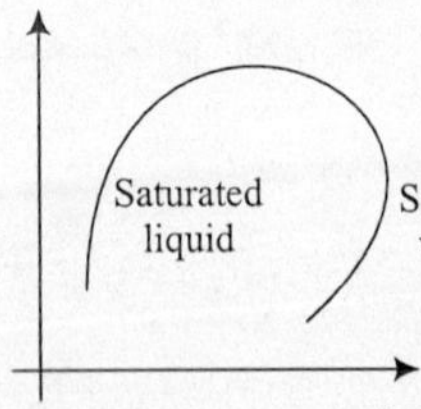

(a) temperature and pressure (b) enthalpy and entropy
(c) pressure and volume (d) pressure and enthalpy

8. With increase of pressure, the latent heat of steam
   (a) remains same (b) increases
   (c) decreases (d) behaves unpredictably
9. Select the right code for the following statements. In the throttling process of wet steam.
   1. The steam pressure and temperature decrease but enthalpy remains constant.
   2. The steam pressure decreases, the temperature increases but enthalpy remains constant.
   3. The entropy, specific volume and dryness fraction increase.
   4. The entropy increases but volume and dryness fraction decrease.

Codes: (a) 1 and 4 (b) 2 and 3 (c) 1 and 3 (d) 2 and 4

Keys

1 (b); 2 (a); 3 (c); 4 (d); 5 (c); 6 (d); 7 (d); 8 (c); 9 (a)

SECTION C POWER CYCLES

1. In the bottoming cycle of cogeneration, low grade waste heat is used for
   (a) processing (b) power generation
   (c) feedwater heating (d) none of above
2. During which of the following processes does heat rejection takes place in Carnot vapour cycle?
   (a) constant volume (b) constant pressure
   (c) constant temperature (d) constant entropy
3. In a steam power plant, feedwater heater is a heat exchanger to preheat feedwater by
   (a) live steam from steam generator
   (b) hot flue gases coming out of the boiler furnace
   (c) hot air from air preheater
   (d) extracting steam from turbine
4. The efficiency of Rankine cycle is lower than the corresponding Carnot cycle because
   (a) the average temperature at which heat is supplied in Rankine cycle is lower than corresponding Carnot cycle
   (b) the Carnot cycle has gas as working substance and Rankine cycle has steam as working substance
   (c) the Rankine cycle efficiency depends upon properties of the working substance whereas Carnot cycle efficiency is independent of the properties of working substance
   (d) the temperature range of Carnot cycle is greater than that for Rankine cycle
5. A steam power plant is shown below.

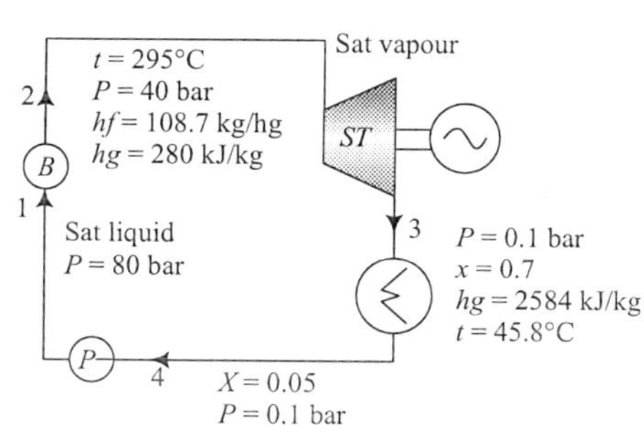

   (a) the cycle violates first and second laws of thermodynamics
   (b) the cycle does not satisfy condition of Clausis inequality

(c) the cycle only violates second law of thermodynamics
(d) the cycle satisfies the Clausius inequality

6. In thermal power plants the deaerator is used mainly to
   (a) remove air from condenser (b) reduce steam pressure
   (c) increase feedwater temperature
   (d) remove dissolved gases from feedwater
7. Select the correct statements. Expansion joints in steam pipelines are installed to:
   1. allow for future expansion of plant
   2. intake stresses away from flanges and fittings
   3. permit expansion of pipes due to temperature rise

Codes: (a) 1, 2 and 3 (b) 1 and 2 (c) 2 and 3 (d) 1 and 3

8. In a Rankine cycle, with the maximum steam temperature being fixed from metallurgical consideration, as the boiler pressure increases
   (a) the condenser load will increase
   (b) the quality of turbine exhaust will decrease
   (c) the quality of turbine exhaust will increase
   (d) the quality of turbine exhaust will remain unchanged
9. Select the correct code. The following power plants use heat recovery boiler (unfired) for steam generator.
   1. Combined cycle power plants
   2. All thermal power plants using coal
   3. Nuclear power plant
   4. Power plants using fluidized bed combustion

   (a) 1 and 2 (b) 3 and 4
   (c) 1 and 3 (d) 2 and 4
10. Select the right code for the following statements.
    The effect of heating of steam in a steam turbine.
    1. It increases the specific output of turbine.
    2. It decreases the cycle efficiency.
    3. It increases blade erosion.
    4. It improves the quality of exit steam.

Codes: (a) 1 and 2 (b) 2 and 3 (c) 3 and 4 (d) 1 and 4

11. The reheat cycle in steam power plant is mainly adopted to
    (a) improve thermal efficiency
    (b) decrease the moisture content in low pressure stages to a safe value
    (c) decrease the capacity of condenser
    (d) recover the waste heat of boiler

12. Select the correct code.

    The efficiency of the vapour power Rankine cycle can be increased by

    1. increasing the temperature of the working fluid at which heat is added.
    2. increasing the pressure of the working fluid at which heat is added.
    3. decreasing the temperature of the working fluid at which heat is rejected.

Codes: (a) 2 and 3 (b) 1 alone (c) 1 and 2 (d) 1, 2 and 3

13. Select the correct codes. The correct sequence of processes in a Rankine cycle is

    1. Constant pressure heat addition
    2. Adiabatic compression
    3. Adiabatic expansion
    4. Constant pressure heat rejection

Codes: (a) 1, 2, 3, 4 (b) 2, 1, 4, 3 (c) 2, 1, 3, 4 (d) 1, 2, 4, 3

14. In a steam power plant, the ratio of the isentropic heat drop in the prime mover to the amount of heat supplied per unit mass of steam is known as

    (a) stage efficiency (b) degree of reaction
    (c) Rankine efficiency (d) relative efficiency

15. The temperature-entropy diagram of a steam turbine power plant, operating on the Rankine cycle with reheat and regenerative feed heating is shown. If m denotes the fraction of steam bled for feed heating, the work developed in the turbine per kg steam entering the turbine at state 5 is

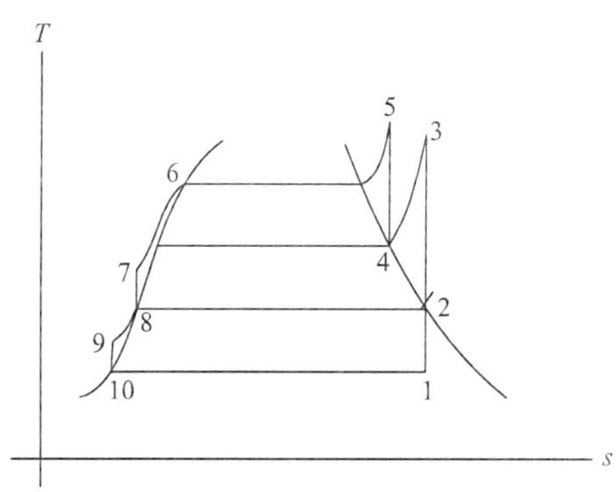

    (a) $(h_5 - h_4) + (1 - m)(h_3 - h_1)$
    (b) $(h_5 - h_4) + (h_3 - h_2) + (1 - m)(h_2 - h_1)$
    (c) $2h_5 - h_4 - h_2 + (1 - m)(h_2 - h_1)$
    (d) $(h_5 - h_4) + (1 - m)(h_3 - h_2)$

16. In a regenerative feed heating cycle, the economic number of the stages of regeneration
    (a) increases as the initial pressure and temperature increase
    (b) decreases as the initial pressure and temperature increase
    (c) is independent of the initial pressure and temperature
    (d) depends only on the condenser pressure

Keys

1 (b); 2 (c); 3 (d); 4 (a); 5 (d); 6 (d); 7 (c); 8 (b); 9 (d);
10 (d); 11 (a); 12 (d); 13 (c); 14 (a); 15 (b)

SECTION D BOILERS

Tick (√) the right answer

1. The calorific value determined by the bomb calorimeter is
   (a) lower calorific value at constant pressure
   (b) higher calorific value at constant pressure
   (c) lower calorific value at constant volume
   (d) higher calorific value at constant volume
2. Economizer is generally placed between
   (a) last superheater/preheater and air preheater
   (b) air preheater and chimney
   (c) electrostatic precipitators
   (d) induced draft fan and forced draft fan
3. Attempering is
   (a) reduction of steam pressure
   (b) reduction of steam temperature
   (c) discharge of flue gases at certain height
   (d) conditioning of disposable fluids for minimum pollution
4. Which one of the following factors has the maximum effect on the equivalent evaporation of a boiler?
   (a) steam generator pressure (b) feedwater inlet temperature
   (c) heating surface of the boiler (d) quality of steam produced
5. The following modes of heat transfer would take place predominantly from boiler furnace to water wall.
   (a) convection (b) conduction
   (c) radiation (d) conduction and convection
6. Blow down is necessary for boiler, because
   1. the boiler water level is lowered rapidly in case it accidently rises too high.
   2. the precipitated sediment or sludge is removed while the boiler is in service.
   3. the concentration of suspended solids in the boiler is controlled.

   The following statements are correct:
   (a) 1, 2 and 3 (b) 1 and 2
   (c) 3 alone (d) 1 and 3
7. Once through boiler is named as such because
   (a) flue gas passes only in one direction
   (b) there is no recirculation of water

(c) air is sent through the same direction

(d) steam is sent only in one direction

8. The output of a boiler is normally stated as:

(a) evaporative capacity in tonnes of steam that can be produced from and at 100°C.

(b) weight of steam actually produced at rated pressure in tonnes per hour.

(c) boiler horsepower

(d) weight of steam produced per kg of fuel

9. Match List I with List II and select the proper code.

List I	List II
A. Bin system	1. Dust collection
B. Cyclone furnace	2. High turbulence
C. Tangential burner	3. High stage recovery
D. Scrubber	4. Pulverized fuel

Codes:	A	B	C	D
(a)	4	3	1	2
(b)	4	3	2	1
(c)	3	4	1	2
(d)	3	4	2	1

10. The boiler accessories would include:

1. Safety valve   2. Steam trap

3. Steam separator   4. Economizer

(a) 1, 2 and 3   (b) 2, 3 and 4

(c) 1 and 4   (d) 1, 2, 3, and 4

11. Match List I with List II and select the right code.

List I (Type of boiler)	List II (Classification of boiler)
A. Babcock and Wilcox	1. Forced circulation
B. Lancashire	2. Fire tube
C. La-mont	3. Water tube
D. Cochran	4. Vertical

Codes:	A	B	C	D
(a)	1	2	3	4
(b)	2	3	4	1
(c)	3	2	1	4
(d)	2	4	1	3

12. Select the correct statements.
    1. Boiler mountings are mainly protective devices
    2. Steam stop valve is an accessory
    3. Feedwater pump is an accessory

Codes: (a) 1, 2 and 3 (b) 1 and 2 (c) 2 and 3 (d) 3 and 1

13. In forced circulation boilers, about 90% of water is remicirculated without evaporation. The circulation ratio is:

    (a) 0.1 (b) 0.9
    (c) 9 (d) 10

14. Given that, h is draught in mm of water, H is chimney height in metres, T_1 is atmospheric temperature in K, the maximum discharge of gases through the chimney is given by

    (a) $h = \dfrac{176.5T_1}{H}$ (b) $h = \dfrac{H}{176.5T_1}$

    (c) $h = \dfrac{1.765H}{T_1}$ (d) $h = \dfrac{176.5H}{T_1}$

15. When solid fuels are burnt, the nitrogen content of the flue gas by volume is about

    (a) 60% (b) 70%
    (c) 80% (d) 90%

16. The excess air required for combustion of pulverized coal is of the order of

    (a) 100 to 150% (b) 30 to 60%
    (c) 15 to 40% (d) 5 to 10%

17. The main reasons for providing water wall enclosure in high pressure boiler furnaces would include the following. Select the right code.
    1. Increasing evaporation rate using convection heat transfer for hot gases.
    2. Increasing evaporation rate using radiation.
    3. Protecting the refractory walls of the furnace.
    4. Increasing water circulation rate.

Codes: (a) 2 and 3 (b) 1 and 3 (c) 1 and 2 (d) 1, 2, 3 and 4

18. In high pressure natural circulation boilers, the flue gases flow through the following boiler accessories:

    1. Superheater 2. Air heater
    3. Economizer 4. I.D.fan

    Select the correct sequence of flue gas flow through the above boiler accessories.

(a) 1, 3, 4 and 2 (b) 3, 1, 4, 2
(c) 3,1, 2, 4 (d) 1, 3, 2, 4

19. In Benson boiler, the correct sequence of entry of water through the following component is:
    1. Radiation evaporator
    2. Economizer
    3. Radiation superheater
    4. Convection superheater

Codes: (a) 1, 2, 3, 4 (b) 1, 2, 4, 3 (c) 2, 1, 3, 4 (d) 2, 1, 4, 3

20. Coal fired power plant boilers manufactured in India generally use.
    (a) pulverized fuel combustion (b) fluidized bed combustion
    (c) circulating fluidized bed combustion
    (d) moving stroker firing system

21. Select the correct code. The boiler mounting are:
    1. Economizer
    2. Feed check valve
    3. Steam trap
    4. Superheater

Codes: (a) 2 alone (b) 1 and 3 (c) 2, 3 and 4 (d) 1, 2, 3, 4

22. Once through boiler will NOT have
    (a) Drum, headers and pumps
    (b) Drum, steam separators and pumps
    (c) Drum, headers and steam separators
    (d) Drum, headers, steam separators and pumps

23. Benson boiler is one of the high pressure boilers having
    (a) one drum
    (b) one water drum and one steam drum
    (c) three drums (d) no drum

24. Which one of the following safety devices is used to protect the boiler when the water level falls below a minimum level?
    (a) water level indicator (b) fusible plug
    (c) blow off cock (d) safety valve

25. Select the right code for the following statements
    1. Pulverized fuel gives high and controlled burning rate
    2. Insufficient air causes excessive smoking of exhaust
    3. Excess air is provided to control the flue gas temperature
    4. Effect of sulphur in fuel is to give high heat transfer rate

Codes: (a) 3 and 4 (b) 2 and 3 (c) 1 and 2 (d) 1 and 4

26. Select the right code for the following statements
    1. Forced circulation is always used in high pressure power boilers
    2. Root blowers are used for cleaning tube surfaces at regular intervals
    3. Electrostatic precipitator is used to remove fly ash from flue gases

Codes: (a) 1, 2 and 3 (b) 2 and 3 (c) 1 and 3 (d) 1 and 2

27. Once-through boilers operate at
    (a) subcritical pressure
    (b) supercritical pressure
    (c) subcritical as well as supercritical pressure
    (d) critical pressure only
28. Select the right code after matching List I with List II

List I (Components)	List II (Function)
A. Steam trap	1. Controls steam flow rate
B. Fusible plug	2. Controls rate of water flow to boiler
C. Blow off cock	3. Puts off furnace fire when water level reaches unsafe limit
D. Feed check valve	4. Removes mud and dirt collected at the bottom of the boiler
	5. Drains off water collected by partial condensation of steam in pipes

Codes:	A	B	C	D
(a)	5	1	4	2
(b)	1	3	5	4
(c)	5	3	4	2
(d)	1	2	5	4

29. Select the correct sequence of flue gas flow in a steam power plant.
    (a) superheater, economizer, air preheater
    (b) economizer, air preheater, superheater
    (c) air preheater, economizer, superheater
    (d) economizer, superheater, air preheater
30. In a fluidized-bed boiler; the following is not correct:
    (a) the combustion temperatures are higher than those in the conventional boilers
    (b) inferior grade of coal can be used without slagging problems
    (c) the formation of NO_x is less than in the conventional boilers
    (d) the volumetric heat release rates are higher than those in the conventional boilers
31. Device used to drain off water from steam pipes without escape of steam is called:
    (a) steam separator (b) steam trap
    (c) pressure reducing valve (d) injector

32. Match List I with List II and select the correct code

	List I (Type of coal)		List II (Coal properties)
A.	Lignite	1.	Artificial fuel derived from coal
B.	Anthracite	2.	Contains inflammable gas (volatile matter) and burns with a flame
C.	Bituminous	3.	Very hard and high heating value
D.	Coke	4.	High ash content and less volatile matter

Codes:	A	B	C	D
(a)	2	3	4	1
(b)	4	1	2	3
(c)	2	1	4	3
(d)	4	3	2	1

33. Match List I with List II and select the right code

	List I (Boiler)		List II (Type/Description)
A.	Lancashire	1.	Horizontal straight tube, fire-tube boiler
B.	Benson	2.	Horizontal straight tube, water-tube boiler
C.	Babcock and Wilcox	3.	Bent tube, water tube boiler
D.	Stirling	4.	High pressure boiler

Codes:	A	B	C	D
(a)	4	2	1	3
(b)	1	4	2	3
(c)	4	2	3	1
(d)	1	4	3	2

34. Select the right code for the following statements regarding fluidized bed combustion boilers.
   1. the combustion temperatures are low, around 900°C
   2. the formation of oxides of nitrogen is low
   3. it removes sulphur from coal during combustion process
   4. it requires high quality of coal as fuel

Codes: (a) 1, 2, 3 and 4 (b) 1, 2 and 3 (c) 2, 3 and 4 (d) 1 and 4

35. The correct gas flow path in a typical large modern natural circulation boiler is
   (a) Combustion chamber, Reheater, Superheater, Economizer, Air preheater, I D Fan, Electrostatic precipitator, stack
   (b) Combustion chamber, Superheater, Reheater, Economizer, Air preheater, Electrostatic precipitator, I D Fan, stack

(c) Combustion chamber, Reheater, Superheater, Air preheater, Economizer, Electrostatic precipitator, I D Fan, stack

(d) Combustion chamber, Superheater, Reheater, Economizer, Air preheater, I D Fan, Electrostatic precipitator, stack

Keys

1 (d); 2 (a); 3 (b); 4 (a); 5 (c); 6 (a); 7 (b); 8 (a); 9 (b);
10 (b); 11 (c); 12 (d); 13 (d); 14 (d); 15 (c); 16 (b); 17 (a); 18 (d);
19 (c); 20 (a); 21 (a); 22 (b); 23 (d); 24 (b); 25 (c); 26 (a); 27 (d);
28 (a); 29 (a); 30 (a); 31 (b); 32 (a); 33 (d); 34 (b); 35 (b)

SECTION E STEAM TURBINE

1. The main aim of compounding steam turbine is to:
   (a) improve efficiency (b) reduce steam consumption
   (c) reduce motor speed (d) reduce turbine size
2. A throttle governed steam turbine develops 20 I.H.P. with 281 kg/hr of steam and 50 I.H.P. with 521 kg/hr of steam. The steam consumption in kg/hr when developing 15 I.H.P will be nearly
   (a) 150 kg/hr (b) 156 kg/hr
   (c) 241 kg/hr (d) 290 kg/hr
3. Governing of steam turbines can be done by the following
   1. Nozzle control
   2. Throttle control
   3. Providing additional valve and passage

   The correct answer will be
   (a) 1, 2 and 3 (b) 1 and 2 only
   (c) 2 and 3 only (d) 1 and 3 only
4. In steam turbine terminology, diaphragm refers to
   (a) separating wall between rotors carrying nozzles
   (b) the ring of guide blades between rotors
   (c) a partition between low and high pressure discs
   (d) the flange connecting the turbine exit to the condenser
5. A three-stage Rateau turbine is designed in such a manner that the first two stages develop equal power will identical velocity diagrams while the third one develops more power with higher blade speed. In such a multistage turbine, the blade ring diameter
   (a) is the same for all the stages
   (b) gradually increases from the first to the third stage
   (c) of the third stage is greater than that of the first two stages
   (d) of the third stage is less than that of the first stages
6. A single stage impulse turbine with a diameter of 120 cm runs at 3000 rpm. If the blade speed ratio is 0.42, then the inlet velocity of steam will be
   (a) 79 m/s (b) 188 m/s
   (c) 450 m/s (d) 900 m/s
7. In an ideal impulse turbine, the
   (a) absolute velocity at the inlet of moving blade is equal to that at the outlet
   (b) relative velocity at the inlet of the moving blade is equal to that at the outlet
   (c) axial velocity at inlet is equal to that at the outlet
   (d) whirl velocity at inlet is equal to that at the outlet

8. For a Parson's reaction turbine if α_1 and α_2 are fixed angles at inlet and exit respectively and β_1 and β_2 are the moving blade angles at entrance and exit respectively then

   (a) $\alpha_1 = \alpha_2$ and $\beta_1 = \beta_2$ (b) $\alpha_1 = \beta_1$ and $\alpha_2 = \beta_2$
   (c) $\alpha_1 < \beta_1$ and $\alpha_2 > \beta_2$ (d) $\alpha_1 = \beta_2$ and $\beta_2 = \alpha_2$

9. The isentropic enthalpy drop in moving blade is two-thirds of the isentropic enthalpy drop in fixed blades of a turbine. The degree of reaction will be:

   (a) 0.4 (b) 0.6
   (c) 0.66 (d) 1.66

10. The efficiency of nozzle governed turbines is affected merely by losses due to

    (a) partial admission (b) throttling
    (c) interstage pressure drop (d) condensation in last stage

11. Which one of the following pairs is correctly matched?

    (a) Stage efficiency $\rightarrow \dfrac{\text{actual enthalpy drop}}{\text{isentropic enthalpy drop}}$

    (b) Nozzle efficiency $\rightarrow \dfrac{\text{work delivered}}{\text{isentropic enthalpy drop}}$

    (c) Diagram efficiency $\rightarrow \dfrac{\text{work delivered by blades}}{\text{isentopic enthalpy drop}}$

    (d) Blade efficiency $\rightarrow \dfrac{\text{work done on moving blades}}{\text{actual enthalpy drop}}$

12. Match List I with List II and select the right code

List I (Turbine)	List II (Classification)
A. Parson's	1. Pressure compounded
B. De Laval	2. Reaction
C. Rateau	3. Simple impulse
D. Curtis	4. Velocity compounded

Codes:	A	B	C	D
(a)	3	2	1	4
(b)	2	3	4	1
(c)	2	3	1	4
(d)	3	2	4	1

13. The outward radial flow turbine in which there are two rotors rotating in opposite directions is known as

(a) 50% reaction radial turbine
(b) Cantilever turbine
(c) Ljungstorm turbine
(d) Pass-out turbine

14. Select the right code from the following statements:
    1. Impulse turbine rotor blades are thick at the centre
    2. Rateau turbine is more efficient than Curtis turbine
    3. Blade velocity coefficient for an impulse turbine is of the order of 60%

Codes: (a) 1, 2 and 3 (b) 1 and 2 (c) 1 and 3 (d) 2 and 3

15. Given that α_1 = nozzle angle; n = number of rows of moving blades in a velocity compounded impulse turbine, the optimum blade speed ratio is

(a) $2 \cos \alpha_1 n$ (b) $\dfrac{n \cos \alpha_1}{2}$

(c) $\dfrac{\cos \alpha_1}{2(n+1)}$ (d) $\dfrac{\cos \alpha_1}{2n}$

16. In a Parson's turbine stage, blade velocity is 320 m/s at the mean radius and the rotor blade exit angle is 30°. For minimum kinetic energy of steam leaving the stage, the steam velocity at the exit of the stator will be:

(a) $640/\sqrt{3}$ m/s (b) 640 m/s

(c) $320/\sqrt{3}$ m/s (d) $160/\sqrt{3}$ m/s

17. For a free vortex design of blade in the rotor of a reaction axial turbine, the specific work along the blade height is
    (a) higher at the blade hub and lower at the blade tip
    (b) constant from hub to tip
    (c) lower at the hub and tip but different from the mean section
    (d) same at the hub and tip but different from the mean section

18. A reaction turbine stage has angles α, β, γ as nozzle angle, inlet blade and outlet blade angle respectively. The expression for maximum efficiency of the turbine is given by

(a) $\dfrac{2\cos^2 \beta}{1+\cos^2 \beta}$ (b) $\dfrac{2\cos^2 \gamma}{1+\cos^2 \gamma}$

(c) $\dfrac{2\cos^2 \alpha}{1+\cos^2 \alpha}$ (d) $\dfrac{\cos(\alpha+\beta)}{\cos^2 \gamma}$

19. Which one of the following sketches represents an impulse turbine blade?

(a)  (b) 

(c)  (d) 

20. For which turbine, the variation of absolute velocity of steam along the length of steam turbine is shown

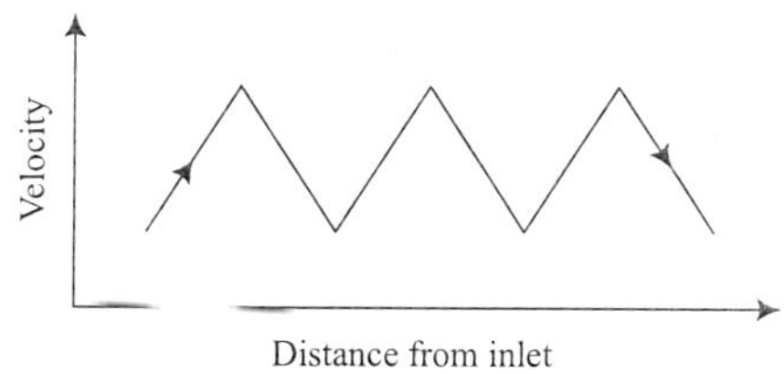

(a) Curtis turbine (b) De Laval turbine
(c) Radial turbine (d) Parson's turbine

21. The correct sequence of turbines in ascending order of efficiency at their design point is
    (a) Rateau, De Laval, Parson's; Curtis
    (b) Curtis, De Laval, Rateau, Parson's
    (c) De Laval, Curtis, Rateau, Parson's
    (d) Parson's, Curtis, Rateau, De Laval

22. Among other things, the poor part-load performance of de Laval turbine is due to the
    (a) formation of shock waves in the nozzle
    (b) formation of expansion waves at the nozzle
    (c) turbulent mixing at the nozzle exit
    (d) increases profile losses in the rotor

23. Match List I with List II and select the correct code

List I (Velocities in velocity diagram of two-stage impulse turbine)	List II (Blade angles)
A. Relative velocity of steam at inlet tip of blade	1. Nozzle angle
B. Absolute velocity of steam at inlet tip of blade	2. Moving blade leading edge angle
C. Relative velocity of steam at outlet tip of blade	3. Moving blade trailing edge angle
D. Absolute velocity of steam at outlet tip of blade	4. Fixed blade leading edge angle

Codes:	A	B	C	D
(a)	1	2	4	3
(b)	2	1	4	3
(c)	2	1	3	4
(d)	1	2	3	4

24. The following relationship between angles of fixed blades (α) and moving blades (β) exists for Parson's turbine
    (a) $\alpha_1 = \alpha_2$ (b) $\alpha_1 = \beta_2$
    (c) $\alpha_2 = \beta_2$ (d) $\beta_1 = \beta_2$
25. Relative velocity of steam at inlet to the rotor for an axial flow turbine stage is 79.0 m/s and relative velocity at rotor exit is 152.0 m/s, the approximate degree of reaction is:
    (a) 0.9 (b) 0.8
    (c) 0.7 (d) 0.6
26. The clearance flow between the blade tips and casing of a steam turbine is
    (a) greater in the reaction turbine than in impulse turbine
    (b) greater in the impulse turbine than in reaction type
    (c) independent of type of turbine
    (d) independent of size of turbine
27. The line *abc* and *cd* in the given graph are known as:

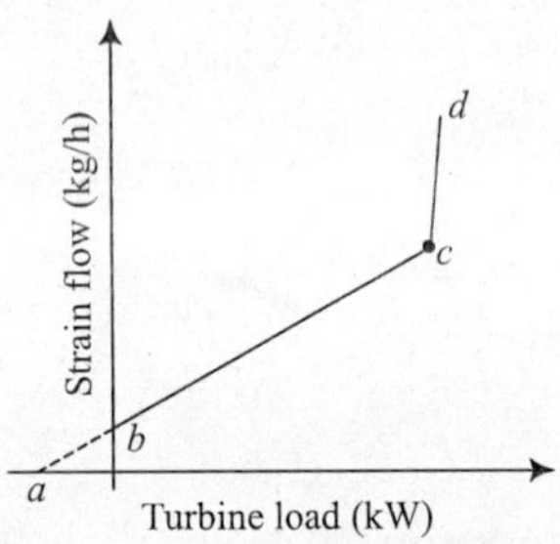

    (a) Wilson's line (b) Willan's line
    (c) S.C. line (d) Throttling line
28. Select the correct statements regarding the nozzle governing of steam turbines
    1. working nozzles receive steam at full pressure
    2. high efficiency is maintained at all loads
    3. stage efficiency suffers due to partial admission
    4. in practice each nozzle of the first stage is governed individually

Codes: (a) 1, 2 and 3 (b) 2, 3 and 4 (c) 1, 3 and 4 (d) 1, 2 and 4

29. Running speeds of steam turbine can be brought down to practical limits by which of the following methods?
    1. By using heavy flywheel
    2. By using a quick response governor
    3. By compounding
    4. By reducing fuel feed to the furnace

Codes: (a) 3 alone (b) 1, 2, 3 and 4 (c) 1, 2 and 4 (d) 2 and 3

30. The net result of pressure-velocity compounding of steam turbine is
    (a) Less number of stages
    (b) Large turbine for a given pressure drop
    (c) Shorter turbine for a given pressure drop
    (d) Lower friction loss
31. If V_b – blade speed, V = absolute velocity of steam entering the blade, α = nozzle angle, then the maximum efficiency of an impulse turbine will be when:
    (a) $V_b = 0.5\ V \cos \alpha$ (b) $V_b = V \cos \alpha$
    (c) $V_b = 0.5\ V^2 \cos \alpha$ (d) $V_b = V^2 \cos \alpha$
32. An impulse turbine produces 50 kW of power when blade mean speed is 400 m/s. What is the rate of change of momentum tangential to the rotor?
    (a) 200 N (b) 175 N
    (c) 150 N (d) 125 N
33. At a particular section of a reaction turbine, the diameter of the blade is 1.8 m, the velocity of flow of steam is 49 m/s and the quantity of steam flow is 5.4 m^3/s. The blade height at the section will be approximately:
    (a) 4 cm (b) 2 cm
    (c) 1 cm (d) 0.5 cm
34. Select the right code. If steam is reheated during the expansion through turbine stages:
    1. erosion of blade will decrease
    2. the overall pressure ratio will increase
    3. the total heat drop will increase

Codes: (a) 1, 2 and 3 (b) 1 and 2 (c) 2 and 3 (d) 1 and 3

35. In an impulse-reaction turbine stage, the heat drop in fixed and moving blades are 15 kJ/kg and 30 kJ/kg respectively. The degree of reaction will be:
    (a) 1/3 (b) 1/2
    (c) 2/3 (d) 3/4
36. If *D* is the diameter of turbine wheel and *U* is its peripheral velocity, then the disc friction loss will be proportional to

(a) $(DU)^3$ (b) D^2U^3
(c) D^3U^2 (d) DU^4

37. The compounding of steam turbine is done to:
(a) improve efficiency (b) reduce turbine speed
(c) increase blade speed ratio (d) reduce axial thrust

38. Select the right code for the following statements:
1. De Laval nozzle is a subsonic nozzle
2. Supersonic nozzle is a converging nozzle
3. Subsonic diffuser is a diverging passage

Codes; (a) 1 and 2 (b) 2 and 3 (c) 1 alone (d) 3 alone

39. Match List I with List II and select the right code.

List I (Different turbine stages)	List II (Type of turbine)
A. 50% reaction stage	1. Rateau
B. Two-stage velocity compounded turbine	2. Parson
C. Single-stage impulse	3. Curtis
D. Two-stage pressure compounded turbine	4. De Laval
	5. Hero

Codes:	A	B	C	D
(a)	5	1	2	3
(b)	5	3	2	1
(c)	2	3	4	1
(d)	3	1	4	2

40. If α is the angle made by absolute velocity of steam at inlet, the maximum efficiency of a Parson's turbine is

(a) $\dfrac{\cos^2\alpha}{2(1+\cos^2\alpha)}$ (b) $\dfrac{2+\cos^2\alpha}{2\cos^2\alpha}$

(c) $\dfrac{2\cos\alpha}{1+\cos^2\alpha}$ (d) $\dfrac{2\cos^2\alpha}{1+\cos^2\alpha}$

41. Partial admission steam turbine refers to the situation where the
(a) steam is admitted partially into the blades through nozzles
(b) nozzles occupy the complete circumference leading into the blade annulus
(c) nozzles do not occupy the complete circumference leading into the blade annulus
(d) steam is admitted partially into the blades directly

42. Select the right code. For a 100% reaction turbine
    1. Change in absolute velocity of steam across the moving blade is zero
    2. Change in absolute velocity of steam across the moving blade is negative
    3. Enthalpy drop in fixed blades is zero.

Codes: (a) 1 alone (b) 2 alone (c) 2 and 3 (d) 1 and 3

43. Point out the wrongly matched pair

	List I	List II
(a)	Internal efficiency of steam turbine	Product of stage efficiency and reheat factor
(b)	Stage efficiency of a turbine	Ratio of adiabatic heat drop to the isentropic heat drop per stage
(c)	Dryness fraction of steam within a stage	Decreases due to reheating
(d)	Steam condensation during expansion through the turbine	Enhances blade erosion

44. Velocity triangle of a reaction turbine stage is shown where

$AB = V_1$ = absolute velocity at rotor blade inlet

$CB - w_1$ = relative velocity at rotor blade inlet

$CE = w_2$ = relative velocity at rotor blade exit

$CD = CB$

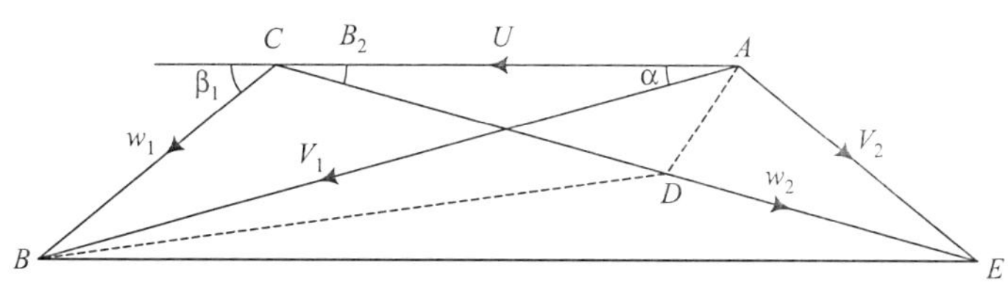

The ratio of reaction force to impulse force is

(a) CE/CB (b) CD/CE

(c) DE/BD (d) AE/AB

45. Select the right code for the following statements.
    1. Throttle governing improves quality of steam in the last few stages.
    2. Internal efficiency of steam is not seriously affected by throttle governing.
    3. Throttle governing is better than nozzle governing.

Codes: (a) 1, 2 and 3 (b) 1 and 3 (c) 2 and 3 (d) 1 and 2

46. Tick the right statement
    (a) Reheat factor is zero if efficiency of the turbine is close to unity.
    (b) Lower the efficiency, higher will be the reheat factor.

(c) Reheat factor is independent of steam condition at turbine inlet.
(d) Availability of reheat is higher at low pressure end.

47. Match List I with List II and select the right code.

List I (Machines)	List II (Features)
A. Steam engine	1. Velocity compounding
B. Impulse turbine	2. Diagram factor
C. Reaction turbine	3. Continuous pressure drop
D. Centrifugal compressor	4. Isentropic efficiency

Codes:	A	B	C	D
(a)	3	4	2	1
(b)	2	1	3	4
(c)	2	4	3	1
(d)	3	1	2	4

48. For maximum blade efficiency of a single-stage impulse turbine, blade speed ratio should be as follows, where α = angle of absolute velocity at inlet.

(a) $\cos 2\alpha$
(b) $\frac{\cos 2\alpha}{2}$
(c) $\frac{\cos\alpha}{2}$
(d) $\frac{2}{\cos\alpha}$

49. The curve A-B-C shows the variation of certain steam parameter in case of a simple impulse turbine. This curve represents the variation of:

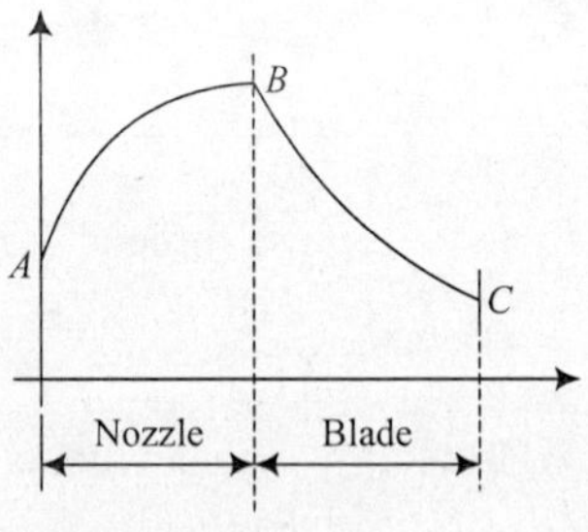

(a) pressure in nozzle and blades
(b) velocity in nozzle and blades
(c) temperature in nozzle and blades
(d) enthalpy in nozzle and blades

50. For a 50% reaction turbine, the efficiency is maximum when the blade speed is equal to (V = absolute velocity of steam; α = angle of V to the tangent of wheel)

(a) $\frac{V\cos\alpha}{2}$ (b) $2\,V\cos\alpha$

(c) $V\cos^2\alpha$ (d) $V\cos\alpha$

51. In a reaction turbine, the heat drop in fixed blade is 8 kJ/kg and total heat drop per stage is 20 kJ/kg. The degree of reaction is:

(a) 40% (b) 66.7%

(c) 60% (d) 25%

52. Match List I with List II and select the right code

List I (Equipment)	List II (Application area)
A. Anticipatory gear	1. Sealing system
B. Labyrinth	2. Steam power plant
C. Invented T-attachment	3. Turbine governing system
D. Deaerator	4. Blade

Codes:	A	B	C	D
(a)	4	2	3	1
(b)	3	1	4	2
(c)	4	1	3	2
(d)	3	2	4	1

53. Employing superheated steam leads to

(a) Increase in erosion of blade

(b) Decrease in erosion of blades

(c) No erosion in blading

(d) No change in erosion of blading

54. Steam enters a De Laval steam turbine with an inlet velocity of 30 m/s and leaves with an outlet velocity of 10 m/s. The work done by 1 kg of steam is

(a) 400 N-m (b) 600 N-m

(c) 800 N-m (d) 1200 N-m

55. In a 50% reaction stage, the absolute velocity angle at inlet is 45°, mean peripheral speed is 75 m/s and the absolute velocity at the exit is axial. The stage specific work is

(a) 2500 m^2/s^2 (b) 3270 m^2/s^2

(c) 4375 m^2/s^2 (d) 5625 m^2/s^2

56. In a reaction turbine stage, enthalpy drop in the stator blade is 4.62 kJ/kg and that in the rotor blade is 2.38 kJ/kg. The degree of reaction of the stage is:

(a) 0.52 (b) 0.43

(c) 0.34 (d) 0.26

57. An emergency governor of a steam turbine trips the turbine when
    1. shaft exceeds 100% of its rated speed
    2. condenser becomes hot due to inadequate cooling water circulation
    3. lubrication system fails
    4. balancing of turbine is not proper

    Select the right code.

Codes: (a) 1, 2 and 3 (b) 2, 3 and 4 (c) 3, 4 and 1 (d) 4, 1 and 2

58. Match List I with List II and select right code.

List I (Blades)	List II (Features)
A. Ceramic blades	1. High creep strength
B. Steam turbine blades	2. Forged and machined
C. Alloy steel blades	3. Precision cast
D. Compressor blades	4. Thick at mid chord
	5. Thin trailing edge

Codes:	A	B	C	D
(a)	2	1	5	4
(b)	3	4	5	1
(c)	2	4	3	5
(d)	3	2	1	5

Keys

1 (c); 2 (c); 3 (a); 4 (a); 5 (c); 6 (c); 7 (b); 8 (d); 9 (a); 10 (b); 11 (a); 12 (c); 13 (c); 14 (b); 15 (d); 16 (a); 17 (c); 18 (c); 19 (a); 20 (d); 21 (c); 22 (d); 23 (c); 24 (b); 25 (d); 26 (b); 27 (b); 28 (a); 29 (d); 30 (c); 31 (a); 32 (d); 33 (b); 34 (d); 35 (c); 36 (c); 37 (b); 38 (d); 39 (a); 40 (d); 41 (a); 42 (b); 43 (b); 44 (b); 45 (a); 46 (c); 47 (b); 48 (c); 49 (b); 50 (d); 51 (c); 52 (b); 53 (b); 54 (a); 55 (d); 56 (c); 57 (a); 58 (a).

SECTION F NOZZLES

Tick (√) the right answer

1. A convergent divergent nozzle is designed for a throat pressure of 2.2 MPa. If the pressure at the exist is 2.3 MPa, then the nature of the flow at the exit will be:

   (a) supersonic (b) sonic
   (c) subsonic (d) not predictable

2. The effect of friction on flow of steam through a nozzle is to:

   (a) decrease the mass flow rate and to increase the wetness at the exit
   (b) increase the mass flow rate and to increase the exit temperature
   (c) decrease the mass flow rate and to decrease the wetness of steam
   (d) increase the exit temperature, without any effect on mass flow rate

3. In a De Laval nozzle expanding superheated steam from 10 bar to 0.1 bar, the pressure at the minimum cross-section will be:

   (a) 3.3 bar (b) 5.46 bar
   (c) 8.2 bar (d) 9.9 bar

4. When compared to stable flow, for supersaturated flow of steam through a nozzle, the available enthalpy drop.

   (a) remains the same (b) increases
   (c) decreases (d) unpredictable

5. The critical pressure ratios for flow of dry saturated steam through a nozzle are respectively

   (a) 0.5279 and 0.550 (b) 0.577 and 0.550
   (c) 0.577 and 0.546 (d) 0.5279 and 0.546

6. The shock-boundary layer interaction takes place in a converging diverging nozzle at:

   (a) converging portion (b) throat
   (c) inlet (d) diverging portion

7. The effect of friction in a steam nozzle is to

   (a) increase velocity and increase dryness friction
   (b) increase velocity and decrease dryness friction
   (c) decrease velocity and increase dryness friction
   (d) decrease velocity and decrease dryness friction

8. Select the right code for the following statements.

   When dry saturated or slightly superheated steam expands through a nozzle

   1. the coefficient of discharge is greater than unity

2. it is dry upto Wilson's line
3. expansion is isentropic throughout

Codes: (a) 1, 2 and 3 (b) 1 and 2 (c) 1 and 3 (d) 2 and 3

9. The total and static pressure at the inlet of a steam nozzle are 186 kPa and 178 kPa respectively. If the total pressure at the exit is 180 kPa and static pressure is 100 kPa, then the loss of energy per unit mass in the nozzle will be:
   (a) 78 kPa (b) 8 kPa
   (c) 6 kPa (d) 2 kPa
10. Under ideal condition, the velocity of steam at outlet of a nozzle for a heat drop of 400 kJ/kg will be approximately:
   (a) 1200 m/s (b) 900 m/s
   (c) 60 m/s (d) sonic velocity
11. In a steam nozzle, inlet pressure of superheated steam is 10 bar. The exit pressure is decreased from 3 bar to 1 bar. The discharge rate will
   (a) remain constant (b) decrease
   (c) increase slightly
   (d) increase or decrease depending on whether the nozzle is convergent or convergent-divergent
12. Select the right code. For supersaturated flow through a steam nozzle, the
   1. enthalpy drop reduces further
   2. exit temperature increases
   3. flow rate increases

Codes: (a) 1, 2 and 3 (b) 1 and 2 (c) 2 and 3 (d) 1 and 3

13. The critical pressure ratio for maximum discharge through a nozzle is given by:

   (a) $\left(\frac{n+1}{2}\right)^{\frac{n}{n-1}}$ (b) $\left(\frac{2}{n+1}\right)^{\frac{n}{n-1}}$

   (c) $\left(\frac{n+1}{2}\right)^{\frac{n-1}{n}}$ (d) $\left(\frac{2}{n+1}\right)^{\frac{n-1}{n}}$

14. Select the right code after matching List I with List II

List I (Names)	List II (Figure)
A. Subsonic nozzle	1. 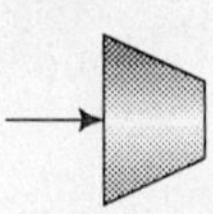
B. Supersonic nozzle	2. 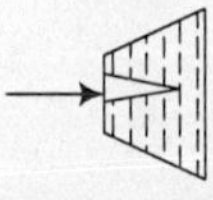

C. Subsonic diffuser

3. 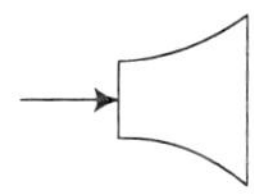

D. Centrifugal compressor

4. 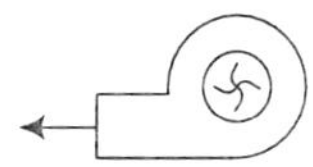

5. 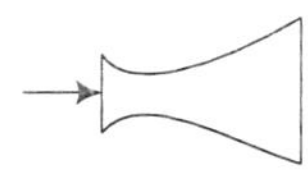

Codes:	A	B	C	D
(a)	3	4	2	5
(b)	1	5	3	4
(c)	3	5	2	4
(d)	1	4	3	5

15. Select the right code for the statement relevant to critical flow through a steam nozzle.
    1. flow rate through the nozzle is minimum
    2. flow rate through the nozzle is maximum
    3. velocity at the throat is supersonic
    4. velocity at the throat is sonic

Codes: (a) 1 alone (b) 1 and 3 (c) 2 and 4 (d) 4 alone

16. Select the correct code for the following statements.

    A convergent-divergent steam nozzle operating under choked condition:
    1. In the convergent portion, steam velocity is less than sonic velocity.
    2. In the convergent portion, steam velocity is greater than sonic velocity.
    3. In the divergent portion, steam velocity is less than sonic velocity.
    4. In the divergent portion, steam velocity is greater than sonic velocity.

Codes: (a) 1 and 3 (b) 1 and 4 (c) 2 and 3 (d) 2 and 4

17. For maximum discharge through a convergent nozzle, the pressure ratio $\frac{p_2}{p_1}$ should be (n = isentropic expansion index)

(a) $\left(\frac{n}{n+1}\right)^{\frac{2}{n+1}}$ (b) $\left(\frac{2}{n+1}\right)^{\frac{n}{n-1}}$

(c) $\left(\frac{n+1}{2}\right)^{\frac{n}{n-1}}$ (d) $\left(\frac{2}{n+1}\right)^{\frac{n}{n+1}}$

Keys

1 (c); 2 (c); 3 (b); 4 (c); 5 (c); 6 (d); 7 (c); 8 (b); 9 (c); 10 (b); 11 (a); 12 (a); 13 (d); 14 (b); 15 (c); 16 (c); 17 (b)

3rd Proof 29/5/09

SECTION G REFRIGERATION CYCLES

Tick (√) the right answer

1. Heat engine E has an efficiency of 0.4 of $Q_2 + Q_4 = 3Q$, the COP of refrigerator R will be:
   (a) 2.5; (b) 3.0;
   (c) 4.0; (d) 5.0.

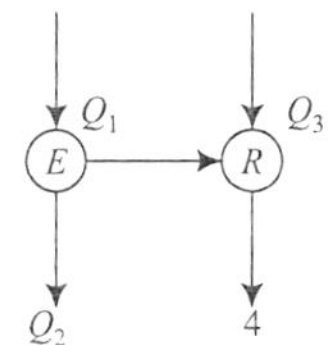

2. When the lower temperature is fixed, COP of a refrigerating machine can be improved by:
   (a) operating the machine at higher speeds;
   (b) operating the machine at lower speeds;
   (c) raising the higher temperature;
   (d) lowering the higher temperature.
3. Which relation is convert?
   (a) $(COP)_{HP} = (COP)_R + 1$ (b) $(COP)_{HP} = (COP)_R - 1$
   (c) $\dfrac{(COP)_{HP}}{(COP)_R} = 1$ (d) $(COP)_{HP} < (COP)_R$
4. A good refrigerant showed have:
   (a) High latent heat of vaporization and low freezing point;
   (b) High operating pressure and low freezing point;
   (c) High specific volume and high latent heat of vaporization;
   (d) Low COP and low freezing point.
5. Which of the following refrigerant has the maximum ozone depletion in the stratosphere?
   (a) ammonia; (b) carbon dioxide;
   (c) sulphur dioxide; (d) fluorine
6. A thermostatic expansion value in a refrigeration system
   (a) ensures the evaporation completely filled with refrigerant of the load;
   (b) is suitable only for constant load system;
   (c) maintains different temperature in evaporation in proportion to load
   (d) none of the above.
7. A heat pump working on a recovered cannot cycle COP of 5. If it works as a refrigerator taking 1 kW of the input, the refrigerating effect will be:
   (a) 1 kW; (b) 2 kW;
   (c) 3 kW; (d) 4 kW.
8. A refrigerating system on reversed Brayton refrigeration cycle is used for maintaining 250 K. If the temperature at the end of constant pressure cool-

ing is 300 K and rise in the temperature of air in the refrigerator is 50 K, then the net work of compression will be (assume air as worthy substance with $Cp = 1$ kJ/kg-k).

(a) 250 kJ/kg; (b) 200 kJ/kg;
(c) 50kJ/kg; (d) 25kJ/kg.

9. In a vapour compression refrigeration plant, the refrigerant leaves the evaporation at 190 kJ/kg and condenser at 65 kJ/kg. For every kg of refrigerant the plant can supply per second, a cooling load of:

(a) 70 kW; (b) 100 kW;
(c) 130 kW; (d) 160 kW.

10. While designing the refrigeration system of an aircraft, the prime consideration is that the

(a) System has high *COP*.
(b) HP/tonne is low
(c) Weight of the refrigerant circulated in the system is low
(d) Weight of refrigeration equipment is low.

11. A value of which maintains a constant degree of superheat at the end of the evaporation coil is called:

(a) automatic expansion value
(b) high side float value,
(c) thermostatic expansion value,
(d) low side float value.

12. Which of the following statements are true for ammonia as a refrigerant?

(1) It has higher compressor discharge temperature compared to fluorocarbon.
(2) It is toxic to mucous membranes.
(3) It requires layer displacement for *TR* compared to fluorocarbons.
(4) It reacts with copper and its alloys.

Select the correct answer using the codes given below:

(a) 1 and 2; (b) 1, 2 and 3;
(c) 1, 2 and 4; (d) 2, 3 and 4.

13. The operating temperatures of a cold storage is –2°C. Heat leakage from the surrounding is 30 kW for the ambient temperature of 40°C. The actual *COP* of the refrigeration plant used is one-fourth that of an ideal plant working between the same temperatures. The power required to drive the plant is:

(a) 1.86 kW; (b) 3.72 kW;
(c) 7.44 kW; (d) 18.6 kW.

14. A heat pump is shown schematic as:

(a) 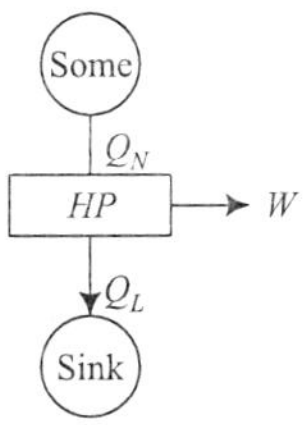

(b) 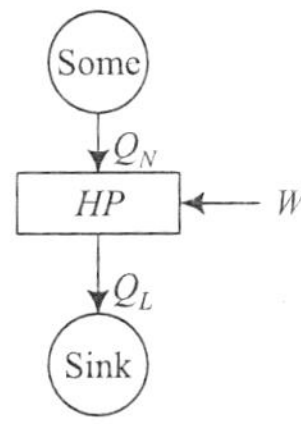

(c) 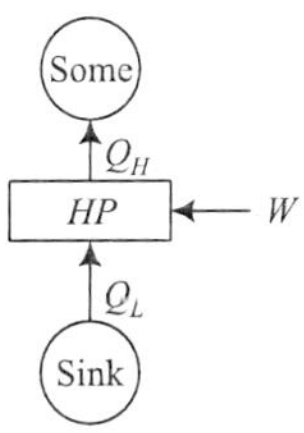

(d) 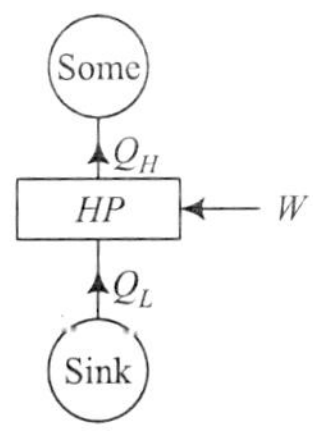

15. In a domestic refrigerator, a capacity tube controls the flow of refrigerant from the
    (a) expansion value to the evaporation
    (b) condenser to the expansion value
    (c) condenser to evaporation
    (d) evaporation to thermostat.
16. The COP of a refrigerator working as heat pump is given by:
    (a) $(COP)_{HP} = (COP)_{Ref} + 2$
    (b) $(COP)_{HP} = (COP)_{Ref} + 1$
    (c) $(COP)_{HP} = (COP)_{Ref} - 1$
    (d) $(COP)_{HP} = (COP)_{Ref}$
17. Match List I (Basic components of Aqua-ammonia refrigeration system) with List II (function of the components in the system) and select the correct answer using the codes given below the table.

List I	List II
A. Generator	1. Dehydration
B. Analyses	2. Removal of vapour from strong aqua-ammonia solution
C. Rectifier	3. Producing dry ammonia vapour by removing traces of water particles completely.
D. Receiver	4. Storage of high pressure liquid ammonia
	5. Formation of liquid ammonia from high pressure vapours.

Codes	A	B	C	D
(a)	3	1	2	5
(b)	5	3	4	2
(c)	1	3	2	5
(d)	2	1	3	4

18. A condenser of a refrigeration system rejects heat at the rate of 120 kW, while its compressor consumes a power of 30 kW. The coefficient of power of the system would be:

(a) $\frac{1}{4}$; (b) $\frac{1}{3}$;

(c) 3; (d) 4.

19. Waste heat can be effectively used in one of the following system:
   (a) vapour compression cycle
   (b) vapour absorption cycle
   (c) air refrigeration cycle
   (d) vortex refrigeration system.
20. In aircraft, air refrigeration cycle is used because of
   (a) low unit weight tonne of refrigeration.
   (b) high heat transfer rate.
   (c) lower temperature at high altitudes.
   (d) higher coefficient of performance.
21. A pressure gauge or discharge side of a refrigerant compressor reads too high. The reason could be:
   1. Lack of cooling water.
   2. Water temperature being high.
   3. Dirty condenser surface
   4. Refrigerant temperature being too high of these reasons.

   (a) 1, 2 and 4 are valid. (b) 1, 2 and 3 are valid

   (c) 2, 3 and 4 are valid (d) 1, 3 and 4 are valid
22. Consider the following statements:

   A decrease in evaporation temperature of a vapour compression machine leads to
   1. an increase in refrigerating effect,
   2. an increase in specific volume of vapour
   3. a decrease in volumetric efficiency of compressor,
   4. an increase in compressor water of these statements:

(a) 1, 3 and 4 are correct
(b) 1, 2 and 3 are correct
(c) 2, 3 and 4 are correct
(d) 2 and 4 are correct

23. Match List I with List II and select the correct answer using the codes given below the lists

List I (Refrigerant)	List II (Principal application)
A. Air	1. Direct contact freezing of food
B. Ammonia	2. Centrifugal compressor system
C. Carbon dioxide	3. Large industrial temperature installation
D. Refrigerant-II	4. Automatic air-conditioners
	5. Aircraft refrigeration

Codes	A	B	C	D
(a)	4	3	1	2
(b)	5	3	1	2
(c)	2	4	3	5
(d)	5	3	2	1

24. In a compression refrigeration system, throttle value is used in place of an expansion because:
(a) It considerably reduces the system weight
(b) It improves the COP, as condenser is small.
(c) The positive work in isentropic expansion of liquid is very small
(d) It leads to significant cost reduction.

25. Match items in List I with those in List II and III and the correct answer.

List I	List II	List III
A. Reversed Carnot engine	1. Condenser	6. Generator
B. Sub cooling	2. Evaporator.	7. Increase in refrigeration effect
C. Super heating	3. Vortex refrigerator	8. Highest COP
D. Constant enthalpy	4. Throttling	9. Adiabatic
	5. Heat pump	10. Dry compression

Codes	A	B	C	D
(a)	3, 10	1, 7	2, 9	4, 6
(b)	5, 8	1, 7	2, 10	4, 9
(c)	4, 10	3, 8	3, 10	1, 6
(d)	2, 7	5, 8	4, 6	1, 9

26. Oil separator is provided in ammonia refrigeration system because
    1. Oil separation in evaporator lead to reduction in heat transfer coefficient
    2. Oil accumulation in the evaporator causes choking of evaporator.
    3. Oil is partially miscible in the refrigerant.
    4. Oil causes choking of expansion device of these statements:

    (a) 1 and 2 are correct. (b) 2 and 4 are correct.
    (c) 2, 3 and 4 are correct (d) 1, 3 and 4 are correct.
27. Moisture showed be removed from refrigerants to avoid:
    1. compressor seal failure,
    2. freezing at the expansion valve,
    3. restriction to refrigerant flow,
    4, corrosion steel parts of these statements.

    (a) 1, 2, 3 and 4 ate correct. (b) 1 and 2 are correct.
    (c) 2, 3 and 4 are correct. (d) 1, 3 and 4 are correct.
28. Which of the following statements are correct?
    1. Practically all common refrigerants have approximated the same COP and power requirement.
    2. Ammonia mixes freely with lubrication oil and this helps in lubrication of compressors.
    3. Dielectric strength of refrigerants is an important property in hermetically sealed compressor units.
    4. Leakage of ammonia can be deleted by halide torch method.

    Groups:

    (a) 1, 2 and 4 (b) 2 and 4
    (c) 1, 3 and 4 (d) 1 and 3
29. The refrigerant used for absorption refrigerators working heat from solar collectors is nature of water and

    (a) Carbon dioxide (b) Sulphur dioxide
    (c) Lithium bromide (d) Freou 12.
30. A refrigerating machine working on reversed Carnot cycle takes out 2 kW per minute of heat from the system while between temperature limits of 300 K and 200 K. COP and power consumed by the cycle be respectively:

    (a) 1 and 1 kW; (b) 1 and 2 kW
    (c) 2 and 1 kW; (d) 2 and 2 kW.
31. In the case of a vapour compression machine, if the condensing temperature of the refrigerant is closer to the critical temperature, then there will be:

1. excessive power consumption
2. high compression
3. large volume flow

The following group of statements is correct:

(a) 1, 2 and 3 (b) 1 and 2
(b) 2 and 3 (d) 1 and 3

32. Hydrogen is essential in an Electrolux refrigeration system because:
    (a) It acts as a catalyst in the evaporation.
    (b) The reaction between hydrogen and ammonia is endothermic in evaporation and exothermic in absorber.
    (c) The cooled hydrogen leaving the heat exchanger cools the refrigerant entering the evaporation.
    (d) It helps in maintaining a low partial pressure for the evaporating ammonia.
33. In an ideal refrigeration (reversed carnot) cycle, the condensor and evaporator temperature are 27°C and –13°C respectively. The *COP* of this cycle will be
    (a) 6.05 (b) 7.5
    (c) 10.5 (d) 15.0
34. A single stage vapour compression refrigeration system be used to produce ultra low temperatures because:
    (a) refrigerant for ultra-low temperatures are not available
    (b) lubricant for ultra-low temperature are not available
    (c) volumetric efficiency will decrease considerably
    (d) heat leakage into the system will be excessive.
35. Vapour absorption refrigeration system works using the
    (a) ability of a substance to get easily condensed or evaporated.
    (b) ability of a vapour to get compressed or expanded
    (c) affinity of a substance for another substance.
    (d) absorbtivity of a substance.
36. Which statement about ammonia absorption system is correct?
    (a) A function of the temperature and pressure of the solutions
    (b) A function of the pressure of the solution irrespective of the temperature.
    (c) A function of the temperature of the solution above
    (d) Independent of the temperature and pressure of the solution.

37. The following group of the statements are correct regarding coefficient of performance of thermoelectric refrigeration as function of:
    1. electrical conductivity of materials
    2. Pelton coefficient
    3. Seebeck coefficient
    4. temperature at cold and hot junction
    5. thermal conductivity of materials

    Groups: (a) 1, 3, 4 and 5; (b) 1, 2, 3 and 5
    (b) 1, 2, 4 and 5; (d) 2, 3, 4 and 5
38. Air cooling is used for compressors whereas water jacketing is adopted for cooling ammonia compressors because:
    (a) latent heat of ammonia is higher than that of felon
    (b) thermal conductivity of water is higher than that gain
    (c) specific heat of water is higher than that of air.
    (d) of the larger superheat horn of ammonia compression cycle.
39. Air refrigeration cycle is used in
    (a) Commercial refrigeration (b) Domestic refrigeration
    (c) Gas liquefaction (d) Air-conditioning
40. The flash chamber in a single stage simple vapour compression cycle
    (a) increases the refrigeration effect
    (b) decreases the refrigeration effect
    (c) increases the work of compression
    (d) has no effect on refrigeration effect
41. In a vapour compression system, a thermometer placed in the liquid line can indicate whether the
    1. refrigerant flow is too low 2. water circulation is adequate
    3. condensor is fouled 4. pump is functioning properly

    The following group of statements is correct
    (a) 1, 2 and 3 (b) 1,2 and 4
    (c) 1, 3 and 4 (d) 2,3 and 4
42. Match List-I with List-II and select the correct answer using the codes given below the lists.

List-I	List-II
A. Bell Column refrigeration	1. Compressor
B. Vapour compression refrigeration	2. Generator
C. Absorption refrigeration	3. Flash chamber
D. Jet refrigeration	4. Expansion cylinder

Codes	A	B	C	D
(a)	1	4	3	2
(b)	4	1	3	2
(c)	1	4	2	3
(d)	4	1	2	3

43. The maximum *COP* for the absorption cycle is given by (T_G = generator temperating. T_c = environment temperature, T_E = refrigerated space temperature)

    (a) $\dfrac{T_E(T_G - T_c)}{T_G(T_c - T_E)}$; (b) $\dfrac{T_G(T_c - T_E)}{T_E(T_G - T_c)}$;

    (c) $\dfrac{T_c(T_G - T_E)}{T_G(T_c - T_E)}$; (d) $\dfrac{T_G(T_c - T_E)}{T_c(T_G - T_E)}$

44. In milk chilling plants, the usual secondary refrigerant

    (a) Ammonia solution; (b) Sodium silicate;

    (c) Glycol (d) Brine

45. The desirable combination of properties for a refrigerant include:

    (a) High specific heat and low specific volume
    (b) High heat transfer coefficient and low latent heat
    (c) High thermal conductivity and low freezing point
    (d) High specific heat and high boiling point

46. A heat pump operating on Carnot cycle pumps heat from a reservoir at 300 K to a reservoir at 600 K. The COP is:

    (a) 1.5; (b) 0.5;

    (c) 2; (d) 1

47. One tonne refrigeration is equivalent to:

    (a) 3.5 kW; (b) 50 kJ/s;

    (c) 1000 J/min; (d) 1000 kJ/min

48. Solar energy can be directly used in:

    (a) vapour compression refrigeration system
    (b) vapour absorption refrigeration system
    (c) air refrigeration system
    (d) jet refrigeration system

49. The most common type of absorption system in use in industrial applications is based on the refrigerant absorbant combination of:

    (a) Air water; (b) Lithium bromide-air

    (c) Carbon dioxide-water; (d) Ammonia-water

50. A good refrigerant should have:
   (a) Large latent heat of vaporization and low operating pressure
   (b) Small latent heat of vaporization and high operating pressures
   (c) Large latent heat of vaporization and large operating pressures
   (d) Small latent heat of vaporization and low operating pressures.
51. When a burnt-out hermetic compressor is replaced by a new one, it is desirable to include in the system a large drier-cum-strainer also. This is to be replaced in
   (a) liquid line; (b) suction line
   (c) hot gas line (d) discharge line
52. The COP of a refrigerator working on a reversed Carnot cycle is 4. The ratio of the highest absolute temperature to the lowest absolute temperature is
   (a) 1.2; (b) 1.25;
   (c) 3.33; (d) 4
53. The correct sequence of the components of a vapour compression refrigerator is:
   (a) Evaporation, compressor, condensor and throttle value,
   (b) Condenser, throttle value, evaporation and compressor
   (c) Compressor, condenser, throttle value and evaporation
   (d) Throttle value, evaporation, compressor and condenser.
54. The enthalpies at the beginning of the compression and at the end of condensation are respectively 185 kJ/kg = 10 kJ/kg and 85 kJ/kg. The COP of the vapour compression refrigeration system is
   (a) 0.25; (b) 5.4;
   (c) 4; (d) 1.35
55. Which of the groups of statements are correct?
   1. Refrigerant NH_3 is used in reciprocating compressors
   2. Refrigerant CO_2 is used in reciprocating compressors
   3. Refrigerant R-11 is used in reciprocating compressors

   Groups:
   (a) 1 and 3; (b) 1 and 2;
   (c) 2 and 3; (d) 1,2 and 3.
56. The COP of a heat pump β_{HP} and COP of a refrigerant β_{Ry} are related as:
   (a) $\beta_{HP} + \beta_{Ry} = 1$; (b) $\beta_{HP} - \beta_{Ry} = 1$;
   (c) $\beta_{Ry} - \beta_{HP} = 1$; (d) $\beta_{HP} - \beta_{Ry} = 0$.

57. The pressure in a horizontal capillary tube of a refrigeration system decreases due to the
    1. frictional resistance offered by the tube wall
    2. acceleration of refrigerant in the tube
    3. heat transfer from the tube wall
    4. decrease in the potential energy

    The following group of statements are correct.

    (a) 1 and 4; (b) 2, 3 and 4;
    (c) 1, 2 and 3; (d) 1 and 2
58. The discharge pressure of the compressor in the refrigeration system goes up due to the
    (a) lower volumetric efficiency of the compressor
    (b) formation of scale in the condenser.
    (c) large size of the condenser
    (d) undercharge of the refrigerant
59. The leakage in a feron-based refrigeration system be detected by using a/an:
    (a) oxy-acety line torch, (b) halide torch
    (c) sulphur torch, (d) blue litmus paper
60. The working temperature in evaporation and condenser coils of a refrigerator are –23°C and 27°C, respectively. The COP of the refrigerator is 0.8 of the maximum COP. For a power input of 1 kW, the refrigeration effect produced will be:
    (a) 4 kW; (b) 5 kW;
    (c) 8 kW; (d) 2.5 kW.
61. For a heat pump working on vapour compression cycle, enthalpy values of the working fluid at the end of heat addition process, at the end of compression process, at the end of the heat rejection process and at the end of isenthalpic expansion process are 195 kJ/kg, 210 kJ/kg, 90 kJ/kg and 90 kJ/kg respectively. The man flow rate is 0.5 kg/s. Then the heating capacity of the heat pump is nearly.
    (a) 7.5 kW; (b) 45 kW;
    (c) 52.5 kW; (d) 60 kW
62. A one tonne capacity water cooler cools water steadily from 35°C to 20°C. The specific heat of water is 4.18 kJ/kg-k, the water flow rate will be nearly.
    (a) 13.33 L/hr; (b) 33.3 L/hr;
    (c) 200 L/hr; (d) 250 L/hr.

63. Match List-I (refrigerant) with List-II (chemical constituent) and select the correct answer using the codes given below the lists:

List-I (Refrigerant)	List-II (Chemical constituent)
A. R-12	1. Trichlorotrifluoroethane (CCl_2FCClF_2)
B. R-22	2. Difluoro mino chloro methane (CNF_2Cl)
C. R-717	3. Ammonia (NH_3)
D. R-313	4. Difluoro dichloro methane (CCl_2F_2)

Codes	A	B	C	D
(a)	3	2	4	1
(b)	4	2	3	1
(c)	3	1	4	2
(d)	4	1	3	2

64. Which of the following group of statements is correct.

Subcooling in the condenser of a refrigeration system is advisable when:

1. Expansion value is at a higher elevation than condenser.
2. There is a large pressure drop in the line connecting condenser to the expansion
3. The refrigeration effect is to be increased
4. The compressor work is to be reduced.

Groups:

(a) 1 and 2; (b) 1, 3 and 4;

(c) 2, 3 and 4; (d) 1, 2 and 3

65. A solar absorption refrigeration system has generator temperature of 87°C, evaporation temperature of –3°C, condenser and absorber temperature of 27°C each, then its maximum possible COP is

(a) 10.0; (b) 9.0;

(c) 1.80; (d) 1.50.

66. Match List-I (Expansion device) with List-II (operation) and select the correct answer using codes given below the lists:

List-I (Expansion device)	List-II (Operation)
A. Float value	1. Constant degree of superheat at evaporation exit pressure
B. Automatic expression value	2. Constant degree of superheat at evaporation inlet pressure
C. Internally equalized thermostatic expansion value	3. Constant level of refrigerant in the evaporation

D.	Externally equalized thermostatic expansion value	4. Constant pressure in the evaporation

Codes	A	B	C	D
(a)	1	2	4	3
(b)	3	2	4	1
(c)	3	4	2	1
(d)	1	4	2	3

67. A standard vapour compression refrigeration cycle consists of the following 4 thermo-dynamic processes in sequence.
    (a) isothermal expansion, isentropic compression, isothermal compression and isentropic expansion.
    (b) constant pressure heat addition, isentropic compression, constant pressure heat rejection and isentropic expansion.
    (c) constant pressure heat addition, isentropic compression, constant pressure heat rejection and isentropic expansion.
    (d) isothermal expansion, constant pressure heat addition, isothermal compression and constant pressure heat rejection.
68. Sub-cooling heat exchanger is used in a refrigeration cycle. The enthalpies at condenser outlet and evaporation are 78 and 182 kJ/kg respectively. The enthalpy of sub-cooled liquid is 68 kJ/kg. The COP of the cycle is
    (a) 3.25 (b) 2.16
    (c) 3.0 (d) 3.5
69. Maximum possible COP of a solar absorption refrigeration system with generator temperature of 360 K, absorber temperature of 300 K, condenser temperature of 300 K and evaporator temperature of 270 K is
    (a) 9; (b) 6;
    (c) 3; (d) 1.5.
70. Ozone depletion by CFCs occurs by breakdown of:
    (a) Chlorine at ones from refrigerant by UV radiation and reaction with ozone in troposphere.
    (b) Fluorine atoms from refrigerant by UV radiation and reaction with ozone in troposphere.
    (c) Chlorine atoms from refrigerant by UV radiation and reaction with ozone in stratosphere.
    (d) Fluorine atoms from refrigerant by UV radiation and reaction with ozone in stratosphere.
71. The sensible bulb of the thermostatic expansion value is located at the
    (a) exit of evaporation; (b) inlet of the evaporation
    (c) exit of conclusion; (d) inlet of condenser.

72. Experimental measurements on a refrigeration system indicate that rate of heat extraction by the evaporation, rate of heat rejection by the compressor body to environment are 70 kW, 90 kW and 5 kW respectively. The power input (in kW) required to operate the system is
   (a) 15; (b) 20;
   (c) 25; (d) 75.
73. The rate of the clearance volume to the displacement volume of a R-12 reciprocating compressor is 0.05. Specific volume at inlet and outlet of compressor are 0.04 and 0.02 m^3/kg respectively volumetric efficiency of the compressor is
   (a) 95.0%; (b) 47.5%;
   (c) 38.0%; (d) 19.0%
74. In a one tonne capacity water cooler, water enters at 30°C at the rate of 200 litres per hour. The outlet temperature of water will be (sp. heat of water = 4.18 kJ/kg-k].
   (a) 3.5°C; (b) 6.3°C;
   (c) 23.7°C; (d) 15°C.
75. A refrigerant working on a reversed Carnot cycle has a *COP* of 4. If it works as a heat pump and consumes 1 kW, the heating effect will be
   (a) 1 kW; (b) 4 kW;
   (c) 5 kW; (d) 6 kW
76. Theoretical maximum *COP* of a vapour absorption system (where T_G = generator temp, T_E = evaporation temp; T_o = environmental temperature is
   (a) $\frac{T_E}{T_G}\left(\frac{T_G - T_o}{T_o - T_E}\right)$; (b) $\frac{T_E}{T_G}\left(\frac{T_o - T_E}{T_G - T_o}\right)$;
   (c) $\frac{T_G}{T_E}\left(\frac{T_G - T_o}{T_o - T_E}\right)$; (d) $\frac{T_G}{T_E}\left(\frac{T_o - T_E}{T_G - T_o}\right)$
77. When a refrigerator system is started from ambient conditions, the evaporation temperature decreases from ambient temperature to design value. The period is known as a pull-down period. The power requirement of compressor during pull down
   (a) decreases continuously (b) increases continuously
   (c) remaining constant (d) increases and then decreases.
78. Oil separation is NOT required in refrigeration system if:
   (a) the refrigerant and oil are immiscible at all pressures and temperatures.
   (b) the refrigerant and oil are immiscible at condensation pressure and temperatures.

(c) the refrigerant and oil are miscible at all pressures and temperatures.

(d) the refrigerant and oil are miscible at condensation and temperature.

Keys

1. (d)	2. (d)	3. (a)	4. (a)	5. (d)	6. (a)	7. (d)	8. (c)	9. (c)
10. (d)	11. (c)	12. (c)	13. (d)	14. (c)	15. (c)	16. (b)	17. (d)	18. (c)
19. (b)	20. (a)	21. (b)	22. (c)	23. (b)	24. (c)	25. (b)	26. ( )	27. (c)
28. (d)	29. (c)	30. (c)	31. (b)	32. (a)	33. (a)	34. (c)	35. (a)	36. (a)
37. (c)	38. (c)	39. (d)	40. (a)	41. ( )	42. (d)	43. (a)	44. (d)	45. (d)
46. (a)	47. (a)	48. (b)	49. (b)	50. (a)	51. (d)	52. (b)	53. (c)	54. (c)
55. (a)	56. (b)	57. (c)	58. (a)	59. (b)	60. (a)	61. (d)	62. (c)	63. (b)
64. ( )	65. (d)	66. (c)	67. (c)	68. (b)	69. (d)	70. (c)	71. (a)	72. (c)
73. (a)	74. (d)	75. (c)	76. (a)	77. (b)	78. (c)			

Index